PLCs & SCADA

PLCs & SCADA

THEORY AND PRACTICE

By

RAJESH MEHRA
Faculty
Department of Electronics and Communication Engineering
National Institute of Technical Teachers, Training and Research, Chandigarh

VIKRANT VIJ
Faculty (JPTC),
Jaypee University of Information Technology, Himachal Pradesh

UNIVERSITY SCIENCE PRESS
(An Imprint of Laxmi Publications Pvt. Ltd.)
An ISO 9001:2015 Company

BENGALURU • CHENNAI • GUWAHATI • HYDERABAD
KOCHI • KOLKATA • LUCKNOW • RANCHI
NEW DELHI

Printed and bound in India

First Edition : 2011; *Edition* : 2025

ISBN *:* 978-93-81159-11-8

Branches

✆	Bengaluru	080-26 75 69 30
✆	Chennai	044-24 34 47 26
✆	Guwahati	0361-254 36 69
✆	Hyderabad	040-27 55 53 83
✆	Kochi	0484-405 13 03
✆	Kolkata	033-40 04 77 79
✆	Lucknow	0522-430 36 13
✆	Ranchi	0651-224 24 64

Published in India by

UNIVERSITY SCIENCE PRESS

(An Imprint of Laxmi Publications Pvt. Ltd.)

An ISO 9001:2015 Company

113, GOLDEN HOUSE, GURUDWARA ROAD, DARYAGANJ,

NEW DELHI - 110002, INDIA

Corporate Office : B-33, Sector-5, Noida,

Gautambuddha Nagar (U.P.) - 201 301, INDIA

Telephone : +91-120-44 46 900

www.laxmipublications.com

info@laxmipublications.com

CONTENTS

PREFACE

I find the great thing in this world is not so much where we stand, as in what direction we are moving.

—Oliver Wendell Holmes

Control engineering has evolved overtime. In the past humans were the main methods for controlling a system. More recently electricity has been used for control and early electrical control was based on relays. The development of low cost computer has brought the most recent revolution, the Programmable Logic Controller (PLC). The advent of the PLC began in the 1970s, and has become the most common choice for manufacturing controls.

Since the development of first logic controllers, the capabilities of programmable logic controllers have grown by leaps and bounds. Likewise, the applications of PLCs have grown with them. In fact, in today's increasingly computer controlled environment, it is almost impossible to find a technical industry that does not use programmable controllers in one form or another. This book, provides a comprehensive theoretical, yet practical, look at all aspects of PLCs and their associated devices and systems.

After finishing the book, the student will most likely be employed in a position designing, programming and maintaining systems using PLCs of some brand or model, or even more likely, many machines with many different brands and models of PLC. The authors desire that this text not only be used to learn programmable logic controllers, but also that this text will become part of the student's personal technical reference library.

The organized material that resulted in this book is our attempt at providing a reasonable degree of balance between rigor, clarity of presentation and at the same time keeping the length of book at manageable level. The principal objective of the book is to provide an introduction to various PLC concepts and to develop a foundation that can be used as basis for research and further study in this field.

Organization of this Book

Chapter 1 is on PLC introduction highlighting evolution of PLCs, Advantages of PLC, PLC components, PLC Programming techniques, PLC applications and manufacturers.

Chapter 2 is on Fundamental concepts of Digital electronics incorporating number systems, conversions, Logic gates and Boolean Algebra.

Chapter 3 is on Ladder Logic concepts incorporating Basic Control system components and their symbols, Ladder logic input, output and programming concepts,Basic and Advanced Ladder Logic Functions. Relay logic and Boolean logic concepts.

Chapter 4 is on File structure and Addressing Formats, highlighting fundamental concepts of SLC 500 BASED SYSTEM, Addressing formats and elements of output and Input, Timer data, Bit data, Counter data and various other associated files.

Chapter 5 is on PLC Project development, including features of RS Logix 500 and RS Linx software, Various installation steps, Micrologix, SLC 5 /03 driver configurations, Various phases in developing a New Project, followed by LADSIM based simulation.

Chapters 6 and 7 are on Instruction set highlighting operation and significance of various Bit, Timer, Counter, Comparison and sequencing Instructions.

Chapter 8 is on PLC Applications highlighting the use of RS Logix software to program and develop some of PLC applications like Liquid level control, Process control, Drink dispenser.

Chapter 9 is PLC & SCADA interface with emphasis on SCADA software Installation, SCADA Project Development, Basics of PLC interfacing with SCADA working with developed projects.

Chapter 10 is on SCADA animations highlighting various SCADA animations like Visibility animation, text animation, Numeric display, Numeric input and Project creation using memory tags.

Chapter 11 is on Alarming and Data logging highlighting the Use of Alarm, Alarm configurations Procedure for Alarm set up, Alarm start up and operation and Concept of Alarm summary and Data logging.

Chapter 12 is on SCADA Supplements highlighting Various SCADA supplement like Event detection, Derived tags, Tens, OLE, Security and its significance in PLC Control.

The book has been written to position itself within the marketplace midway between a number of texts in this subject area which may be regarded as comprehensive, and a number of other texts which are rather brief and tend to be merely primers. Reference texts can be too detailed and large for a newcomer to the topic and the primer type of text can lack information and be rather bland for many readers. With this in mind the book is written to appeal to two broad ranges of readers:

1. Students of electronics/instrumentation engineering undergraduate and postgraduate courses who are studying Programmable Logic Controllers.
2. Professional engineer who has basic knowledge of operation and control of PLC's and for the new engineer embarking on an exciting career in electronics design and control.

Utmost care was exercised to eliminate errors; Effort has been made, as far as possible, to make each chapter self contained. Suggestions for the improvement of this book shall be gratefully acknowledged.

—Authors

Credit Where Credit is Due

On completing a book of this nature, it becomes clear to the author(s) that the work would not have been possible without the support of others.

Becoming an electronics engineer and, consequently, writing this book was spurred on by our early interest in electronics and computers that was fully supported and encouraged by our families. Whether it was the attic turned into a laboratory, a trip to the electronic supply store, or accompaniment to the science fair, our families have always been behind us. We wish to express gratitude to ***Our loving parents*** for their support and blessings throughout our life.

More directly, several friends and colleagues were kind enough to review the initial draft, provide feedback on the content, and bring to our attention details that required correction and clarification. We are grateful to our friends and colleagues for their suggestions during the course of our work.

We are thankful to our mentors for teaching us the essence of learning and keeping up with emerging technology.

—Authors

1

PROGRAMMABLE LOGIC CONTROLLERS

LEARNING OBJECTIVES

After reading this chapter, you will be able to explain:

1. Evolution of PLCs, programmable logic controller basics, and its advantages
2. PLC architecture and system components
3. PLC programming techniques
4. PLCs application list

1.1 INTRODUCTION

For automation in industries, many manufacturing processes demand a sequence of operations, which are to be performed repetitively. Early automation systems were mechanical in design, timing and sequencing being effected by gears and cams. Slowly these design concepts were replaced by electrical drives, which were controlled by relays. The relay provided many benefits since its operation was well established, it was immune to electrical interference and able to handle typical industrial input and output signals without special interfacing. The electromagnetic relays have provided control mechanism over a long period of time. But the relay does have a number of limitations:

1. It suffers from contact wear and reliability problems.
2. Being single input, multi-output, it is often completely unsuited for controlling sequencing in industrial processes.
3. The relays are hardwired *i.e.*, relays are wired to perform specific functions. When the system needs to be modified the relays have to be rewired. In large systems this would mean replacing complete control panels because it is not economical to rewire the old panels.
4. Hardwired systems are very expensive to produce due to the high labour content and modification is difficult and time consuming.

In order to control ever-increasing configurations of systems, engineers turned to computers. The computer however was not suited to the industrial environment and the use of the computer on the factory floor, was not possible, unless costly interference filtering was used. The other major problem with computers was the need to program it using a high level language. Engineers who understood a process and its control using relay ladder logic could not readily convey this to a computer programmer. So, a system similar to electromechanical relays using logical programming to control the process called PLC found favourable.

1.2 PROGRAMMABLE LOGIC CONTROLLERS

A programmable Logic Controller shown in Fig. 1.1 is a solid-state device, designed to operate in noisy industrial environments and perform all the logic functions previously achieved using electromechanical relays, drum switches, mechanical timers and counters. Microprocessor technology now provides a more reliable and maintenance free alternative to the relay. It contains software relays in place of actual relays or conventional relays. Solid-state relays based on silicon technology with no moving parts may be used to switch on the large loads to control a process. Hence, this replaces the electromechanical relays or hardwired solid-state logic blocks.

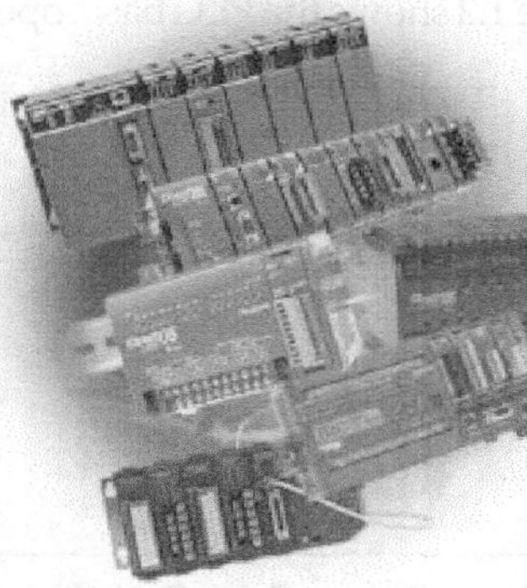

Fig. 1.1 Programmable Logic Controllers

Programmable logic controllers are sometimes also called programmable controllers as it consists of microprocessor based systems. They make technologically advanced control systems having the inherent flexibility and advantages of other programmable controllers, but with an important difference of simplicity. It is a dedicated computer that can be programmed to perform controlling functions in harsh industrial environment.

Today's industrial control problems are much more sophisticated and may need modification in their control strategy from time-to-time. A PLC can easily handle those kinds of problems because it gives required flexibility whereas hardwired devices would be very complicated and provides negligible flexibility towards changes in control strategies. PLCs have replaced old electromechanical relay type systems and have provided with noise free reliable control systems. PLCs have many advantages:

1. It uses simple programming language *e.g.*, ladder diagrams. The program can also be transferred to PLC installed on the factory floor by a programming device.
2. The outputs may be a mixture of conventional relays and solid-state relays.
3. They are designed for an electrically noisy environment no extra filtering is required.
4. They are smaller, faster and more reliable than hardwired systems.
5. All devices like drum switches, timers, and counters are implemented using software programs. No such physical devices exist.
6. PLCs have incorporated object-oriented programming tools and multiple languages based on the IEC 1131-3 standard.
7. Small PLCs have been provided with powerful instructions, which extend the area of application for these small controllers.

8. If system requirements call for flexibility or future growth, a programmable controller brings returns that outweigh any initial cost advantage of a relay control system. Even in a case where no flexibility or future expansion is required, a large system can benefit tremendously from the troubleshooting and maintenance aids provided by a PLC.

1.3 BASIC OPERATION

PLC forms the nucleus of control system. It interfaces input devices to capture the data and actuates control through interfaced output devices. Input parameters are acquired after weak signals from sensors which are conditioned and digitized before use with PLCs. The digital outputs can be converted into analog equivalents if required. Figure 1.2 shows the PLC basic operation to control a machine or plant.

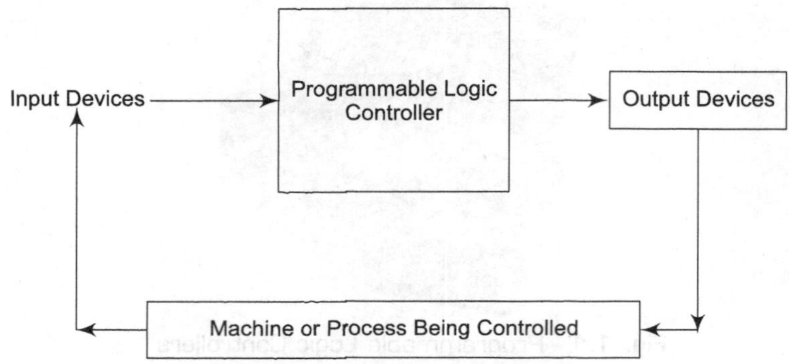

Fig. 1.2 Basic PLC Operation

The input devices are sensors that monitor the machine or the process being controlled. The status of these sensors which can be ON or OFF is fed to the PLC. Depending upon the input status of these sensors the outputs of the PLC may be switched on to energize motors, relays, valves etc. to control the machine or process.

During its operation, the PLC (CPU) completes three processes: (1) it **reads**, or accepts, the input data from the field devices via the input interfaces, (2) it **executes**, or performs, the control program stored in the memory system, and (3) it **writes**, or updates, the output devices via the output interfaces. This process of sequentially reading the inputs, executing the program in memory, and updating the outputs is known as **scanning.** Figure 1.3 illustrates a graphical representation of a scan.

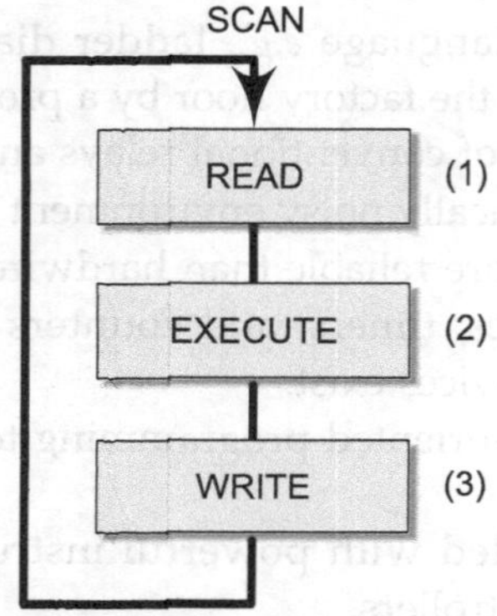

Fig. 1.3 Illustration of a Scan

Although not generally considered a part of the controller, the **programming device**, usually a personal computer or a manufacturer's mini-programmer unit, is required to enter the control program into memory.

1.4 PLC ARCHITECTURE AND COMPONENTS

The PLC, like a computer, employs a microprocessor chip to do the processing and memory chips to store the program. The PLC consists of:

1. A processor
2. Memory
3. Inputs
4. Outputs
5. Programming devices
6. Power supply

The inputs and outputs are connected via proper interfaces. Each input and output has its own address. An interface is a medium where two systems come together and interact or communicate, *i.e.,* communication between the plant being controlled and the PLC. Figure 1.4 shows the basic architecture of a PLC. It contains a processor *e.g.,* a microprocessor chip and memory chips. It also contains all the input and output interfacing. The programming devices can be either handheld or desktop which are remote from the PLC.

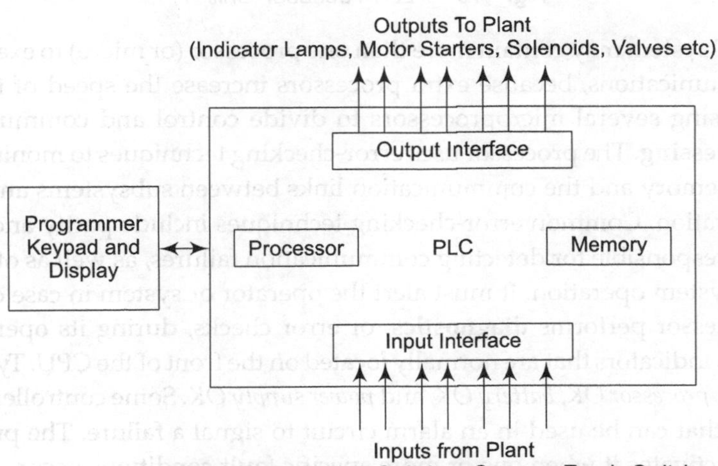

Fig. 1.4 PLC Architecture

1.4.1 Processor

Very small **microprocessors** (or micros)—integrated circuits with tremendous computing and control capability—provide the intelligence of today's programmable controllers. The processor unit contains a microprocessor chip *e.g.,* Zilog Z80, 8085, 8086 or any 16 or 32-bit microprocessor as shown in Fig. 1.5 and necessary circuits to communicate with the devices inside the PLC as well as outside the PLC. This word length affects the speed at which the processor performs most operations. For example, a 32-bit microprocessor can manipulate data faster than a 16-bit microprocessor, since it manipulates twice as much data in one operation. Word length correlates with the capability and degree of sophistication of the controller (*i.e.,* the larger the word length, the more sophisticated the controller). The processor can initiate the operation of counters, sequences, timers and the control relays although no actual physical devices exist. These operations are performed using software routines within the processor unit. The processor can make all the decisions necessary to control a process. It monitors the ON or OFF states of the inputs, scans and solves the logic of the user program and controls the state of the output devices to make them ON or OFF.

Fig. 1.5 PLC Processor Unit

The CPU of a PLC system may contain more than one processor (or micro) to execute the system's duties and/or communications, because extra processors increase the speed of these operations. This approach of using several microprocessors to divide control and communication tasks is known as **multiprocessing**. The processor uses error-checking techniques to monitor the functional status of both the memory and the communication links between subsystems and peripherals, as well as its own operation. Common error-checking techniques include parity and checksum.

The processor is responsible for detecting communication failures, as well as other failures, that may occur during system operation. It must alert the operator or system in case of a malfunction. To do this, the processor performs **diagnostics**, or error checks, during its operation and sends status information to indicators that are normally located on the front of the CPU. Typical diagnostics includes *memory OK*, *processor OK*, *battery OK*, and *power supply OK*. Some controllers possess a set of fault relay contacts that can be used in an alarm circuit to signal a failure. The processor controls the fault relay and activates it when one or more specific fault conditions occur.

1.4.2 Memory

The **memory** system is the area in the PLC's CPU, where all of the sequences of instructions, or *programs*, are stored and executed by the processor to provide the desired control of field devices. PLC memory as shown in Fig. 1.6 contains RAM and EPROM in varying amounts depending upon the design and use of PLC. PLC memory, generally, is sub-divided into five major areas as shown in Table 1.1. Memory capacity is a vital concern when considering a PLC application. Specifying the right amount of memory can save the costs of hardware and time associated with adding additional memory capacity later. Memory capacity is nonexpandable in small controllers (less than 64 I/O capacity) and expandable in larger PLCs. Small PLCs have a fixed amount of memory because the available memory is usually more than enough to provide program storage for small applications. Larger controllers allow memory expandability, since the scope of their applications and the number of their I/O devices have less definition.

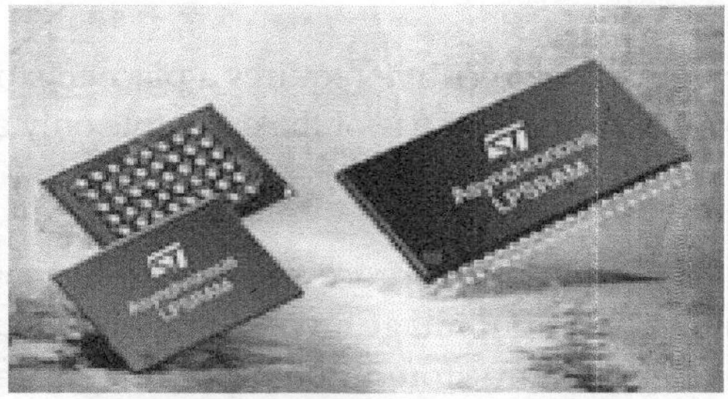

Fig. 1.6 PLC memory Unit

These five types of memories are:

1. Executive memory
2. System memory
3. I/O Status memory
4. Data memory
5. User memory

Executive Memory

The **executive memory** is a collection of permanently stored programs that are considered part of the PLC itself. These supervisory programs direct all system activities, such as execution of the control program and communication with peripheral devices. Executive programs are nothing but operating system programs which are placed in EPROM and these rarely need any change later on. The functions of executive programs are:

(*a*) It instructs microprocessor to examine each user instruction.
(*b*) It converts high-level instruction into machine-level instruction.
(*c*) If the user program requires data from other area of memory, the executive instructs microprocessor to collect those requested data.
(*d*) It provides all system statuses and operating conditions to the microprocessor.
(*e*) It reads all input devices.
(*f*) It solves the user logic program.
(*g*) It updates all output devices accordingly.

In short, the executive program is a special machine level program that literally runs the PLC. It also runs the diagnostic programs to check the proper working and all its components. Read-only memory (ROM) is designed to permanently store a fixed program that is not alterable under ordinary circumstances. It gets its name from the fact that its contents can be examined, or *read*, but not altered once information has been stored. This contrasts with memory types that can be read from and written to. By nature, ROMs are generally immune to alteration due to electrical noise or loss of power. Executive programs are often stored in ROM.

Table 1.1 PLC Memory Types

Memory Name	Type	Function
Executive Memory	EEPROM	It is used to monitor the program
System Memory	RAM	It is used for execution of program
I/O Status Memory	RAM	It is used to check the status of inputs and outputs
Data Memory	RAM	It is used to store the data
User Memory	RAM	It is used to store program codes

System Memory

While performing various functions, the executive requires a place to store intermediate results and information. The system memory is allotted for use of the executive only and not available to user.

I/O Status Memory

During each scan, the executive instructs microprocessor to read the correct statuses of the inputs, and then that information is stored in 'input image area'. After solving user logic program, various output device status are generated and are stored in 'output image area'. This area of memory is always available for user for monitoring and use. User can modify those statuses only through programming.

Data Memory

Random-access memory does not retain its contents if power is lost; therefore, it is a volatile type of memory. Random-access memory normally uses a battery backup to sustain its contents in the event of a power outage.

This RAM area is allocated for timers, counters, math and data functions to store:

(*a*) Preset and accumulated values for timer/counter.
(*b*) Data and result for math instruction.
(*c*) Miscellaneous data and information to be used by data manipulation functions.

User Memory

This is allocated for storage of user program. The executive has to scan this area to find the user instructions. If ASCII messages or some special subroutines are there, then processor can allocate a portion of this area for storage of those special programs.

1.4.3 PLC Inputs

The input interface is the circuitry, which connects the PLC input terminals to the processor. Typical input signals to the PLC are provided from:

1. Sensors
2. Push Buttons
3. Toggle Switches

Sensors

These are used to detect the presence of a component as shown in Fig. 1.7.

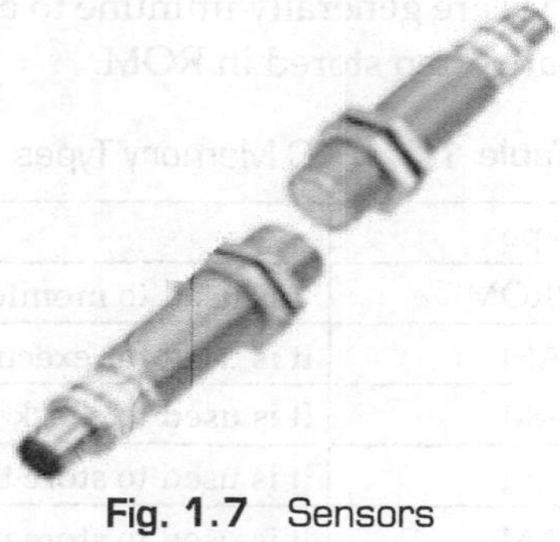

Fig. 1.7 Sensors

Push Buttons

These are used to start or stop a process and are shown in Fig. 1.8

Fig. 1.8 Push Buttons

Toggle Switches

These are used to give a constant high or low output as shown in Fig. 1.9. These switches may also provide inputs to the PLC. All of these inputs however are either present or not present. The PLC is only concerned with ON or OFF signals at its input.

Fig. 1.9 Toggle Switch

The interface is used to convert input signals AC or DC from these input devices to low level DC voltage signals that is used in the processor. The interface may also have to provide electrical isolation to prevent noise and voltage spikes on the input lines from distorting the input signals. The isolation is provided using Opto-Couplers or Opto-Isolators.

1.4.4 PLC Outputs

The output interface is an electronic circuitry, which connects the processor to the PLC output terminals. These output terminals are connected to the plant to control the different outputs like:

1. Indicator lamps
2. Motor starter
3. Valves
4. Hooters

Indicator Lamps

These are used to indicate the start or stop of a process and are shown in Fig. 1.10.

Fig. 1.10 Indicator Lamps

Motor Starter

It is used to change the state of a motor from ON to OFF or *vice-versa*. It is shown in Fig. 1.11 given below.

Fig. 1.11 Motor Starter

Valves

These are used to change the state of a valve from ON to OFF or *vice-versa*. These are shown in Fig. 1.12.

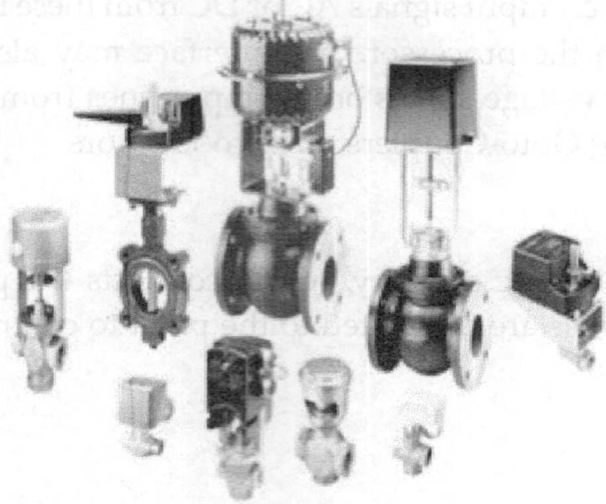

Fig. 1.12 Valves

Hooters

These are used to indicate alarms occurred during a process. These are shown in Fig. 1.13.

Fig. 1.13 Hooters

1.4.5 Programming Device

Although the way to enter the control program into the PLC has changed since the first PLCs came onto the market, PLC manufacturers have always maintained an easy human interface for program entry. The programming device is used to program the PLC which is remote from the PLC and may be connected to enter a user program and disconnected afterwards. Most PLCs are programmed using very similar instructions. The only difference may be the mechanics associated with entering the program into the PLC, which may vary from manufacturer to manufacturer. The user program may then be run without the programming device. Therefore, one device may serve several PLC's. There are 2 types of programming devices: ***Desktop or Laptop and Handheld Programmer.***

Desktop or Laptop

Common usage of the Personal Computer (PC) in our daily lives has led to the practical elimination of dedicated PLC programming devices. Due to the personal computer's general-purpose architecture and standard operating system, most PLC manufacturers and other independent suppliers provide the necessary PC software to implement ladder program entry, editing, documentation, and real-time monitoring of the PLC's control program. A PC can also communicate with a programmable controller through the RS-232C serial port, thus serving either as the data handler and supervisor of the PLC control.

Typical Desktop or Laptop consists of a CRT (Cathode Ray Tube), keyboard, and all the circuitry to interface with the processor. They are portable, designed to be moved from site to site, and be operated in areas with high levels of electrical noise, high humidity and temperature. These are basically an industrial IBM computer as shown in Fig. 1.14. In addition to programming and data collection activities, PC software that provides ladder programming capability often includes PLC documentation options. This documentation capability allows the programmer to define the purpose and function of each I/O address that is used in a PLC program.

Fig. 1.14 Laptop as PLC Programming Device

Handheld Programmers

Mini-programmers, also known as *handheld* or *manual programmers*, are an inexpensive and portable way to program small PLCs (up to 128 I/O). These are smaller and lighter but have limited display facilities as shown in Fig. 1.15. They are useful in situations where the installation requires constant changes in the circuit. The display is either LED's, which only allows one line or element of the ladder to be observed at time, or an LCD display, which allows overall rungs to be observed, and some also have a scroll facility. Instead of handheld units, some controllers have built-in mini-programmers. In some instances, these built-in programmers are detachable from the PLC. Even though they are used mainly for editing and inputting control programs, mini-programmers can also be useful tools for starting up, changing, and monitoring the control logic.

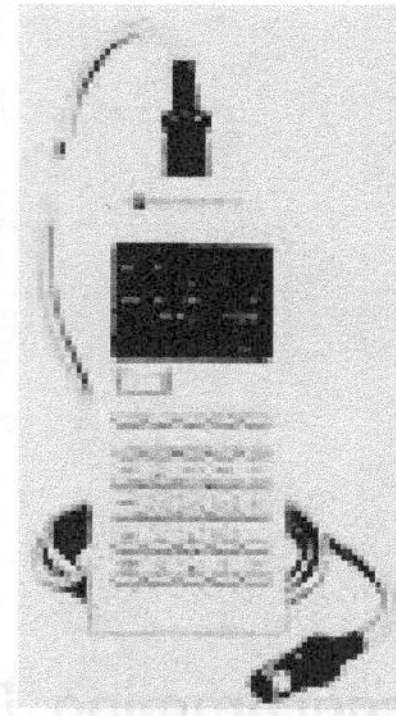

Fig. 1.15 Handheld Programmer as PLC Programming Device

1.4.6 Power Supply

The system power supply plays a major role in the total system operation. In fact, it can be considered the "first-line manager" of system reliability and integrity. The power supply converts the field power into a form, more suitable for the PLC typically +5 V DC, +12 V /–12 V DC. etc. The power supply must be capable of providing multiple voltage levels for circuits inside the PLC and also for the field devices outside the PLC. Since industrial facilities normally experience fluctuations in line voltage and frequency, a PLC power supply must be able to tolerate a 10 to 15% variation in line voltage conditions. Line voltage variations in some plants can eventually become disruptive and may result in frequent loss of production. Normally, in such a case, a constant voltage transformer is installed to stabilize line conditions. A **constant voltage transformer** compensates for voltage changes at its input (the primary) to maintain a steady voltage to its output (the secondary). When operated at less than the rated load, the transformer can be expected to maintain approximately ±1% output voltage regulation with an input voltage variation of as much as 15%.Often, a programmable controller will be installed in an area where the AC line is stable; however, surrounding equipment may generate considerable amounts of Electromagnetic Interference (EMI). Such an installation can result in intermittent misoperation of the controller, especially if the controller is not electrically isolated (on a separate AC power source) from the equipment generating the EMI. Placing the controller on a separate **isolation transformer** from the potential EMI generators will increase system reliability.

The power supply may be categorized as:

Clean Power Supply: This should be free from noise and voltage level fluctuations. This is to be used by processor.

Normal Power Supply: This is not as clean as the above one. This should provide all necessary voltage levels to cater the need of I/O devices.

1.5 PROGRAMMING LANGUAGES

As PLCs have developed and expanded, programming languages have developed with them. Programming languages allow the user to enter a control program into a PLC using an established syntax. Today's advanced languages have new, more versatile instructions, which initiate control program actions.

Logically microprocessor based system and PLC include similar blocks. But physically they are quite different as they are used in different applications and environment. The PLC based systems are useful for embedded applications and designed to work in hostile noisy environment. On the other hand microprocessor based systems are used for general purpose applications. PLC uses simple programming languages which are easy to understand whereas microprocessor uses assembly language which is difficult as compared to PLC languages. PLCs can be programmed with different type of languages and some of them are:

(*a*) Ladder diagram
(*b*) Boolean
(*c*) Grafcet

1.5.1 Ladder Diagram

The programmable controller was developed for ease of programming using existing relay ladder symbols and expressions to represent the program logic needed to control the machine or process. The ladder diagram language is based on three common instructions which are:

(*i*) Normally open instruction
(*ii*) Normally close instruction
(*iii*) Output energize instruction

Normally Open (NO)

The ladder diagram in Fig. 1.16 shows the working of NO as input connected to an output energize instruction. In this ladder diagram when the input connected to NO instruction is logically low then output connected to output energize instruction will be low and when input is high then output will be high.

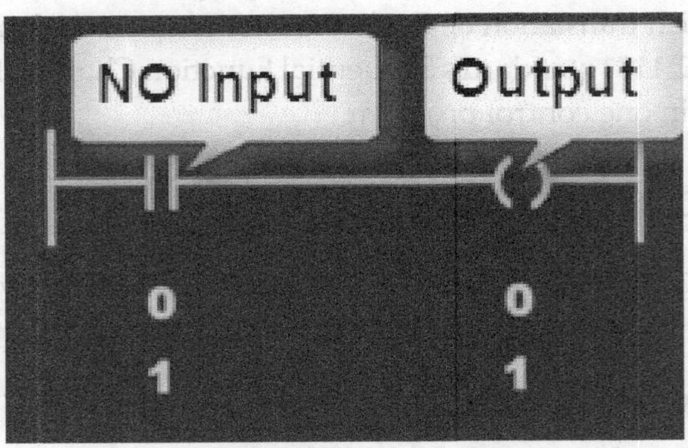

Fig. 1.16 NO Operation

Normally Closed (NC)

The ladder diagram in Fig. 1.17 shows the working of NC as input. In this ladder diagram when the input is logically low then the output will be high and when input is high then output will be low. So it functions as a NOT gate.

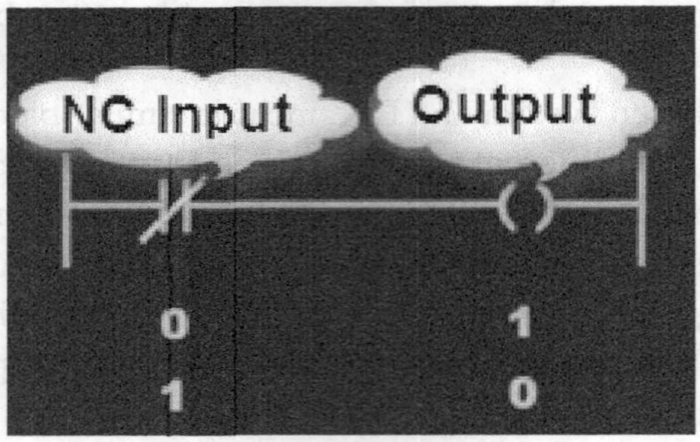

Fig. 1.17 NC Operation

Output Energize

It is a simple output instruction which will be high when rung conditions are true. It will be low when rung conditions are false.

1.5.2 Boolean

Some PLC manufacturers use Boolean language, also called *Boolean mnemonics*, to program a controller. The Boolean language uses Boolean algebra syntax (see Chapter 3) to enter and explain the control logic. That is, it uses the AND, OR, and NOT logic functions to implement the control circuits in the control program. The Boolean language is primarily just a way of entering the control program into a PLC, rather than an actual instruction-oriented language. When displayed on the programming monitor, the Boolean language is usually viewed as a ladder circuit instead of as the Boolean commands that define the instruction.

1.5.3 Grafcet

Grafcet (Graphe Fonctionnel de Commande Étape Transition) is a symbolic, graphic language, which originated in France, that represents the control program as steps or stages in the machine or process. In fact, the English translation of Grafcet means "step transition function charts." Grafcet is the foundation for the IEC 1131 standard's Sequential Function Charts (SFCs), which allow several PLC languages to be used in one control program.

1.6 PLC APPLICATIONS AND MANUFACTURES

Since its inception, the PLC has been successfully applied in virtually every segment of industry, including steel mills, paper plants, food-processing plants, chemical plants, and power plants.

1.6.1 PLC Applications

PLCs made by different manufacturers can be used for wide variety of application wherever we need automation. Some of the industries employing PLCs including their typical applications are listed below.

1. **Automotive :** Internal combustion engine monitoring, monitoring automotive production machines, power steering, valve assembly and testing.
2. **Chemical and Petrochemical :** Ammonia and ethylene processing, dye processing used in the textile industry, chemical batching, fan control, gas transmission and distribution. Programmable controllers monitor and regulate pressures and flows of gas transmission

and distribution systems. Data is gathered and measured in the field and transmitted to the PLC system.

3. **Glass Processing :** Annealing Lehr control, glass batching, cullet weighing.
4. **Manufacturing/Machining :** Production machines, transfer line machines, wire machine.

1.6.2 PLC Manufacturers

At present, PLC made by different manufacturers are available in the market. Some of which are ALLEN BRADLEY, SIEMENS, SATICON, SQUARE D, MERLIN GERIN, MEM, MODICON, GEC GEM, AEG, SPRECHER & SCHUL, MITSUBISHI, TEXAS. All of these manufacturers are leader in the field of automation and control. Each offers a full line of programmable controllers plus other associated equipments. PLCs offered by Allen Bradley, Siemens and Modicon are widely accepted in most of the industries. So for remaining part of this book we will discuss programming of Allen Bradley based SLC 500 series using RS Logix 500 software and SCADA programming using RS View 32 software. We will also use Demo version of LADSIM software to simulate some PLC programs in order to make PLC programming concepts more clear to the user.

SUMMARY

Every aspect of industry-from power generation to automobile painting to food packaging-uses programmable controllers to expand and enhance production. Programmable controllers have many definitions. However, PLCs can be thought of in simple terms as industrial computers with specially designed architecture in both their central units (the PLC itself) and their interfacing circuitry to field devices (input/output connections to the real world). Hardware advances and software advances have led to more powerful PLCs. The processor and the power supply are important parts of the central processing unit. Although the way to enter the control program into the PLC has changed since the first PLCs came onto the market, PLC manufacturers have always maintained an easy human interface for program entry. This means that users do not have to spend much time for learning how to enter a program, but rather they can spend their time for programming and solving the control problems. The programming languages used in programmable controllers have been evolving since the inception of the PLC in the late 1960s. The three types of programming languages used in PLCs are: ladder, Boolean and Grafcet, PLCs perform a great variety of control tasks, from repetitive ON/OFF control of simple machines to sophisticated manufacturing and process control facilitating its use in various industrial applications.

REVIEW QUESTIONS

1. Compare PLC based systems with relay based systems.
2. Define PLCs and explain its operation with the help of diagrams.
3. Discuss PLC architecture with the help of diagrams and explain its each block in detail.
4. Differentiate between PLC based and microprocessor based systems.
5. Write down some languages mostly used for PLC programming.
6. Explain the function of NC and NO in relation with ladder diagrams.
7. Discuss various applications of PLC.
8. List PLCs made by different manufacturers.

2

Numeric Systems and Logic Concepts

LEARNING OBJECTIVES

After reading this chapter, you will be able to explain:

1. Fundamental concepts of digital electronics
2. Various number systems and conversions
3. Gray code, excess-3 and BCD codes.
4. Boolean algebra, logic gates and IC 7400 series of logic family

2.1 INTRODUCTION

In this chapter, we will explain the number systems and digital codes that are most often used in programmable controller applications. We will first introduce the four number systems most frequently used during input/output address assignment and programming: binary, octal, decimal, and hexadecimal. Then, we will discuss the Binary Coded Decimal (BCD) and gray codes. Since these codes and systems are the foundation of the logic behind PLCs, a basic knowledge of them will help you understand how PLCs work. To understand programmable controllers and their applications, you must first understand the logic concepts behind them. In this chapter, we will discuss basic logic functions, and show you how, with just these three functions, you can make control decisions ranging from very simple to very complex. We will also introduce you to the fundamentals of Boolean algebra and its associated operators.

2.1.1 Analogue Vs Digital

An analogue signal is a continuous or discrete—time signal whose amplitude is continuous in value between a lower and upper limit, but may be either a continuous time or discrete time. A digital signal is a continuous or discrete-time signal with discrete values between a lower and upper limit. When an analogue signal is sampled and converted to digital, this is undertaken using an Analogue-to-Digital Converter (ADC).When a digital signal is converted back to analogue, this is undertaken using Digital-to-Analogue Converter (DAC).

2.1.2 History of Digital Logic

Early electronic circuits were analogue, and before the advent of digital logic, signal processing was undertaken using analogue electronic circuits. The invention of the semiconductor transistor in 1947 at Bell laboratories and the improvements in transistor characteristics and fabrication during the 1950s led to the introduction of linear (analogue) ICs and the first Transistor-Transistor Logic (TTL) digital logic IC in the early 1960s, closely followed by Complementary Metal Oxide Semiconductor (CMOS) ICs. The early devices incorporated a small number of logic gates. However, rapid growth in the ability to fabricate an increasing number of logic gates in a single IC led to the microprocessor in the early 1970s. This, with the ability to create memory ICs with ever increasing capacities, laid the foundation for the rapid expansion in the computer industry and the types of complex digital systems based on the computer architecture that we have available today. The last fifty years have seen a revolution in the electronics industry.

Fundamentally, a digital circuit will be categorized into one of three general types, each of which is created and fabricated within an integrated circuit:

- **Combinational logic,** in which the response of the circuit is based on a Boolean logic expression of the input only and the circuit responds immediately to a change in the input.
- **Sequential logic,** in which the response of the circuit is based on the current state of the circuit and the sometimes the current input. This may be asynchronous or synchronous. In synchronous sequential logic, the logic changes state whenever an external clock control signal is applied. In asynchronous sequential logic, the logic changes state on changes of the input data (the circuit does not utilize a clock control signal).
- **Memory,** in which digital values can be stored and retrieved some time later. For a user, memory can be either read-only (ROM) or random-access (RAM). In ROM, the data stored in the memory are initially placed in the memory and can only be read by the user. Data cannot normally be altered in the circuit application. In RAM (also referred to as read-write memory, RWM), the user can write data to the memory and read the data back from the memory.

2.2 NUMBER SYSTEMS

There are several number systems which we normally use, such as decimal, binary, octal, hexadecimal, etc. Amongst them we are most familiar with the decimal number system. These systems are classified according to the values of the **base or radix** of the number system.

The number system having the value of the base as 10 is called a decimal number system, whereas that with a base of 2 is called a binary number system. Likewise, the number systems having base 8 and 16 are called octal and hexadecimal number systems respectively.

With a decimal system we have 10 different digits, which are 0, 1, 2, 3, 4, 5, 6, 7, 8, and 9. But a binary system has only 2 different digits-0 and 1. Hence, a binary number cannot have any digit other than 0 or 1. So to deal with a binary number system is quite easier than a decimal system.

In an octal number system there are 8 digits-0, 1, 2, 3, 4, 5, 6, and 7. Hence, any octal number cannot have any digit greater than 7. Similarly, a hexadecimal number system has 16 digits-0 to 9 and the rest of the six digits are specified by letter symbols as A, B, C, D, E, and F. Here A, B, C, D, E, and F represent decimal 10, 11, 12, 13, 14, and 15 respectively. Octal and hexadecimal codes are useful to write assembly level language.

In general, we can express any number in any base or radix "X." OR "r". Any number with base X, having n digits to the left and m digits to the right of the decimal point, can be expressed as:

$$a_n X^{n-1} + a_{n-1} X^{n-2} + a_{n-2} X^{n-3} + \dots + a_2 X^1 + a_1 X^0 + b_1 X^{-1} + b_2 X^{-2} + \dots + b_m X^{-m}$$

where a_n is the digit in the nth position. The coefficient a_n is termed as the MSD or most significant digit and b_m is termed as the LSD or the least significant digit.

2.2.1 Decimal Number System

The invention of decimal number systems has been the most important factor in the development of science and technology. The term decimal comes from the Latin word for "ten." The decimal number system uses positional number representation, which means that the value of each digit is determined by its position in a number. The base (also called radix) of a number system is the number of symbols that the system contains. The decimal system has ten symbols: 0, 1, 2, 3, 4, 5, 6, 7, 8, 9; in other words it has a base of 10. Each position in the decimal system is 10 times more significant than the previous position.

2.2.2 Binary Number System

The binary number system is a radix-2 number system with '0' and '1' as the two independent digits. All larger binary numbers are represented in terms of '0' and '1'. The procedure for writing higher order binary numbers after '1' is similar to the case of the decimal number system. For example, the first 16 numbers in the binary number system would be 0, 1, 10, 11, 100, 101, 110, 111, 1000, 1001, 1010, 1011, 1100, 1101, 1110, and 1111. The next number after 1111 is 10000, which is the lowest binary number with five digits. Starting from the binary point, the place values of different digits in a mixed binary number are 2^0, 2^1, 2^2 and so on (for the integer part) and 2^{-1}, 2^{-2}, 2^{-3} and so on (for the fractional part).

Bit is an abbreviation of the term 'binary digit' and is the smallest unit of information. It is either '0' or '1'. A *byte* is a string of eight bits. The byte is the basic unit of data operated upon as a single unit in computers. A *computer word* is again a string of bits whose size, called the 'word length' or 'word size', is fixed for a specified computer, although it may vary from computer to computer. The word length may equal one byte, two bytes, four bytes or be even larger. The *1's complement* of a binary number is obtained by complementing all its bits, *i.e.*, by replacing 0s with 1s and 1s with 0s.

2.2.3 Octal Number Systems

Digital systems operate only on binary numbers. Since binary numbers are often very long, two shorthand notations, octal and hexadecimal, are used for representing large binary numbers. The octal number system uses a base or radix of 8; thus it has digits from 0 to 7. As in the decimal and binary systems, the positional value of each digit in a sequence of numbers is definitely fixed. Each position in an octal number is a power of 8, and each position is 8 times more significant than the previous position.

In the octal number system, we have the 7's and 8's complements. The *7's complement* of a given octal number is obtained by subtracting each octal digit from 7. For example, the 7's complement of $(562)_8$ would be $(215)_8$. The *8's complement* is obtained by adding '1' to the 7's complement. The 8's complement of $(562)_8$ would be $(216)_8$.

2.2.4 Hexadecimal Number System

The hexadecimal numbering system has a base 16; that is, there are 16 symbols. The decimal digits 0 to 9 are used as the first ten digits as in the decimal system, followed by the letters A, B, C, D, E, and F, which represent the values 10, 11, 12, 13, 14, and 15, respectively.

The 15's and 16's complements are defined with respect to the hexadecimal number system. The *15's complement* is obtained by subtracting each hex digit from 15. For example, the 15's complement of $(3BF)_{16}$ would be $(C40)_{16}$. The *16's complement* is obtained by adding '1' to the 15's complement. The 16's complement of $(2AE)_{16}$ would be $(D52)_{16}$.

2.3 NUMBER SYSTEM CONVERSIONS

It is often required to convert a number in a particular number system to any other number system, *e.g.*, it may be required to convert a decimal number to binary or octal or hexadecimal. The reverse is also true, *i.e.*, a binary number may be converted into decimal and so on. The methods of interconversions are now discussed.

2.3.1 Decimal-to-Binary Conversion

To convert a number in decimal to a number in binary we have to divide the decimal number by 2 repeatedly, until the quotient of zero is obtained. This method of repeated division by 2 is called the 'double-dabble' method. The remainders are noted down for each of the division steps. Then the column of the remainder is read in reverse order *i.e.*, from bottom to top order.

For the fractional part, it is found by successively multiplying the fractional part of the decimal number by 2 and recording the carry until the result of multiplication is '0'. The carry sequence written in forward order constitutes the binary equivalent of the fractional part of the decimal number. If the result of multiplication does not seem to be heading towards zero in the case of the fractional part, the process may be continued only until the requisite number of equivalent bits has been obtained. We try to show the method with an example.

Example: *We will find the binary equivalent of* $(13.375)_{10}$.

- The integer part = 13

Divisor	Dividend	Remainder
2	13	–
2	6	1
2	3	0
2	1	1
–	0	1

- The binary equivalent of $(13)_{10}$ is therefore $(1101)_2$
- The fractional part =.375
- $0.375 \times 2 = 0.75$ with a carry of 0
- $0.75 \times 2 = 0.5$ with a carry of 1
- $0.5 \times 2 = 0$ with a carry of 1
- The binary equivalent of $(0.375)_{10} = (.011)_2$
- Therefore, the binary equivalent of $(13.375)_{10} = (1101.011)_2$

2.3.2 Decimal-to-Octal Conversion

Similarly, to convert a number in decimal to a number in octal we have to divide the decimal number by 8 repeatedly, until the quotient of zero is obtained. This method of repeated division by 8 is called 'octal-dabble.' The remainders are noted down for each of the division steps. Then the column of the remainder is read from bottom to top order, just as in the case of the double-dabble method.

Example: *We will find the octal equivalent of* $(73.75)_{10}$

- The integer part = 73

Divisor	Dividend	Remainder
8	73	–
8	9	1
8	1	1
–	0	1

- The octal equivalent of $(73)_{10} = (111)_8$
- The fractional part = 0.75
- 0.75 × 8 = 0 with a carry of 6
- The octal equivalent of $(0.75)_{10} = (.6)_8$
- Therefore, the octal equivalent of $(73.75)_{10} = (111.6)_8$

2.3.3 Decimal-to-Hexadecimal Conversion

The same steps are repeated to convert a number in decimal to a number in hexadecimal. Only here we have to divide the decimal number by 16 repeatedly, until the quotient of zero is obtained. This method of repeated division by 16 is called 'hex-dabble.' The remainders are noted down for each of the division steps. Then the column of the remainder is read from bottom to top order as in the two previous cases. We try to discuss the method with an example.

Example: *Determine the hexadecimal equivalent of* $(82.25)_{10}$

- The integer part = 82

Divisor	Dividend	Remainder
16	82	–
16	5	2
–	0	5

- The hexadecimal equivalent of $(82)_{10} = (52)_{16}$
- The fractional part = 0.25
- 0.25 × 16 = 0 with a carry of 4
- Therefore, the hexadecimal equivalent of $(82.25)_{10} = (52.4)_{16}$

2.3.4 Binary-to-Decimal Conversion

We have to keep in mind that each of the binary, octal, or hexadecimal number system is a positional number system, *i.e.*, each of the digits in the number systems discussed above has a positional weight as in the case of the decimal system.

Example: *Convert* $(10110)_2$ *into a decimal number.*

The binary number given is 1 0 1 1 0

Positional weights ***4 3 2 1 0***

The positional weights for each of the digits are written in italics above each digit. Hence the decimal equivalent number is given as:

$$1 \times 2^4 + 0 \times 2^3 + 1 \times 2^2 + 1 \times 2^1 + 0 \times 2^0$$

$$= 16 + 0 + 4 + 2 + 0$$

$$= (22)_{10}.$$

2.3.5 Hex-Binary and Binary-Hex Conversions

A hexadecimal number can be converted into its binary equivalent by replacing each hex digit with its four-bit binary equivalent. We take the four-bit equivalent because the base of the hexadecimal number system is 16 and it is the fourth power of the base of the binary number system. A given binary number can be converted into an equivalent hexadecimal number by splitting the integer and fractional parts into groups of four bits, starting from the binary point on both sides. The 0s can be added to complete the outside groups if needed.

Example: *Convert* $(11010110)_2$ *into an equivalent hexadecimal number.*

The binary number given is 11010110

Starting with LSB and grouping 4 bits **1101 0110**
Hexadecimal equivalent D 6
Hence the hexadecimal equivalent number is $(D6)_{16}$.

Example: *Convert $(29C)_{16}$ into an equivalent binary number.*
The hexadecimal number given is 2 9 C
4-bit binary equivalent **0010 1001 1100**
Hence the equivalent binary number is $(001010011100)_2$.

2.3.6 Hex-Octal and Octal-Hex Conversions

For hexadecimal-octal conversion, the given hex number is firstly converted into its binary equivalent which is further converted into its octal equivalent. An alternative approach is firstly to convert the given hexadecimal number into its decimal equivalent and then convert the decimal number into an equivalent octal number. The former method is definitely more convenient and straightforward.

For octal-hexadecimal conversion, the octal number may first be converted into an equivalent binary number and then the binary number transformed into its hex equivalent. The other option is firstly to convert the given octal number into its decimal equivalent and then convert the decimal number into its hex equivalent. The former approach is definitely the preferred one.

Example: *Convert $(A72E)_{16}$ into octal equivalent*
Given hexadecimal number is A 7 2 E
Binary equivalent is 1010 0111 0010 1110
= 1010011100101110
Forming groups of 3 bits from the LSB 001 010 011 100 101 110
Octal equivalent 1 2 3 4 5 6
Hence the octal equivalent of $(A72E)_{16}$ is $(123456)_8$.

Example: *Convert $(36.532)_8$ into an equivalent hexadecimal number*
Given octal number is 3 6 5 3 2
Binary equivalent is 011 110 101 011 010
= 011110.101011010
Forming groups of 4 bits 0001 1110. 1010 1101
Hexadecimal equivalent 1 E A D
Hence the hexadecimal equivalent of $(36.532)_8$ is $(1E.AD)_{16}$.

2.4 BINARY ARITHMETIC

Arithmetic operations using binary numbers are far simpler than the corresponding operations using decimal numbers due to the very elementary rules of addition and multiplication.

2.4.1 Binary Addition

The rules of binary addition are:

$0 + 0 = 0$

$0 + 1 = 1$

$1 + 0 = 1$

$1 + 1 = 0$ (carry 1)

As in decimal addition, the least significant bits of the addend and the augend are added first. The result is the sum, possibly including a carry. The carry bit is added to the sum of the digits of the next column. The process continues until the bits of the most significant column are summed.

Example: *Consider the addition of the decimal numbers 27 and 28 in binary*

Decimal	Binary	
27	11011	Addend
+ 28	+ 11100	Augend
55	110111	Sum
	11000	Carry

To verify that the sum is correct, we convert 110111 to decimal:

$$1 \times 2^5 + 1 \times 2^4 + 0 \times 2^3 + 1 \times 2^2 + 1 \times 2^1 + 1 \times 2^0$$
$$= 32 + 16 + 0 + 4 + 2 + 1$$
$$= 55$$

2.4.2 Binary Subtraction

Binary subtraction is carried out by following the same method as in the decimal system. Each digit in the subtrahend is deducted from the corresponding digit in the minuend to obtain the difference. When the minuend digit is less than the subtrahend digit, then the radix number (*i.e.*, 2) is added to the minuend, and a borrow 1 is added to the next subtrahend digit. The rules applied to the binary subtraction are:

$0 - 0 = 0$
$0 - 1 = 1$ (borrow 1)
$1 - 0 = 1$
$1 - 1 = 0$

Example: *Consider the subtraction of 21 from 27 in binary*

Decimal	Binary	
27	11011	Minuend
– 21	– 10101	Subtrahend
6	010110	Difference
	00100	Borrow

It can easily be verified that the difference $(00110)_2$ corresponds to decimal 6.

Example: *Consider the subtraction of 22 from 17.*

In this case, the subtrahend is greater than the minuend. Therefore the result will be negative

Decimal	Binary	
17	10001	
– 22	– 10110	
– 5	– 00101	Difference
	00001	Borrow

2.4.3 Binary Multiplication

Binary multiplication is performed in the same way as decimal multiplication, by multiplying, then shifting one place to the left, and finally adding the partial products. Since the multiplier can only be 0 or 1, the partial product is either zero or equal to the multiplicand.

The rules of multiplication are:

$$0 \cdot 0 = 0$$
$$0 \cdot 1 = 0$$
$$1 \cdot 0 = 0$$
$$1 \cdot 1 = 1$$

Example: *Consider the multiplication of the decimal numbers 67 by 13 in binary:*

Decimal	Binary	
67	1000011	Multiplicand
× 13	× 1101	Multiplier
871	1000011	First partial product
	0000000	Second partial product
	1000011	Third partial product
	1000011	Fourth partial product
	1101100111	Final product

2.4.4 Binary Division

Binary division is very similar to standard decimal division. However, division is simpler in binary because when one checks to see how many times the divisor fits into the dividend, there are only two possibilities, 0 or 1.

Example: *Consider the division of* $(101110)_2$ *i.e.,* (46_{10}) *by* $(111)_2$ *i.e.,* (7_{10})

```
               0001     Quotient
Divisor 111 | 101110    Dividend
               0111
               ----
               100
```

Since the divisor, 111, is greater than the first three bits of the dividend, the first three quotient bits are 0. The divisor is less than the first four bits of the dividend; therefore the division is possible, and the fourth quotient bit is 1. The difference is less than the divisor, so we bring down the net bit of the dividend:

```
        00011
  111 | 101110
        0111
        ----
        1001
         111
        ----
          10
```

The difference is less than the divisor, so the next bit of the dividend is brought down:

```
        000110
  111 | 101110
        0111
        ----
        1001
         111
        ----
         100   Remainder
```

In this case the dividend is less than the divisor; hence the next quotient bit is 0 and the division is complete. The decimal conversion yields 46/7 = 6 with remainder 4, which is correct.

To conclude, all arithmetic operations in digital systems are carried out in modified forms; in fact, they use only addition as their basic operation.

2.5 SIGNED NUMBERS

So far, the number representations we considered have not carried sign information. Such unsigned numbers have a magnitude significance only. Normally a prefix + or – may be placed to the left of the magnitude to indicate whether a number is positive or negative. This type of representation, known as sign-magnitude representation, is used in the decimal system. In binary systems, an additional bit known as sign bit, is added to the left of the most significant bit to define the sign of a number. A 1 is used to represent – and a 0 to represent +. Table 2.1 shows 3 bit equivalent in terms of signed and unsigned equivalents.

Table 2.1 Signed and Unsigned Numbers

	Decimal Equivalent	
Binary	Signed	Unsigned
000	+ 0	0
001	+ 1	1
010	+ 2	2
011	+ 3	3
100	– 0	4
101	– 1	5
110	– 2	6
111	– 3	7

Although the sign-magnitude representation is convenient for computing the negative of a number, a problem occurs when two numbers of opposite signs have to be added. To illustrate, let us find the sum of $+(2)_{10}$ and $-(6)_{10}$.

$$+2_{10} = 0010$$
$$-6_{10} = 1110$$
$$\overline{\quad 10000 \quad}$$

The addition produced a sum that has 5 bits, exceeding the capability of the number system (sign + 3 bits); this results in an overflow. Also, the sum is wrong; it should be – 4 (*i.e.*, 1100) instead of 0.

An alternative representation of negative numbers, known as the complement form, simplifies the computations involving signed numbers. The complement representation enjoys the advantage that no sign computation is necessary. There are two types of complement representations: diminished radix complement and radix complement.

2.5.1 Diminished Radix Complement

In the decimal system (X = 10) the complement of a number is determined by subtracting the number from (X = 1), *i.e.*, 9. Hence the process is called finding the 9's complement. For example, 9's complement of 5 (9 – 5) = 4.

In binary notation ($r = 2$), the diminished radix complement is known as the 1's complement. A positive number in 1's complement is represented in the same way as in sign magnitude representation. The 1's complement representation of a negative number x is derived by subtracting the binary value of x from the binary representation of $(2^n - 1)$, where n is the number of bits in the binary value of x.

Example: *Compute the 1's complement form of – 43.*

The binary value of 43 = 00101011. Since $n = 8$, the binary representation of $2^8 - 1$ is

$$\begin{array}{r} 2^8 = 100000000 \\ -1 \\ \hline 2^8 - 1 = \quad 11111111 \end{array}$$

Hence the 1's complement form of – 43 is

$$\begin{array}{r} 2^8 - 1 = \quad 11111111 \\ -43 = -00101011 \\ \hline 11010100 \end{array}$$

Notice that the 1's complement form of –43 is a number that has a 0 in every position that +43 has a 1, and *vice-versa*. Thus the 1's complement of any binary number can be obtained by complementing each bit in the number.

The advantage of using 1's complement numbers is that they permit us to perform subtraction by actually using the addition operation. It means that, in digital systems, addition and subtraction can be carried out by using the same circuitry. The addition operation for two 1's complement numbers consists of the following steps :

(*i*) Add the two numbers including the sign bits.

(*ii*) If a carry bit is produced by the leftmost bits (*i.e.*, the sign bits), add it to the result. This is called end-around carry.

Example: *Consider addition of –7 to –5*

$$\begin{array}{r} \text{1's complement form of} - 7 = 11000 \\ \text{1's complement form of} - 5 = \underline{11010} \\ \text{Sum } 110010 \\ \text{Carry} \quad \rightarrow 1 \\ \text{1's complement sum } \quad 10011 \end{array}$$

The result is a negative number. By complementing the magnitude bits, we get 11100, that is, –12 Thus the sum of –7 and –5 is –12, which is correct.

Table 2.2 Comparison of 3 bit Signed, Unsigned and 1's Complement

	Decimal Equivalent		
Binary	Unsigned	Sign-Magnitude	1's Complement
000	0	+ 0	+ 0
001	1	+ 1	+ 1
010	2	+ 2	+ 2
011	3	+ 3	+ 3
100	4	– 0	– 3
101	5	– 1	– 2
110	6	– 2	– 1
111	7	– 3	– 0

The subtraction operation for two 1's complement numbers can be carried out as follows:

(*i*) Derive the 1's complement of the subtrahend and add it to the minuend.

(*ii*) If a carry bit is produced by the leftmost bits (*i.e.*, the sign bits), add it to the result.

Example: *Consider subtraction of +21 from +35:*

```
                 Minuend = +35 =  0100011
              Subtrahend = +21 =  1101010  (in 1's complement form)
                           Sum = 10001101
                          Carry ↗ ↘     1
      1's complement sum =  0001110
```

The sign is correct, and the decimal equivalent + 14 is also correct.

In nutshell, one can conclude that the 1's complement number system has the advantage that addition and subtraction are actually one operation. It is important to highlight here that unfortunately, there is still a dual representation for 0. With 3-bit numbers, for example, 000 is positive zero and 111 is negative zero.

2.5.2 Radix Complement

In the decimal system, the radix complement is the 10's complement. The 10's complement of a number is the difference between 10 and the number. For example, 10's complement of 5, is 10 – 5 = 5

In binary number system, the radix complement is called the 2's complement. The 2's complement representation of a positive number is the same as in sign-magnitude form. The 2's complement form of a negative number can be obtained by adding 1 to the 1's complement representation of the number.

Example: *Consider 2's complement representation of –43*

```
+43 = 0101011    (sign-magnitude form)
    = 1010100    (1's complement form)
           +1    (add 1)
      -------
      1010101    2's complement form
```

A distinct advantage of 2's complement form is that, unlike 1's complement form, there is a unique representation of 0. Moreover, addition and subtraction can be treated as the same operation as in 1's complement; however, the carry bit can be ignored and the result is always in correct 2's complement notation. Thus addition and subtraction are easier in 2's complement than in 1's complement.

2.6 BINARY CODES

As we have discussed, digital circuits use binary signals but are required to handle data which may be alphabetic, numeric, or special characters. Hence the signals that are available in some other form other than binary have to be converted into suitable binary form before they can be processed further by digital circuits. This means that in whatever format the information may be available it must be converted into binary format. To achieve this, a process of coding is required where each letter, special character, or numeral is coded in a unique combination of 0s and 1s using a coding scheme known as *code*.

In digital systems a variety of codes are used to serve different purposes, such as data entry, arithmetic operation, error detection and correction, etc. Selection of a particular code depends on the requirement. Even in a single digital system a number of different codes may be used for different operations and it may even be necessary to convert data from one type of code to another. For conversion of data, code converter circuits are required, which will be discussed in due time.

2.6.1 Binary Coded Decimal or 8421 Code

The Binary Coded Decimal (BCD) is a type of binary code used to represent a given decimal number in an equivalent binary form. BCD-to-decimal and decimal-to-BCD conversions are very easy and straightforward.

The BCD equivalent of a decimal number is written by replacing each decimal digit in the integer and fractional parts with its four-bit binary equivalent. As an example, the BCD equivalent of $(23.15)_{10}$ is written as $(0010\ 0011.0001\ 0101)_{BCD}$. The BCD code described above is more precisely known as the 8421 BCD code, with 8, 4, 2 and 1 representing the weights of different bits in the four-bit groups, starting from MSB and proceeding towards LSB. This feature makes it a weighted code, which means that each bit in the four-bit group representing a given decimal digit has an assigned weight. Other weighted BCD codes include the 4221 BCD and 5421 BCD codes. Again, 4, 2, 2 and 1 in the 4221 BCD code and 5, 4, 2 and 1 in the 5421 BCD code represent weights of the relevant bits. Since the 8421 code is the most popular of all the BCD codes, it is simply referred to as the BCD code.

2.6.1.1 BCD-to-Binary Conversion

A given BCD number can be converted into an equivalent binary number by first writing its decimal equivalent and then converting it into its binary equivalent. The first step is straightforward, and the second step was explained in the previous sections.

Example: *Find the binary equivalent of the BCD number 0010 1001.0111 0101*

- BCD number: 0010 1001.0111 0101.
- Corresponding decimal number: 29.75.
- The binary equivalent of 29.75 can be determined to be 11101 for the integer part and .11 for the fractional part.
- Therefore, $(0010\ 1001.0111\ 0101)_{BCD} = (11101.11)_2$.

Table 2.3 BCD Codes

Decimal	8421 BCD code	4221 BCD code	5421 BCD code
0	0000	0000	0000
1	0001	0001	0001
2	0010	0010	0010
3	0011	0011	0011
4	0100	1000	0100
5	0101	0111	1000
6	0110	1100	1001
7	0111	1101	1010
8	1000	1110	1011
9	1001	1111	1100

2.6.1.2 Binary-to-BCD Conversion

The process of binary-to-BCD conversion is the same as the process of BCD-to-binary conversion executed in reverse order. A given binary number can be converted into an equivalent BCD number by first determining its decimal equivalent and then writing the corresponding BCD equivalent.

Example: *Find the BCD equivalent of the binary number 10101011.101*

- The decimal equivalent of this binary number can be determined to be 171.625.
- The BCD equivalent can then be written as 0001 0111 0001.0110 0010 0101.

2.6.2 Excess-3 Codes

The excess-3 code is another important code. It is particularly significant for arithmetic operations as it overcomes the shortcomings encountered while using the 8421 BCD code to add two decimal digits whose sum exceeds 9. The excess-3 code has no such limitation, and it considerably simplifies arithmetic operations.

The excess-3 code for a given decimal number is determined by adding '3' to each decimal digit in the given number and then replacing each digit of the newly found decimal number by its four-bit binary equivalent. It may be mentioned here that, if the addition of '3' to a digit produces a carry, as is the case with the digits 7, 8 and 9, that carry should not be taken forward. The result of addition should be taken as a single entity and subsequently replaced with its excess-3 code equivalent.

Example: *Find the excess-3 code for the decimal number 597*

- The addition of '3' to each digit yields the three new digits/numbers '8', '12' and '10'.
- The corresponding four-bit binary equivalents are 1000, 1100 and 1010 respectively.
- The excess-3 code for 597 is therefore given by: 1000 1100 1010 = 100011001010.

One significant feature that makes this code attractive for performing arithmetic operations is that the complement of the excess-3 code of a given decimal number yields the excess-3 code for 9's complement of the decimal number. As adding 9's complement of a decimal number B to a decimal number A achieves A – B, the excess-3 code can be used effectively for both addition and subtraction of decimal numbers.

Table 2.4 Excess-3 Codes of Decimal Numbers

Decimal number	Excess-3 code	Decimal number	Excess-3 code
0	0011	5	1000
1	0100	6	1001
2	0101	7	1010
3	0110	8	1011
4	0111	9	1100

2.6.3 Gray Codes

The Gray code was designed by Frank Gray at Bell labs and patented in 1953. It is an unweighted binary code in which two successive values differ only by 1 bit. Owing to this feature, the maximum error that can creep into a system using the binary Gray code to encode data is much less than the worst-case error encountered in the case of straight binary encoding. Table 2.5 lists the binary and Gray code equivalents of decimal numbers 0–15.

An examination of the four-bit Gray code numbers, as listed in Table 2.5, shows that the last entry rolls over to the first entry. That is, the last and the first entry also differ by only 1 bit. This is known as the *cyclic property* of the Gray code. Although there can be more than one Gray code for a given word length, the term was first applied to a specific binary code for non-negative integers and called the *binary-reflected Gray code* or simply the Gray code.

There are various ways by which Gray codes with a given number of bits can be remembered. One such way is to remember that the least significant bit follows a repetitive pattern of '2' (11, 00, 11, ...), the next higher adjacent bit follows a pattern of '4' (1111, 0000, 1111, ...) and so on. We can also generate the n-bit Gray code recursively by prefixing a '0' to the Gray code for n–1 bits to obtain the first 2^{n-1} numbers, and then prefixing '1' to the reflected Gray code for n–1 bits to obtain the remaining 2^{n-1} numbers. The reflected Gray code is nothing but the code written in reverse order.

Table 2.5 Gray Codes

Decimal	Binary	Gray	Decimal	Binary	Gray
0	0000	0000	8	1000	1100
1	0001	0001	9	1001	1101
2	0010	0011	10	1010	1111
3	0011	0010	11	1011	1110
4	0100	0110	12	1100	1010
5	0101	0111	13	1101	1011
6	0110	0101	14	1110	1001
7	0111	0100	15	1111	1000

2.6.3.1 Binary-Gray Code Conversion

A given binary number can be converted into its Gray code equivalent by going through the following steps:

1. Begin with the Most Significant Bit (MSB) of the binary number. The MSB of the Gray code equivalent is the same as the MSB of the given binary number.
2. The second most significant bit, adjacent to the MSB, in the Gray code number is obtained by adding the MSB and the second MSB of the binary number and ignoring the carry, if any. That is, if the MSB and the bit adjacent to it are both '1', then the corresponding Gray code bit would be a '0'.
3. The third most significant bit, adjacent to the second MSB, in the Gray code number is obtained by adding the second MSB and the third MSB in the binary number and ignoring the carry, if any.
4. The process continues until we obtain the LSB of the Gray code number by the addition of the LSB and the next higher adjacent bit of the binary number.

Example: *Convert $(1011)_2$ into its Gray code equivalent*

Binary 1011
Gray code 1- - -
Binary 1011
Gray code 11- -
Binary 1011
Gray code 111-
Binary 1011
Gray code 1110

2.6.3.2 Gray Code-Binary Conversion

A given Gray code number can be converted into its binary equivalent by going through the following steps:

1. Begin with the Most Significant Bit (MSB). The MSB of the binary number is the same as the MSB of the Gray code number.
2. The bit next to the MSB (the second MSB) in the binary number is obtained by adding the MSB in the binary number to the second MSB in the Gray code number and disregarding the carry, if any.
3. The third MSB in the binary number is obtained by adding the second MSB in the binary number to the third MSB in the Gray code number. Again, carry, if any, is to be ignored.
4. The process continues until we obtain the LSB of the binary number.

Example: *Convert Gray code number 1110 into its binary equivalent*

Gray code 1110
Binary 1- - -
Gray code 1110
Binary 10- -
Gray code 1110
Binary 101-
Gray code 1110
Binary 1011

2.6.3.3 Applications of Gray Codes

1. The Gray code is used in the transmission of digital signals as it minimizes the occurrence of errors.
2. The Gray code is preferred over the straight binary code in angle-measuring devices.
3. The Gray code is used for labelling the axes of Karnaugh maps, a graphical technique used for minimization of Boolean expressions.
4. The use of Gray codes to address program memory in computers minimizes power consumption. This is due to fewer address lines changing state with advances in the program counter.
5. Gray codes are also very useful in genetic algorithms since mutations in the code allow for mostly incremental changes. However, occasionally a one-bit change can result in a big leap, thus leading to new properties.

2.7 BOOLEAN ALGEBRA

George Boole, in 1854, developed a system of mathematical logic, which we now call *Boolean algebra*. Based on Boole's idea, Claude Shannon, in 1938, showed that circuits built with binary switches can easily be described using Boolean algebra.

Boolean algebra may be defined for a set A (which could be finite or infinite) in terms of two binary operations + and, are called the inclusive OR and AND, respectively; they should not be confused with the addition and multiplication operations in conventional algebra. The operations in Boolean algebra are based on the following axioms or postulates, known as Huntington's postulates:

Postulate 1. If $x, y \in$ A, then

$$x + y \in \text{A}; \quad x \cdot y \in \text{A}$$

This is called the closure property.

Postulate 2. If $x, y \in$ A, then

$$x + y = y + x; \quad x \cdot y = y \cdot x$$

that is, + and · operations are commutative.

Postulate 3. If $x, y, z \in A$, then

$$x + (y \cdot z) = (x + y) \cdot (x + z)$$

$$x \cdot (y + z) = (x \cdot y) + (x \cdot z)$$

that is, + and · operations are distributive.

Postulate 4. Identity elements, denoted as 0 and 1, must exist such that $x + 0 = x$ and $x \cdot 1 = x$ for all elements of A.

Postulate 5. For every element x in A there exists an element $\overline{x}$, called the complement of x, such that

$$x + \overline{x} = 1 \quad x \cdot \overline{x} = 0$$

2.7.1 Boolean Theorems

Several theorems used for the manipulation of Boolean expressions are given below:

Theorem 1. The identity elements 0 and 1 are unique

Theorem 2. The Idempotent Laws

$$x + x = x$$

$$x \cdot x = x$$

Theorem 3.

$$x + 1 = x$$

$$x \cdot 0 = 0$$

Theorem 4. The Absorption Law

$$x + xy = x$$

$$x(x + y) = x$$

Theorem 5. Every element in set A has a unique complement

Theorem 6. The Involution Theorem

$$\overline{(\overline{x})} = x$$

Theorem 7. Demorgan' Theorem

$$\overline{(x + y)} = \overline{x} \cdot \overline{y}$$

$$\overline{xy} = \overline{x} + \overline{y}$$

2.7.2 Venn Diagrams

A Venn diagram is a useful illustration to visualize the relationship among the variables of a Boolean expression. The diagram consists of a rectangle as shown in Fig. 2.1, inside two overlapping circles are drawn, which represent two variables. Each circle is labelled by a variable. We consider that the area inside the circle belongs to the named variable and area outside circle does not belong to that variable. For example, if the variables are A and B, then A = 1 for inside the circle A, A = 0 for outside of circle A, and B = 1 for inside the circle B, B = 0 for outside of circle B. Now for two overlapping circles as shown in Fig. 2.1, four distinct areas are available inside the rectangle area belonging to A only or AB′, area belonging to B only or A′B, area belonging to both A and B *i.e.*, A.B and area belonging to neither A or B *i.e.*, A′B′.

Example: *Represent Distributive law with venn diagram*

Distributive law states that A(B + C) = AB + AC. In Fig. 2.2, three variables A, B and C are used. It is possible to demonstrate eight distinct areas available for three variables in a Venn diagram. For this particular example, the distributive law is explained by showing the area intersecting the circle A with area enclosed by B or C is the same area belonging to AB or AC.

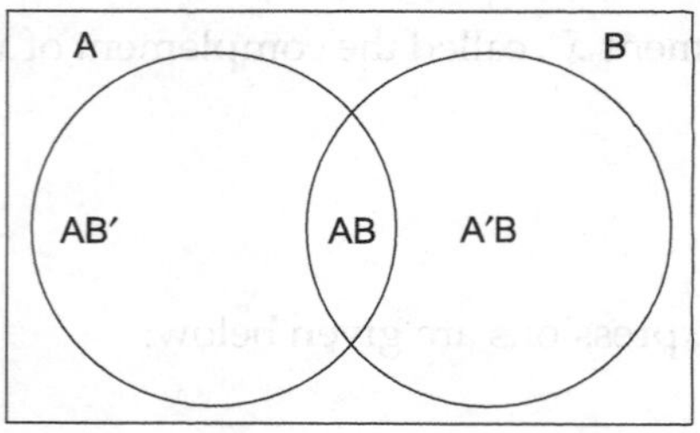

Fig. 2.1 Illustration of Venn Diagram

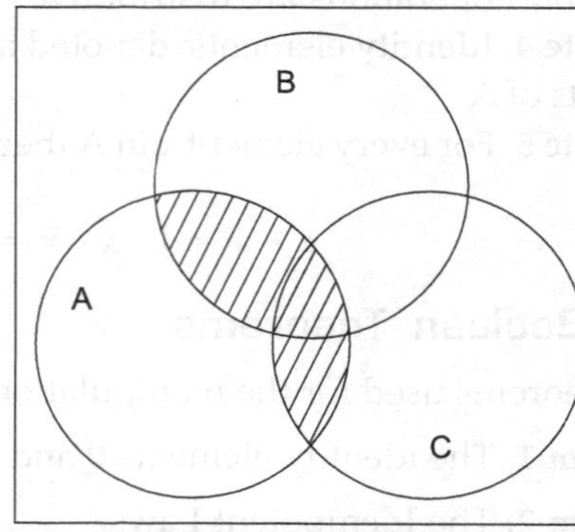

Fig. 2.2 Distributive Law

2.7.3 Simplification of Boolean Expressions

Minimization of the number of literals and the number of terms leads to less complex circuits as well as less number of gates, which should be a designer's aim. There are several methods to minimize the Boolean function. In this section, minimization of complex algebraic expressions will be shown with the help of postulates and theorems of Boolean algebra.

Example: *Simplify the Boolean function F = AB + BC + B′C*

$$\begin{aligned} F &= AB + BC + B'C \\ &= AB + C(B + B') \\ &= AB + C \end{aligned}$$

Example: *Simplify the Boolean function F = A′B′C + A′BC + AB′*

$$\begin{aligned} F &= A'B'C + A'BC + AB' \\ &= A'C\,(B' + B) + AB' \\ &= A'C + AB' \end{aligned}$$

Example: *Simplify the Boolean function F = A + A′B*

$$\begin{aligned} F &= A + A'B \\ &= (A + A')\,(A + B) \\ &= A + B \end{aligned}$$

Example: *Simplify the Boolean function F = XYZ + XY′Z + XYZ′.*

$$\begin{aligned} F &= XYZ + XY'Z + XYZ' \\ &= XZ\,(Y + Y') + XY\,(Z + Z') \\ &= XZ + XY \\ &= X\,(Y + Z) \end{aligned}$$

Example: *Simplify the Boolean function F = ((XY′ + XYZ)′ + X(Y + XY′))′.*

$$\begin{aligned} F &= ((XY' + XYZ) + X(Y + XY'))' \\ &= ((X(Y' + YZ))' + XY + XY')' \end{aligned}$$

$$= ((X(Y'Z+Y'+YZ))'+X(Y+Y'))' \qquad (Y'=Y'(Z+1)=Y'Z+Y')$$
$$= ((X(Y'+Z))'+X)'$$
$$= (X'+(Y'+Z)'+X)'$$
$$= (1+YZ')'$$
$$= 1'$$
$$= 0$$

2.8 LOGIC GATES

Logic gates are electronic circuits that can be used to implement the most elementary logic expressions, also known as Boolean expressions. The logic gate is the most basic building block of combinational logic. There are three basic logic gates, namely the OR gate, the AND gate and the NOT gate. Other logic gates that are derived from these basic gates are the NAND gate, the NOR gate, the EXCLUSIVE-OR gate and the EXCLUSIVE-NOR gate. This section deals with logic gates.

2.8.1 Positive and Negative Logic

As discussed, the binary variables, can have either of the two states, *i.e.*, the logic '0' state or the logic '1' state. These logic states in digital systems such as computers, for instance, are represented by two different voltage levels or two different current levels.

Positive Logic

If the more positive of the two voltage or current levels represents a logic '1' and the less positive of the two levels represents a logic '0', then the logic system is referred to as a *positive logic system*.

Negative Logic

If the more positive of the two voltage or current levels represents a logic '0' and the less positive of the two levels represents a logic '1', then the logic system is referred to as a *negative logic system*.

Example: *If the two voltage levels are 0 V and +5 V, then in the positive logic system the 0 V represents a logic '0' and the +5 V represents a logic '1'. In the negative logic system, 0 V represents a logic '1' and +5 V represents a logic '0'.*

2.8.2 Truth Table

The operation of the AND, OR and NOT logic operators can be formally described by using a truth table as shown in Fig. 2.3. A truth table is a two-dimensional array, where there is one column for each input and one column for each output (a circuit may have more than one output). Since we are dealing with binary values, each input can be either a 0 or a 1. We simply enumerate all possible combinations of 0's and 1's for all the inputs. Usually, we want to write these input values in the normal binary counting order. With two inputs, there are 2^2 combinations giving us the four rows in the table. The values in the output column are determined from applying the corresponding input values to the functional operator.

x	*y*	F
0	0	0
0	1	0
1	0	0
1	1	1

x	*y*	F
0	0	0
0	1	1
1	0	1
1	1	1

x	F
0	1
1	0

Fig. 2.3 Truth Tables of AND, OR, NOT Logical Operations

Using a truth table is one method to formally describe the operation of a circuit or function. The truth table for any given logic expression (no matter how complex it is) can always be derived.

2.8.3 OR Gate

An OR gate performs an ORing operation on two or more than two logic variables. The OR operation on two independent logic variables A and B is written as Y = A + B and reads as Y equals A OR B and not as A plus B. An OR gate is a logic circuit with two or more inputs and one output. The output of an OR gate is LOW only when all of its inputs are LOW. For all other possible input combinations, the output is HIGH.

Figure 2.4 shows the circuit symbol and the truth table of a two-input OR gate. The operation of a two-input OR gate is explained by the logic expression

$$Y = A + B$$

This is represented by switches in parallel, as shown in Fig. 2.5

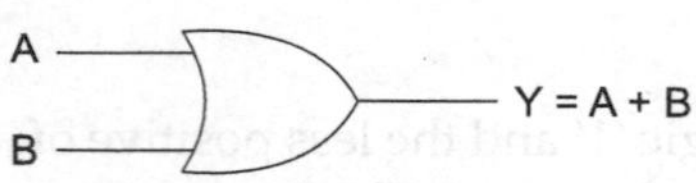

A	B	Y
0	0	0
0	1	1
1	0	1
1	1	1

Fig. 2.4 Two Input OR Gate

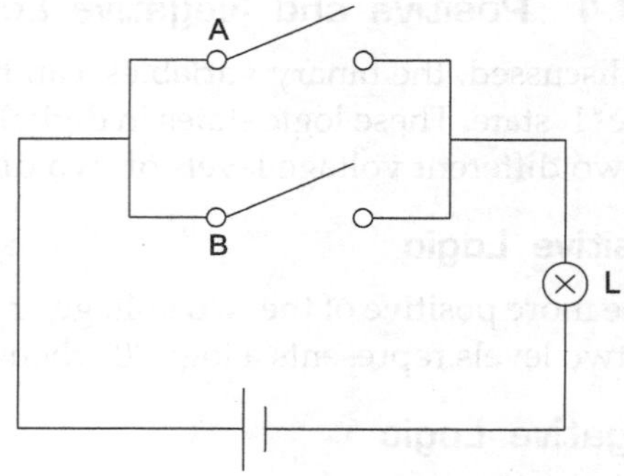

Fig. 2.5 OR Operation

Example: *Design and -implement a four-input OR gate using two-input OR gates only*

Figure 2.6 (*a*) shows one possible arrangement of two-input OR gates that simulates a four-input OR gate. A, B, C and D are logic inputs and Y3 is the output. Figure 2.6 (*b*) shows another possible arrangement. In the case of Fig. 2.6(*a*), the output of OR gate 1 is Y1 = (A + B). The second OR gate produces the output Y2 = (Y1 + C) = (A + B + C). Similarly, the output of OR gate 3 is Y3 = (Y2 + D) = (A + B + C + D). In the case of Fig. 2.6(*b*), the output of OR gate 1 is Y1 = (A + B). The second OR gate produces the output Y2 = (C + D). Output Y3 of the third OR gate is given by (Y1 + Y2) = (A + B + C + D).

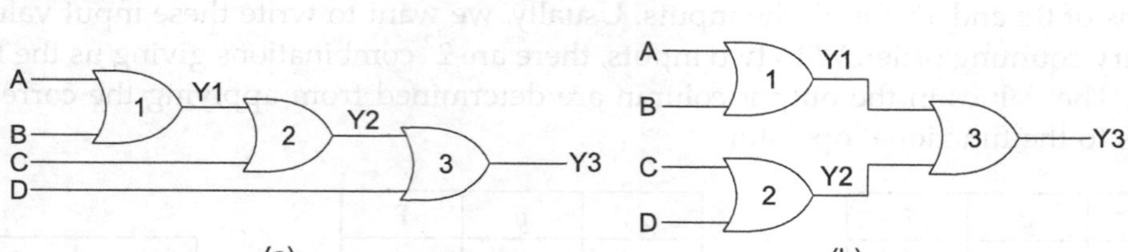

Fig. 2.6 Four Input OR Gate Using two Input or Gates Only

2.8.4 AND Gate

An AND gate is a logic circuit having two or more inputs and one output. The output of an AND gate is HIGH only when all of its inputs are in the HIGH state. In all other cases, the output is LOW.

In other words, output of the AND gate is a logic '1' only when all of its inputs are in logic '1' state. In all other cases, the output is logic '0'. The logic symbol and truth table of a two-input AND gate are shown in Figs. 2.7 (*a*) and (*b*) respectively.

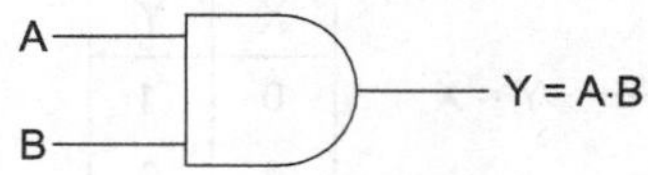

(a)

A	B	Y
0	0	0
0	1	0
1	0	0
1	1	1

(b)

Fig. 2.7 AND Gate

The circuit in Fig. 2.8 requires that both switches A and B, which are normally open, must be closed before the light (L) comes on. In terms of Boolean algebra, a closed switch may correspond to a 1 and an open switch to a 0. Similarly, the light OFF and ON conditions may be represented by 0 and 1, respectively. Thus, an AND gate is equivalent to two switches in series.

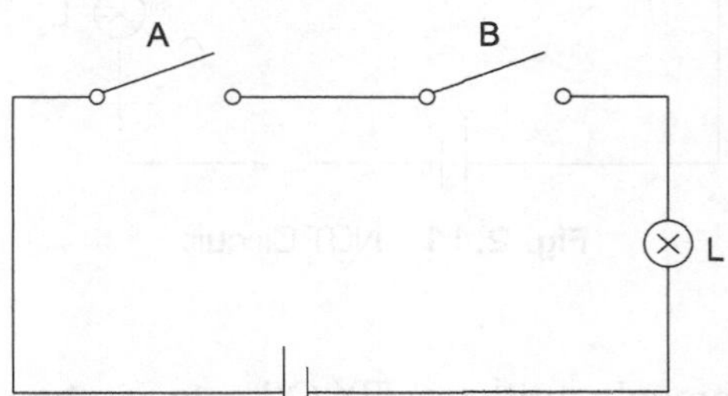

Fig. 2.8 AND Circuit

Example: *Implement a four input AND gate with two inputs AND gates only.*

The hardware implementation of a four-input AND gate using two-input AND gates is shown in Fig. 2.9. The output of AND gate 1 is Y1 = (A·B).The second AND gate produces an output Y2 given by Y2 = Y1·C = A·B·C. Similarly, the output of AND gate 3 is Y = Y2·D = A·B·C·D, which gives us the desired result.

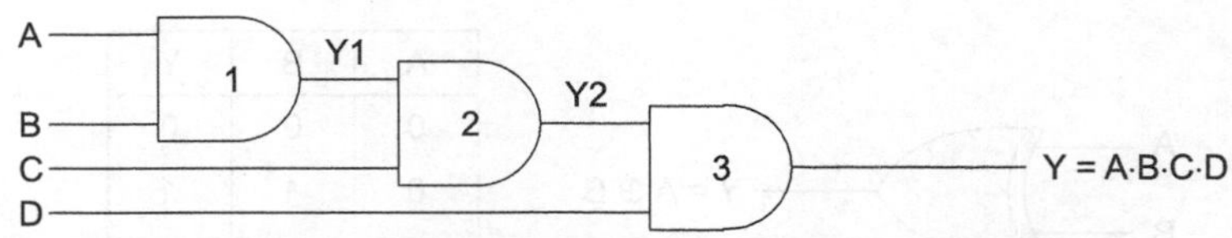

Fig. 2.9 Four Input AND Gate Using two Input AND Gates

2.8.5 NOT Gate

A NOT gate is a one-input, one-output logic circuit whose output is always the complement of the input. That is, a LOW input produces a HIGH output, and *vice-versa.* In other words, a logic '0' at the input produces a logic '1' at the output, and *vice-versa.* It is also known as a 'complementing circuit' or an 'inverting circuit'. Figure 2.10 shows the circuit symbol and the truth table.

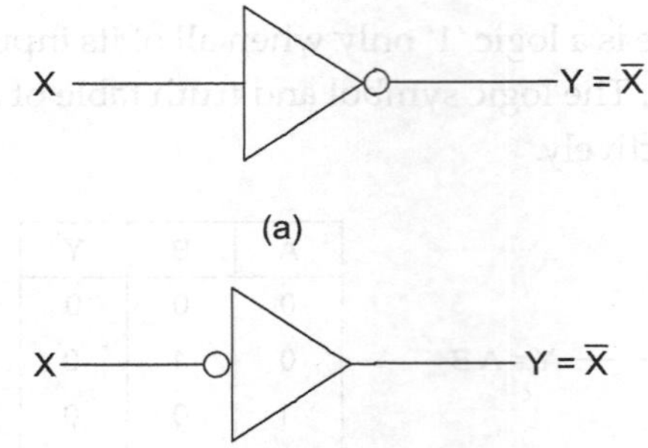

(a)

(b)

X	Y
0	1
1	0

Fig. 2.10 NOT Gate

The circuit representation of NOT gate is shown in Fig. 2.11.

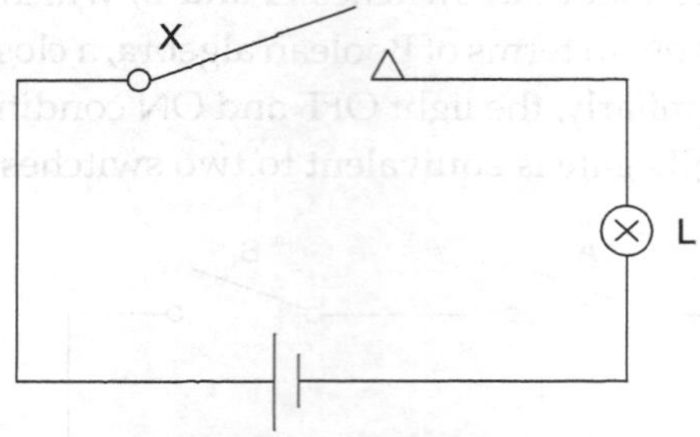

Fig. 2.11 NOT Circuit

2.8.6 EX-OR Gate

The EXCLUSIVE-OR gate, commonly written as EX-OR gate, is a two-input, one-output gate. The output of an EX-OR gate is a logic '1' when the inputs are unlike and a logic '0' when the inputs are like. Figures 2.12(*a*) and (*b*) respectively show the logic symbol and truth table of a two-input EX-OR gate. Although EX-OR gates are available in integrated circuit form only as two-input gates, unlike other gates which are available in multiple inputs also, multiple-input EX-OR logic functions can be implemented using more than one two-input gates.

The output of a two-input EX-OR gate is expressed by

$$Y = (A \oplus B)$$

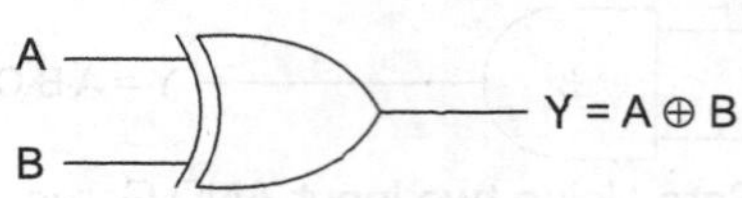

A	B	Y
0	0	0
0	1	1
1	0	1
1	1	0

(a) (b)

Fig. 2.12 EX-OR Gate

An EX-OR gate can be formed from a combination of AND, OR and NOT gates as shown in Fig. 2.13.

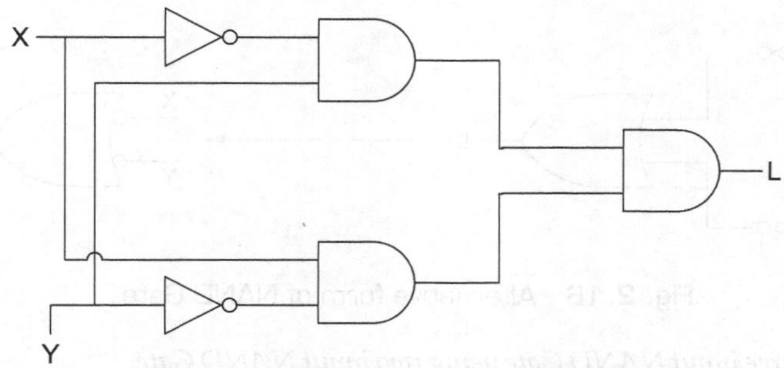

Fig. 2.13 Realization of EX-OR Gate

Example : *Design and implement a 3 input EX-OR gate using two inputs EX-OR gates only.*

Three input EX-OR gate using two inputs EX-OR gates can be realized as shown in Fig. 2.14. The output Y1 is given by A ⊕ B. The final output Y is given by Y = (Y1 ⊕ C) = (A ⊕ B) ⊕ C = A ⊕ B ⊕ C.

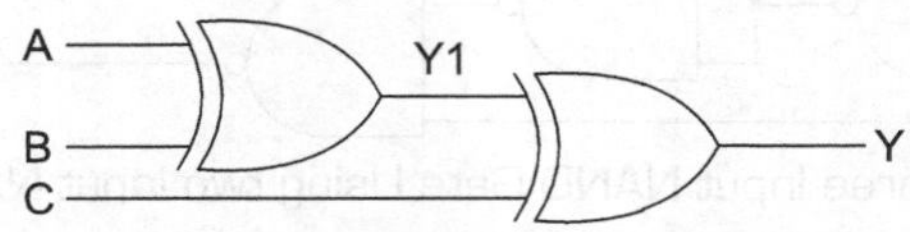

Fig. 2.14 Realization of 3 Input EX-OR Gate Using two Inputs EX-OR Gates

2.8.7 NAND Gate

A NAND gate is formed by combining a NOT gate with an AND gate (Fig. 2.15). The truth table for the NAND gate is shown in Table 2.6. It can be seen from the truth table that the output of the NAND gate will be 1 if at least one of the inputs is 0. The output is 0 if and only if the inputs are 1; that is,

$$L = \overline{X \cdot Y}$$

By Demorgan's Law

$$L = \bar{X} + \bar{Y}$$

Thus a NAND gate may also be formed by combining two NOT gates with an OR gate, as shown in Fig. 2.16.

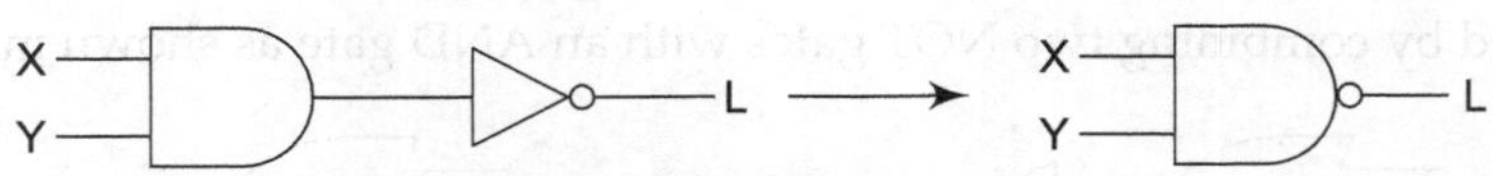

Fig. 2.15 NAND Gate

Table 2.6 Truth Table of NAND Gate

X	Y	L
0	0	1
0	1	1
1	0	1
1	1	0

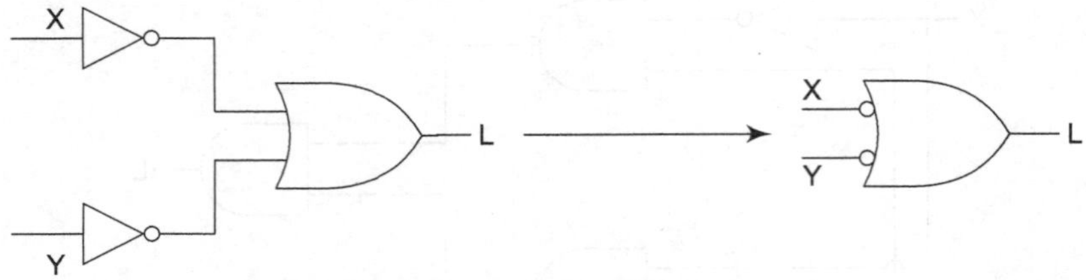

Fig. 2.16 Alternative form of NAND Gate

Example: *Design a three input NAND Gate using two input NAND Gate.*

The first step is to get a two-input AND from a two-input NAND. The output of the two-input AND gate and the third input then feed the inputs of another two-input NAND to get the desired output. The realized circuit is shown in Fig. 2.17.

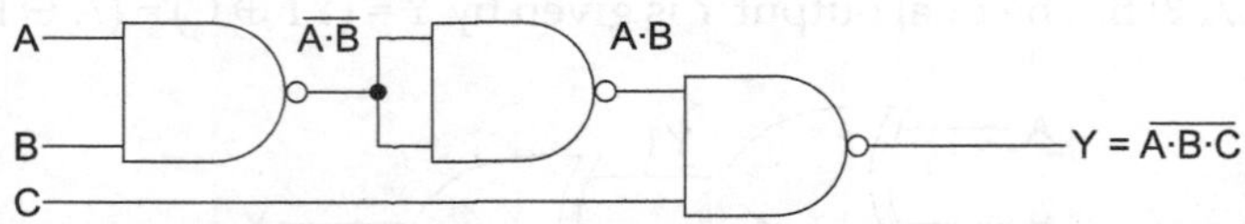

Fig. 2.17 Three Input NAND Gate Using two Input NAND Gates

Example: *Design a NOT gate using NAND Gate.*

A NOT gate can be obtained by shorting the inputs of NAND gate. This is because when all inputs to a NAND are at logic '0' level the output is a logic '1', and when all inputs to a NAND are at logic '1' level the output is a logic '0'. Figure 2.18 shows the implementation of design.

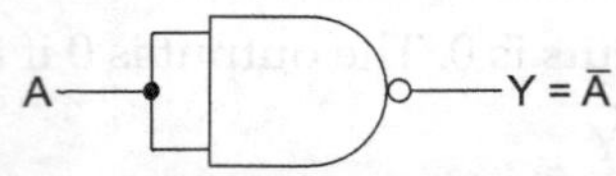

Fig. 2.18 NOT Gate Using NAND Gates

2.8.8 NOR Gate

A NOR gate is formed by combining a NOT gate with an OR gate (Fig. 2.19). The truth table for the NOR gate is shown in Table 2.7. It can be seen from the truth table that if a 1 appears at any input, the output will be 0. The output is 1 if and only if both inputs are 0 (*i.e.*, L = XY). Thus a NOR gate may also be formed by combining two NOT gates with an AND gate as shown in Fig. 2.20.

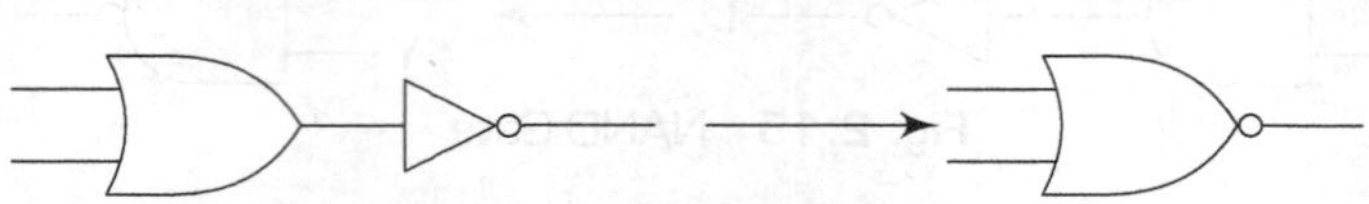

Fig. 2.19 NOR Gate

Table 2.7 Truth Table of NOR Gate

X	Y	L
0	0	1
0	1	1
1	0	0
1	1	0

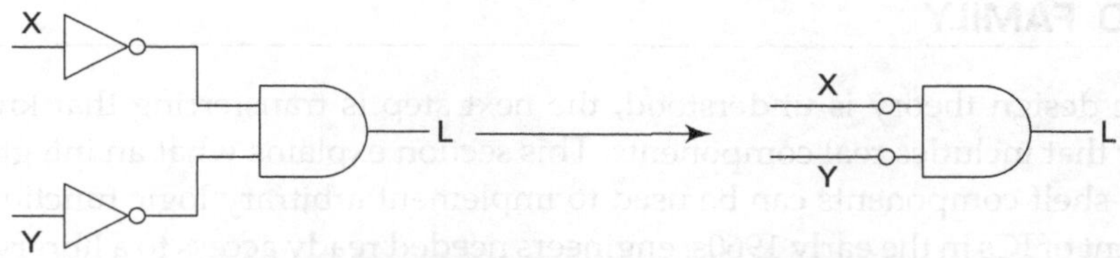

Fig. 2.20 Alternative form of NOR Gate

2.8.9 EX-NOR Gate

An EX-NOR gate can be formed by combining a NOT gate with an EX-OR gate (Fig. 2.21). The truth table for an EX-NOR gate is shown in Table 2.8. It can be seen from the truth table that the output of the EX-NOR gate is 1 only when both the inputs are either 0 or 1. For this reason an EX-NOR gate is also known as an equivalence or coincidence gate. It is represented by ⊙

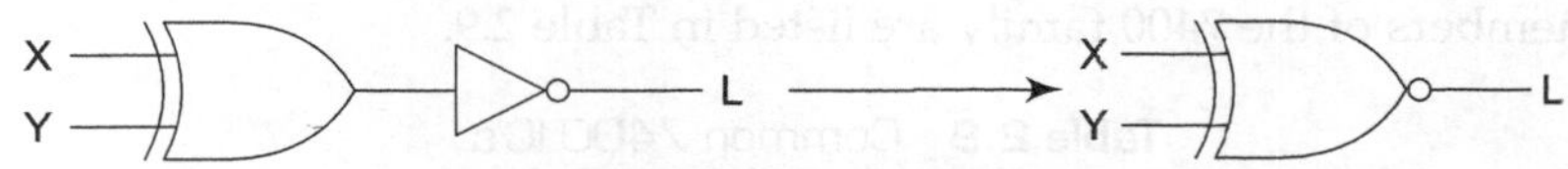

Fig. 2.21 EX-NOR Gate

Table 2.8 Truth Table of EX-NOR Gate

X	Y	L
0	0	1
0	1	0
1	0	0
1	1	1

Example : *Design a NOT gate using EX-NOR gates.*

It is evident from the truth table(Table 2.8) of a two-input EX-NOR gate that, if one of the inputs is permanently tied to a logic '0' level and the other input is treated as the input, then it behaves as a NOT circuit between input and output. Figure 2.22 shows the implementation of design.

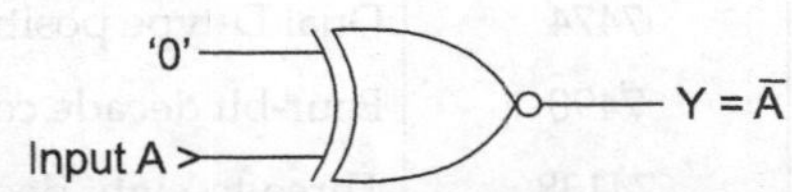

Fig. 2.22 NOT Gate Using EX-NOR Gate

2.8.10 Universal Gates

NOR and NAND gates have the property that they individually can be used to hardware-implement a logic circuit corresponding to any given Boolean expression. That is, it is possible to use either only NAND gates or only NOR gates to implement any Boolean expression. This is so because a combination of NAND gates or a combination of NOR gates can be used to perform functions of any of the basic logic gates. It is for this reason that NAND and NOR gates are universal gates. The realization of NOT gate using NAND and NOR gate have been already discussed in preceding examples.

2.8.11 Applications of Logic Gates

- An OR gate can be used in all those situations where the occurrence of anyone or more than one event needs to be detected and needs to be acted upon.
- An AND gate is commonly used as an ENABLE or INHIBIT gate to allow or disallow passage of data from one point in the circuit to another.
- EX-OR and EX-NOR logic gates are commonly used in parity generation and checking circuits.

2.9 IC 7400 FAMILY

Once basic logic design theory is understood, the next step is transferring that knowledge to a practical context that includes real components. This section explains what an integrated circuit is and how off-the-shelf components can be used to implement arbitrary logic functions.

With the advent of ICs in the early 1960s, engineers needed ready access to a library of basic logic gates so that these gates could be wired together on circuit boards and turned into useful products. Rather than having to design a custom microchip for each new project, semiconductor companies began to recognize a market for standard, off-the-shelf logic ICs. In 1963 and 1964, sylvania and texas instruments began shipment of the 7400-series discrete logic family and unknowingly started a *de facto* industry standard that lasts to this day and shows no signs of disappearing anytime soon.

Using the 7400 family, an engineer can select logic gates, flip-flops, counters, and buffers in individual packages and wire them together as desired to solve a specific problem. Some of the most common members of the 7400 family are listed in Table 2.9.

Table 2.9 Common 7400 IC's

Part Number	Function	Number of Pins
7400	Quad two-input NAND gates	14
7402	Quad two-input NOR gates	14
7404	Hex inverters	14
7408	Quad two-input AND gates	14
7432	Quad two-input OR gates	14
7447	BCD to seven-segment display decoder/driver	16
7474	Dual D-type positive edge triggered flip-flops	14
7490	Four-bit decade counter	14
74138	Three-to-eight decoder	16
74153	Dual 4-to-1 multiplexer	16
74157	Quad 2-to-1 multiplexers	16
74160	Four-bit binary synchronous counter	16
74164	Eight-bit parallel out serial shift registers	16
74174	Quad D-type flip-flops with complementary outputs	16
74193	Four-bit synchronous up/down binary counter	16
74245	Octal bus transceivers with tri-state outputs	20
74373	Octal D-type transparent latch	20
74374	Octal D-type flip-flops	20

These are just a few of the full set of 7400 family members. Many 7400 parts are no longer used, because their specific function is rarely required as a separate chip in modern digital electronics designs. However, the parts listed above, and many others that are not listed, are still readily available today and are commonly found in a broad range of digital designs ranging from low-end to high-tech devices. 7400-series logic has been available in DIPs for a long time, as well as (more recently) SOICs and other high-density surface mount packages. All flavors of basic logic gates are available with varying numbers of inputs. For example, there are 2-, 3- and 4-input AND gates and 2-, 3-, 4-, 8-, 12-, and 13-input NAND gates. There are numerous varieties of flip-flops, counters, multiplexers, shift registers, and bus transceivers. Flip-flops exist with and without complementary outputs, preset/ clear inputs, and independent clocks. Counters are available in 4-bit blocks that can both increment and decrement and count to either 15 (binary counter) or 9 (decade counter) before restarting the count at 0. Shift registers exist in all permutations of serial and parallel inputs and outputs. Bus transceivers in 4- and 8-bit increments exist with different types of output enables and capabilities to function in unidirectional or bidirectional modes. Bus transceivers enable the creation and expansion of tri-state buses on which multiple devices can communicate.

Familiarity with the 7400 series proves very useful no matter what type of digital system you are designing. For low-end systems, 7400-series logic may be the only type of IC at your disposal to solve a wide range of problems. At the high end, many people are often surprised to see a small 14- pin 7400-series IC soldered to a circuit board alongside a fancy 32-bit microprocessor running at 100 MHz. The fact is that the basic logic functions that the 7400 series offers are staples that have direct applications at all levels of digital systems design. It is time well spent to become familiar with the extensive capabilities of the simple yet powerful 7400 family. Manufacturers' logic data books, either in print or on line, are invaluable references.

SUMMARY

Understanding the concept of number systems is fundamental to the operation of PLC. Conversion from one number system to another and similarly conversion from one code to another is discussed at length in this chapters. In digital systems a variety of codes are used to serve different purposes, such as data entry, arithmetic operation, error detection and correction, etc. Selection of a particular code depends on the requirement. Even in a single digital system a number of different codes may be used for different operations and it may even be necessary to convert data from one type of code to another. Binary arithmetic, covering different methods of performing arithmetic operations on binary numbers is discussed next. The switching functions can be expressed with Boolean equations. Complex Boolean equations can be simplified by a new kind of algebra, which is popularly called Switching Algebra or Boolean Algebra, invented by the mathematician George Boole in 1854. Boolean Algebra deals with the rules by which logical operations are carried out.

REVIEW QUESTIONS

1. Convert the following decimal numbers to binary:
(*a*) 623 (*b*) 73.17 (*c*) 53.45 (*d*) 2.575

2. Convert the following binary numbers to decimal:
(*a*) 10110110 (*b*) 110000101 (*c*) 100.1101 (*d*) 1.001101

3. Convert the following binary numbers to hexadecimal and octal:
(*a*) 100101010011 (*b*) 001011101111 (*c*) 1011.111010101101 (*d*) 1111.100000011110

4. Convert the following octal numbers to binary and hexadecimal.
(*a*) 1026 (*b*) 1023

5. Perform the BCD addition of the following numbers:
(*a*) 234 + 146, (*b*) 67 + 39, and (*c*) 9234 + 4542

6. How do you distinguish between positive and negative logic systems? Prove that an OR gate in a positive logic system is an AND gate in a negative logic system.
7. Give brief statements that would help one remember the truth table of AND, NAND, OR, NOR, EX-OR and EX-NOR logic gate functions, irrespective of the number of inputs used.
8. Why are NAND and NOR gates called universal gates? Justify your answer with the help of examples.
9. Simplify the following expressions using the postulates and theorems of Boolean algebra.
   (*a*) $ab + \bar{a}c + \bar{a}b\bar{c}$
   (*b*) $(a + \bar{c}) + abc + ac\bar{d} + cd$
   (*c*) $\overline{(\bar{a}(b + \bar{c}))}(a + \bar{b} + \bar{c})\overline{(abc)}$
   (*d*) $(a + b)\,(a + c)b$
   (*e*) $\bar{a}\bar{b}\,(ac + \bar{b}) + (a + b)(a\bar{b}\bar{c} + \bar{a}bc)$
10. Derive the truth tables for the following gates:
   (*a*) 3-input NAND
   (*b*) 3-input NOR

3

INTRODUCTION TO LADDER LOGIC

LEARNING OBJECTIVES

After reading this chapter, you will be able to explain:

1. Basic Control system components and their symbols
2. Ladder logic input, output and programming concepts
3. Basic and advanced ladder logic functions
4. Relay logic and Boolean logic concepts

3.1 INTRODUCTION

Machine control design is a unique area of engineering that requires the knowledge of certain specific and unique diagramming techniques called ladder diagramming. Although there are similarities between control diagrams and electronic diagrams, many of the component symbols and layout formats are different. This chapter provides a study of the fundamentals of developing, drawing and understanding ladder diagrams. We will begin with a description of some of the fundamental components used in ladder diagrams. The basic symbols will then be used in a study of Boolean logic as applied to relay diagrams. Ladder logic is the main programming method used for PLCs. Ladder logic has been developed to mimic relay logic.

3.2 BASIC COMPONENTS AND THEIR SYMBOLS

In this section we will study some of the fundamental components used in electrical machine controls and their ladder diagram symbols.

3.2.1 Control Transformers

For safety reasons, machine controls are low voltage components. In order to make large powerful machines efficient and cost effective and reduce line current, most are powered by high voltages (240 V AC, 480 V AC, or more). This means that owing to safety reasons the line voltage must be reduced to less value for the controls. This is done using a **control transformer**. Figure 3.1 shows the electrical diagram symbol for a control transformer. The most obvious peculiarity here is that the symbol is rotated 90° with the primaries on top and secondary on the bottom. This is done to make it easier to draw the remainder of the ladder diagram. Notice that the transformer has two primary windings. These are usually each rated at 240 V AC. By connecting them in parallel, we obtain a 240 V AC primary, and by connecting them in series, we have a 480 V AC primary. The secondary windings are generally rated at 120 V AC, 48 V AC or 24 V AC. By offering control transformers with dual primaries, transformer manufacturers can reduce the number of transformer types in their product line, make their transformers more versatile, and make them less expensive.

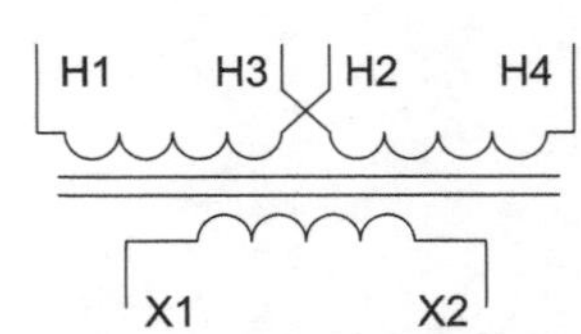

Fig. 3.1 Control Transformer

3.2.2 Fuses

Control circuits are always fuse protected. This prevents damage to the control transformer in the event of a short in the control circuitry. The electrical symbol for a fuse is shown in Fig. 3.2. The fuse used in control circuits is generally a slo-blow fuse and must be rated at a current that is less than or equal to the rated secondary current of the control transformer, and it must be connected in series with the transformer secondary. Most control transformers can be purchased with a fuse block (fuse holder) for the secondary fuse mounted on the transformer.

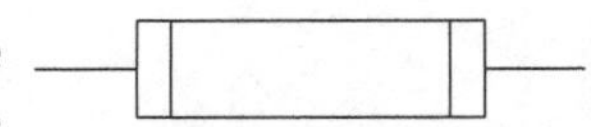
Fig. 3.2 Fuse

3.2.3 Switches

Switch in an important component of machine control. There are two important uses of switch. First, switches are used for operator input to send instructions to the control circuit. Second, switches may be installed on the moving parts of a machine to provide automatic feedback to the control system. Some of important and/or commonly used variants are discussed below.

Pushbutton

The most common switch is the pushbutton. It is widely used in automotive and electronic equipment applications. There are two types of pushbutton, the momentary and maintained. The **momentary** pushbutton switch is activated when the button is pressed, and deactivated when the button is released. The deactivation is done using an internal spring. The **maintained** pushbutton activates when pressed, but remains activated when it is released. Then to deactivate it, it must be pressed a second time. For this reason, it is also called a push-push switch.

The contacts on switches can be of two types. These are normally open (N/O) and normally closed (N/C). Whenever a switch is in its deactivated position, the N/O contacts will be open (non-conducting) and the N/C contacts will be closed (conducting).

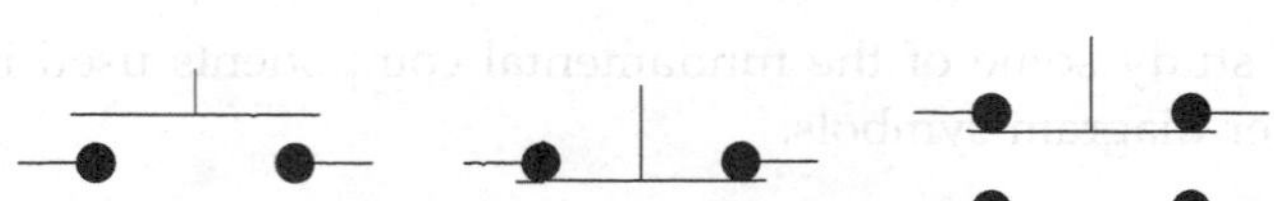
Fig. 3.3 Momentary Push Button Switches

Figure 3.4 illustrates the schematic symbol for the maintained pushbutton. One can interpret that it is the symbol for the momentary pushbutton with a "see-saw" mechanism added to hold in the switch actuator until it is pressed a second time. The maintained switch can have as many contacts of either type as desired.

Selector Switches

A selector switch is also known as a rotary switch. Selector switches use the same symbol as a momentary pushbutton, except a lever is added to the top of the actuator, as shown in Fig. 3.5.

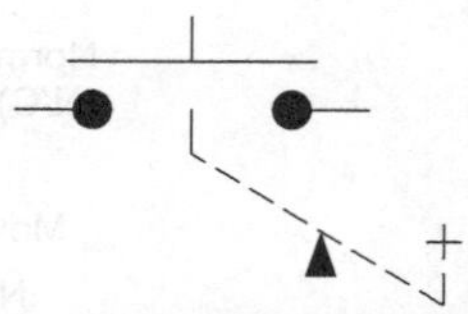

Fig. 3.4 Maintained Switch

The switch on the left is open when the selector is turned to the left and closed when turned to the right. The switch on the right side has two sets of contacts. The top contacts are closed when the switch selector is turned to the left position and open when the selector is turned to the right. The bottom set of contacts work exactly opposite. There is no electrical connection between the top and bottom pairs of contacts.

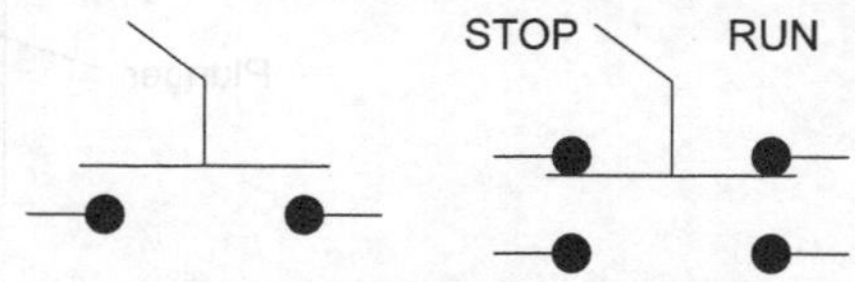

Fig. 3.5 Selector Switch

Limit Switches

Limit switches are usually not operator accessible. Instead they are activated by moving parts on the machine. They are usually mechanical switches, but can also be light activated or magnetically activated. They are sometimes called cam switches because many are operated by a camming action when a moving part passes by the switch. The symbols for both types of limit switches are shown in Fig. 3.6.

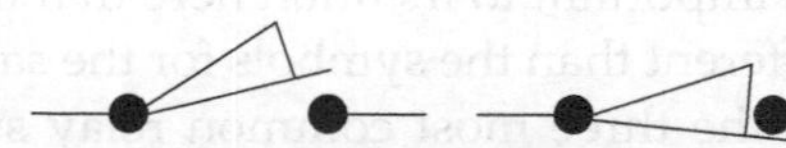

Fig. 3.6 Limit Switch

3.2.4 Indicator Lamps

All control panels include indicator lamps. They are used to indicate the present operating status of the machine. Indicators are drawn as a circle with "light rays" extending on the diagonals as shown in Fig. 3.7.

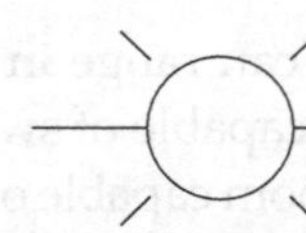

Fig. 3.7 Indicator Lamp

Commonly used colours are usually red, green, or amber, but other colours are also available. Red lamps are reserved for safety critical indicators. Green usually indicates safe conditions (power to the motor is OFF, brakes are ON, etc.). Amber indicates conditions that are important but not dangerous.

3.2.5 Relays

Early electrical control systems were composed of mainly relays and switches. A relay is an electromagnetic device composed of a core with an electromagnet coil and movable or fixed contacts. The movable contacts (and conductor that connects them) are mounted via an insulator to a plunger which moves within a bobbin. A coil of copper wire is wound on the bobbin to create an electromagnet. A spring holds the plunger up and away from the electromagnet. When the electromagnet is energized by passing an electric current through the coil, the magnetic field pulls the plunger into the core, which pulls the movable contacts downward. Two fixed pairs of contacts are mounted to the relay frame on electrical insulators so that when the movable contacts are not being pulled toward the core (the coil is de-energized) they physically touch the upper fixed pair of contacts and, when being pulled toward the coil, touches the lower pair of fixed contacts. There can be several sets of contacts mounted to the relay frame. The contacts energize and de-energize as a result of applying power to the relay coil.

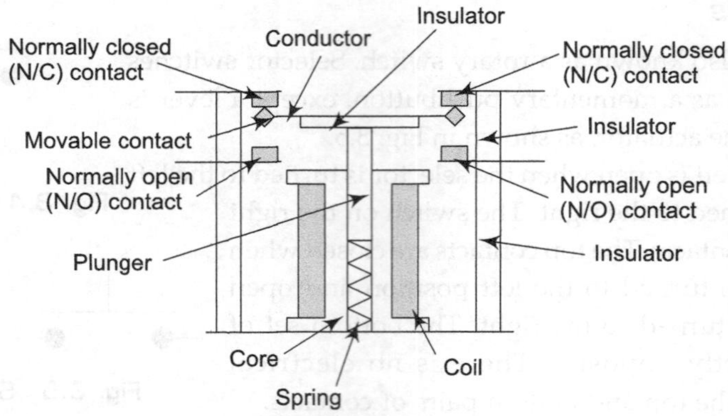

Fig. 3.8 Relay

It is important to mention here that many of the schematic symbols used in electrical diagrams are different than the symbols for the same types of components in electronic diagrams. Figure 3.9 shows the three most common relay symbols used in electrical machine diagrams. These three symbols are a normally open contact, normally closed contact and coil.

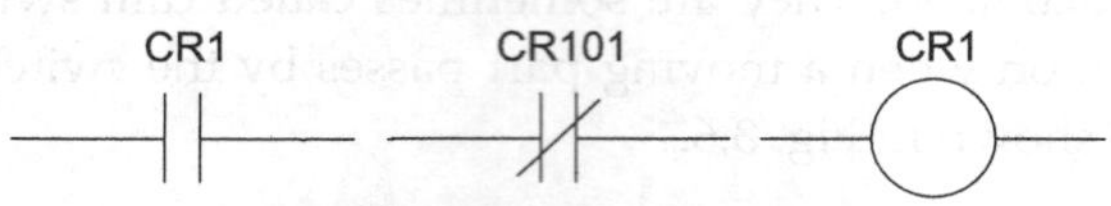

Fig. 3.9 Relay Symbols

Relays can range in size from extremely small reed relays in 14 pin DIP integrated circuit-style packages capable of switching a few tenths of an ampere at less than 100 volts to large contactors the size of a room capable of switching thousands of amperes at thousands of volts. However, for electrical machine diagrams, the schematic symbol for a relay is the same regardless of the relay's size.

Time Delay Relays

It is possible to construct a relay with a built-in time delay device that causes the relay to either switch on after a time delay, or to switch off after a time delay. These types of relays are called **time delay relays,** or TDR's. The schematic symbols for a TDR coil and contacts are the same as for a conventional relay, except that the coil symbol has the letters "TDR" or "TR" written inside, or next to the coil symbol.

Delay-On Timer (TON) Relay

When an on-timer is installed in a circuit, the user adjusts the control on the relay for the desired time delay. This time setting is called the **preset**. Figure 3.10 shows a timing diagram of a delay-on time delay relay. Notice on the top waveform that we are simply turning on power to the relay's coil and some undetermined time later, turning it off. When the coil is energized, the internal timer in the relay begins running (this can be either a motor driven mechanical timer or an electronic timer). When the time value contained in the timer reaches the preset value, the relay energizes. Delay-on relays are useful for delaying turn-on events.

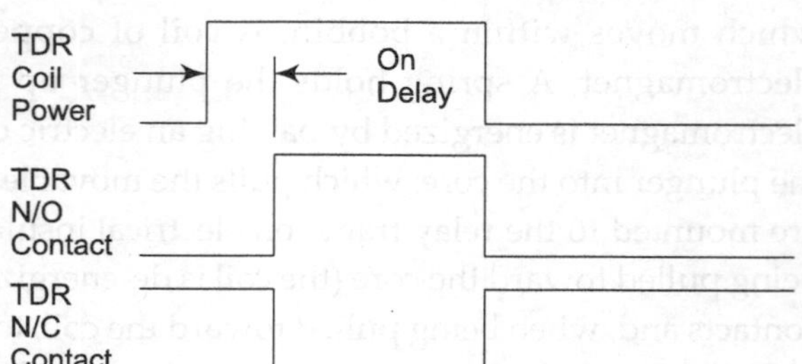

Fig. 3.10 Delay on Timer Relay

Delay-Off Timer (TOF) Relay

In this case, at the instant power is applied to the relay coil, the contacts activate – that is, the N/O contacts close, and the N/C contacts open. The time delay occurs when the relay is switched off. After power is removed from the relay coil, the contacts stay activated until the relay times-out. Delay-off time delay relays are excellent for applications requiring time to be stretched.

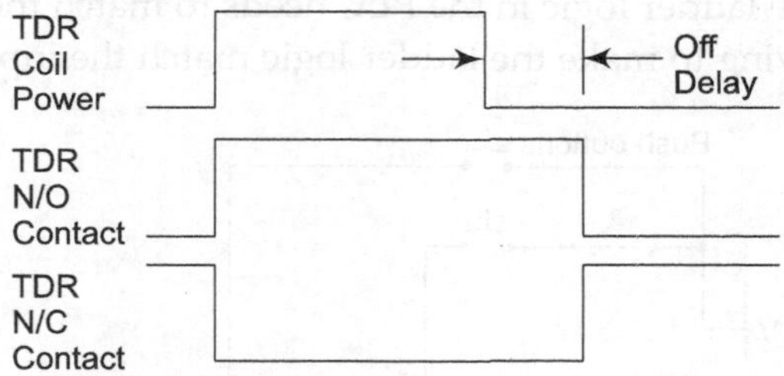

Fig. 3.11 Delay off Timer Relay

3.3 FUNDAMENTALS OF LADDER DIAGRAMS

Ladder logic is the main programming method used for PLCs. As mentioned before, ladder logic has been developed to mimic relay logic. The decision to use the relay logic diagrams was a strategic one. By selecting ladder logic as the main programming method, the amount of retraining needed for engineers and trades people was greatly reduced.

All electrical machine diagrams are drawn using a standard format. This format is called the **ladder diagram**. Beginning with the control transformer, we add a protective fuse on the left side. As mentioned earlier, in many cases the fuse is part of the transformer itself. From the transformer/fuse combination, horizontal lines are drawn to both sides and then drawn vertically down the page as shown in Fig. 3.12. These vertical lines are called **power rails** or simply **rails or uprights**. The voltage difference between the two rails is equal to the transformer secondary voltage, so any component connected between the two rails will be powered. The right side of the control transformer secondary is grounded to the frame of the machine (earth ground). The reason for this is that, without this ground, should the transformer short internally from primary to secondary, it could apply potentially lethal line voltages to the controls. With the ground, an internal transformer short will cause a fuse to blow or circuit breaker to trip farther "upstream" on the line voltage side of the transformer which will shutdown power to the controls. The wires are numbered. In our diagram, the left rail is wire number 2 and the right rail is wire number 1.

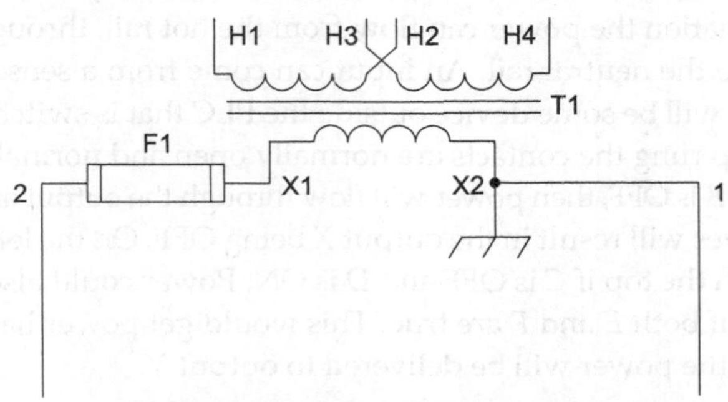

Fig. 3.12 Basic Control Circuit

Figure 3.13 shows a more complete representation of the PLC. Here there are two inputs from push buttons. We can imagine the inputs as activating 24 V DC relay coils in the PLC. This in turn drives an output relay that switches 115 V AC, that will turn on a light. Note, in actual PLCs inputs are never relays, but outputs are often relays. The ladder logic in the PLC is actually a computer program that the user can enter and change. Notice that both of the input push buttons are normally open, but the ladder logic inside the PLC has one normally open contact, and one normally closed contact. Do not think that the ladder logic in the PLC needs to match the inputs or outputs. Many beginners will get caught trying to make the ladder logic match the input types.

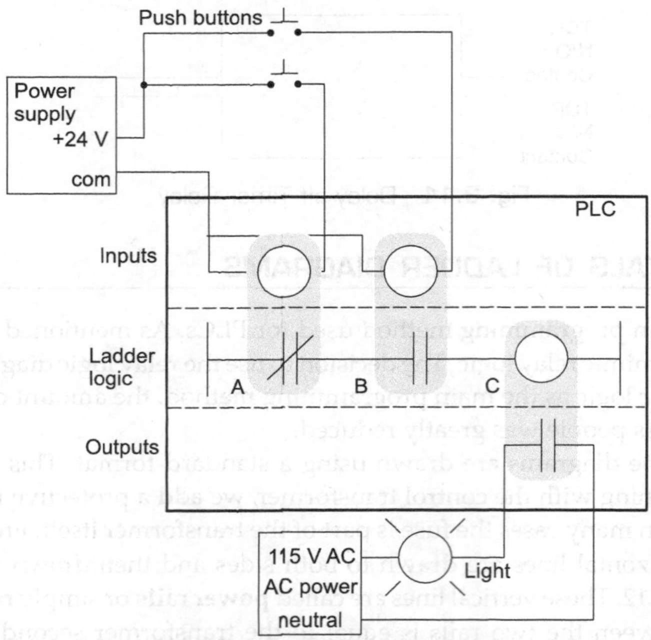

Fig. 3.13 A PLC Illustrated with Relays

3.3.1 Programming Concepts

Earlier programming methods involve programming PLCs with a technique that was based on relay logic wiring schematics. An example of ladder logic can be seen in Fig. 3.14. To interpret this diagram imagine that the power is on the vertical line on the left hand side, we call this the hot rail. On the right hand side is the neutral rail. In the figure there are two rungs, and on each rung there are combinations of inputs (two vertical lines) and outputs (circles). If the inputs are opened or closed in the right combination the power can flow from the hot rail, through the inputs, to power the outputs, and finally to the neutral rail. An input can come from a sensor, switch, or any other type of sensor. An output will be some device outside the PLC that is switched ON or OFF, such as lights or motors. In the top rung the contacts are normally open and normally closed which means if input *A* is ON and input *B* is OFF, then power will flow through the output and activate it. Any other combination of input values will result in the output *X* being OFF. On the left most part of the rung, power could flow through the top if *C* is OFF and *D* is ON. Power could also (and simultaneously) flow through the bottom if both *E* and *F* are true. This would get power half way across the rung, and then if *G* or *H* is true the power will be delivered to output *Y*.

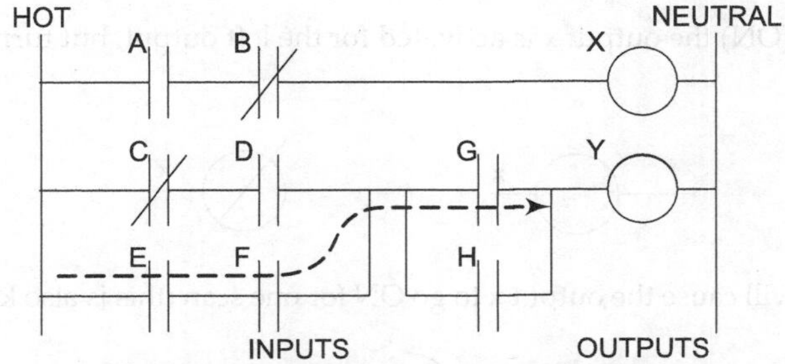

Fig. 3.14 A Simple Ladder Logic Diagram

There are other methods for programming PLCs. One of the earliest techniques involved mnemonic instructions. These instructions can be derived directly from the ladder logic diagrams and entered into the PLC through a simple programming terminal. Structured text programming has been developed as a more modern programming language. It is quite similar to languages such as BASIC.

3.3.2 Ladder Logic Inputs

PLC inputs are easily represented in ladder logic. In Fig. 3.15 there are three types of inputs shown. The first two are normally open and normally closed inputs. The *IIT* (Immediate Input) function allows inputs to be read after the input scan, while the ladder logic is being scanned. This allows ladder logic to examine input values more often than once every cycle.

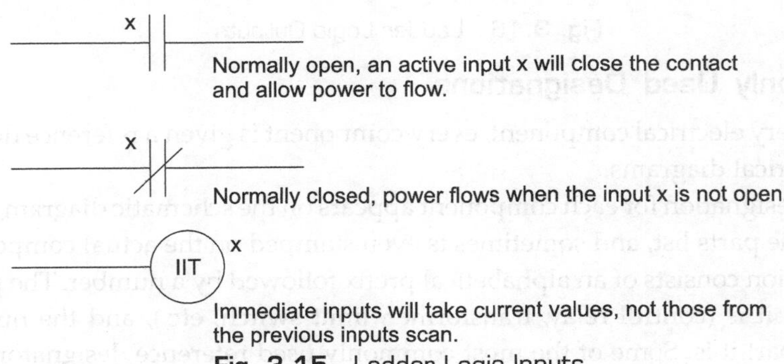

Fig. 3.15 Ladder Logic Inputs

3.3.3 Ladder Logic Outputs

In ladder logic there are multiple types of outputs, but these are not consistently available on all PLCs. Some of the outputs will be externally connected to devices outside the PLC, but it is also possible to use internal memory locations in the PLC. Six types of outputs are shown in Fig. 3.16 The first is a normal output, when energized the output will turn on, and energize an output. The circle with a diagonal line through it is a normally ON output. When energized the output will turn off. This type of output is not available on all PLC types. When initially energized the *OSR* (One Shot Relay) instruction will turn on for one scan, but then be OFF for all scans after, until it is turned off. The *L* (latch) and *U* (unlatch) instructions can be used to lock outputs ON. When an *L* output is energized the output will turn on indefinitely, even when the output coil is de energized. The output can only be turned off using a *U* output. The last instruction is the *IOT* (Immediate Output) that will allow outputs to be updated without having to wait for the ladder logic scan to be completed.

When power is applied (ON) the output x is activated for the left output, but turned off for the output ON the right.

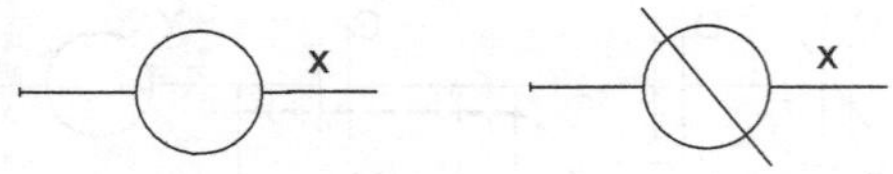

An input transition ON will cause the output x to go ON for one scan (this is also known as a one shot relay)

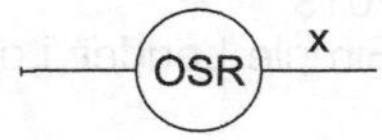

When the L coil energized, *x* will be toggled ON, it will stay ON until the U coil is energized. This is like a flip-flop and stays set even when the PCL is turned off.

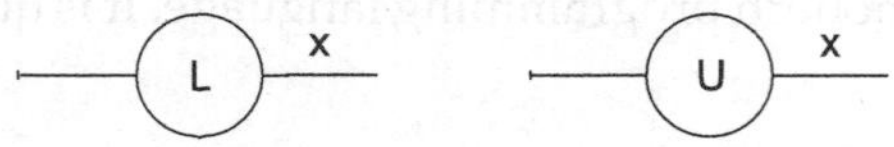

Some PLC's will allow immediate outputs that do not wait for the program scan to end before setting an output. (**Note:** This instruction will only update the outputs using the output table, other instruction must change the individual outputs.)

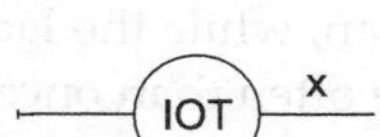

Fig. 3.16 Ladder Logic Outputs

3.3.4 Commonly Used Designations

To easily locate every electrical component, every component is given a reference designator that is to be used in electrical diagrams.

The reference designation for each component appears on the schematic diagram, the mechanical layout diagram, the parts list, and sometimes is even stamped on the actual component itself. The reference designation consists of an alphabetical prefix followed by a number. The prefix identifies what kind of part it is (control relay, transformer, limit switch, etc.), and the number indicates which particular part it is. Some of the most commonly used reference designator prefixes are as follows:

T–Transformer
CR–Control Relay
R–Resistor
C–Capacitor
LS–Limit Switch
PB–Pushbutton
S–Switch
SS–Selector Switch
TDR or TR–Time Delay Relay
M–Motor
L–Indicator Lamp or Line Phase

F–Fuse

CB–Circuit Breaker

OL–Overload Switch or Overload Contact

Some automation equipment and machine tool manufacturers use a reversed component numbering scheme that starts with the number and ends with the alphabetical designation.

3.3.5 Machine Control Terminology

Machine control systems make use of some of words that have special meanings. For safety and standardization purposes, the use of these words is explicit and can have no other meaning. They are generally used when naming control circuits, labelling switch positions on control panels, and describing modes of operation of the machine. A list of some of the most important terms is given below.

1. **ON:** This is a machine state in which power is applied to the machine and to the machine control circuits. The machine is ready to **RUN**. This is also sometimes call the **STANDBY** state.
2. **OFF:** Electrically, the opposite of **ON**. Power is removed from the machine and the machine control circuits.
3. **RUN:** A state in which the machine is cycling or performing the task for which it is designed. This state can only be started by pressing **RUN** switches.
4. **STOP:** The state in which the machine is **ON** but not **RUNNING**. If the machine is **RUNNING**, pressing the **STOP** switch will cause **RUN**NING to cease.
5. **JOG:** A condition in which the machine can be "nudged" a small amount to allow for the accurate positioning of raw material while the operator is holding the material. The machine controls must be designed so that the machine cannot automatically go from the **JOG** condition to the **RUN** condition while the operator is holding the raw material.
6. **CYCLE:** A mode of operation in which the machine **RUNS** for one complete operation and then automatically **STOPS**. Holding down the **CYCLE** button will not cause the machine to **run** more than one cycle. In order to have the machine execute another **CYCLE**, the **CYCLE** button must be released and pressed again. This mode is sometimes called **SINGLE CYCLE**.

3.4 LADDER LOGIC FUNCTIONS

Ladder logic input contacts and output coils allow simple logical decisions. Functions extend basic ladder logic to allow other types of control. For example, the addition of timers and counters allowed event based control. This section will discuss data handling and numerical logic.

3.4.1 Data Handling

1. **Move Functions :** There are two basic types of move functions;
   MOV(value,destination) – moves a value to a memory location
   MVM(value,mask,destination) – moves a value to a memory location, but with a mask to select specific bits.
   The simple MOV will take a value from one location in memory and place it in another memory location

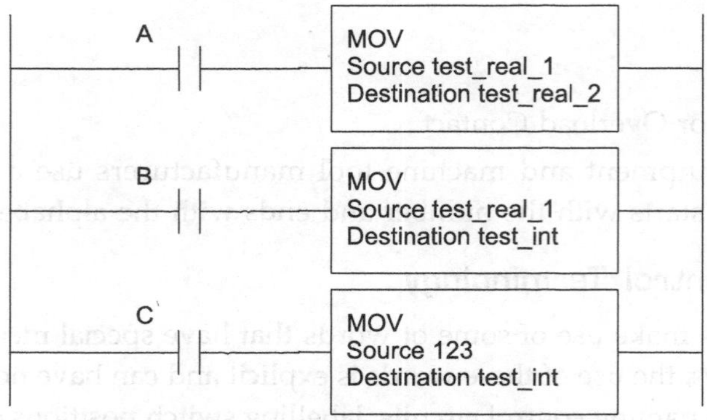

Fig. 3.17 Example of Move Function

If *A*, *B* and *C* are all true, then the value in *test_real_2* will change before the next instruction starts. This is different than the input and output scans that only happen before and after the logic scan.

2. **Mathematical Functions :** Mathematical functions will retrieve one or more values, perform an operation and store the result in memory. A list of arithmetic functions is given below.
   ADD (value,value,destination) – add two values
   SUB (value,value,destination) – subtract
   MUL (value,value,destination) – multiply
   DIV (value,value,destination) – divide
   NEG (value,destination) – reverse sign from positive/negative
   CLR (value) – clear the memory location
   A list of more advanced functions includes basic trigonometry functions, exponents, logarithms and a square root function. The last function *CPT* will accept an expression and perform a complex calculation.
   ACS (value, destination) – inverse cosine
   COS (value, destination) – cosine
   ASN (value, destination) – inverse sine
   SIN (value, destination) – sine
   ATN (value, destination) – inverse tangent
   TAN (value, destination) – tangent
   XPY (value, value, destination) – X to the power of Y
   LN (value, destination) – natural log
   LOG (value, destination) – base 10 log
   SQR (value, destination) – square root
   CPT (destination, expression) – does a calculation
   Figure 3.18 illustrates how an equation can be implemented using CPT function.

$$A = \sqrt{\ln B + e^{C} \operatorname{acos}(D)}$$

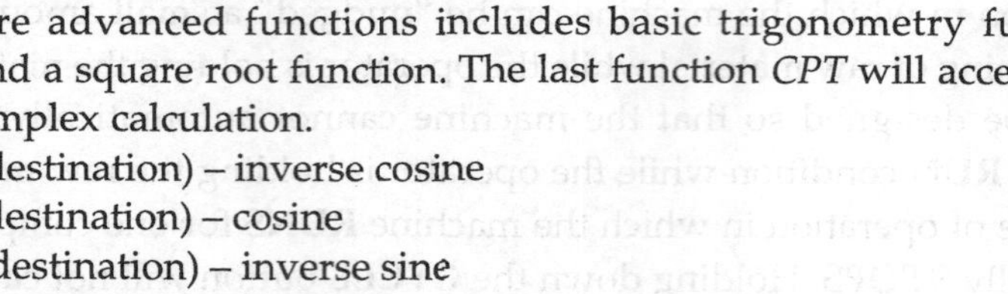

Fig. 3.18 Illustration of Use of Compute Function

3. **Array Data Functions :** Arrays allow us to store multiple data values. In a PLC this will be a sequential series of numbers in integer, floating point, or other memory.
Functions are available that allow statistical calculations. These functions are listed in Fig. 3.19. When *A* becomes true the average (AVE) conversion will start at memory location *weight[0]* and average a total of *4* values. The control word *weight_control* is used to keep track of the progress of the operation, and to determine when the operation is complete. These operations, and the others, are edge triggered. The operation may require multiple scans to be completed. When the operation is done the average will be stored in *weight_avg* and the *weight_control.DN* bit will be turned on.

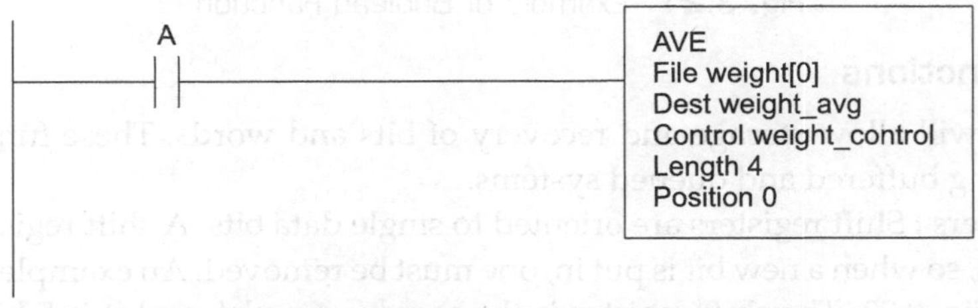

AVE(start value, destination, control, length) – average of values
STD(start value, destination, control, length) – standard deviation of values
SRT(start value, control, length) – sort a list of values

Fig. 3.19 Statistic Functions

Other list of array functions includes block operation functions. Some of these are listed below.
COP (start value,destination,length) – copies a block of values
FAL(control,length,mode,destination,expression) – will perform basic math operations to multiple values.
FSC (control, length, mode, expression) – will do a comparison to multiple values
FLL (value, destination, length) – copies a single value to a block of memory

3.4.2 Logical Functions

1. **Comparison functions :** Comparison functions are shown in Fig. 3.20. Previous function blocks were outputs, these replace input contacts. The example shows an EQU (equal) function that compares two floating point numbers. If the numbers are equal, the output bit *light* is true, otherwise it is false. Other types of equality functions are also listed below.
EQU (value, value) – equal
NEQ (value, value) – not equal
LES (value, value) – less than
LEQ (value, value) – less than or equal
GRT (value, value) – greater than
GEQ (value, value) – greater than or equal
CMP (expression) – compares two values for equality
MEQ (value, mask, threshold) – compare for equality using a mask
LIM (low limit, value, high limit) – check for a value between limits

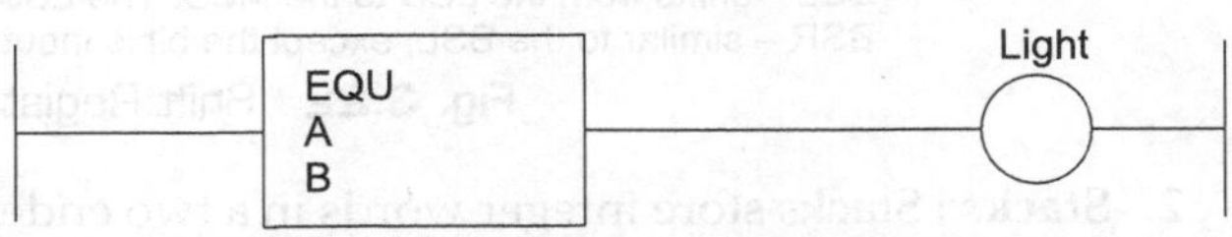

Fig. 3.20 Example of Comparison Function

2. **Boolean functions :** These functions obtain data words from bit memory, perform an boolean operation, and store the results in a new location in bit memory. These functions are all oriented to word level operations. The ability to perform Boolean operations allows logical operations on more than a single bit. A typical example of AND function is shown in Fig. 3.21. Some of commonly used Boolean functions are listed below.

AND (value, value, destination) – Binary and function
OR (value, value, destination) – Binary or function
NOT (value, value, destination) – Binary not function
XOR (value, value, destination) – Binary exclusive or function

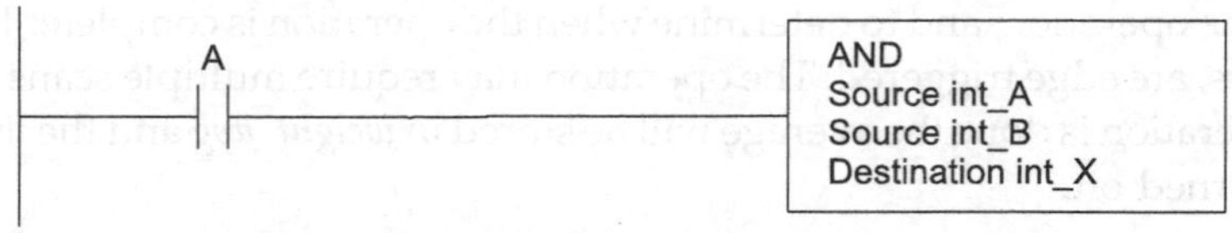

Fig. 3.21 Example of Boolean Function

3.4.3 List Functions

The list functions will allow storage and recovery of bits and words. These functions are useful when implementing buffered and queued systems.

1. **Shift Registers :** Shift registers are oriented to single data bits. A shift register can only hold so many bits, so when a new bit is put in, one must be removed. An example of a shift register is given in Fig. 3.22. The shift register is the word 'example', and it is 5 bits long. When *A* becomes true the bits all shift right to the least significant bit. When they shift a new bit is needed, and it is taken from *new_bit*. The bit that is shifted out, on the right hand side, is moved to the control word UL (unload) bit *c.UL* This function will not complete in a single ladder logic scan, so the control word *c* is used.

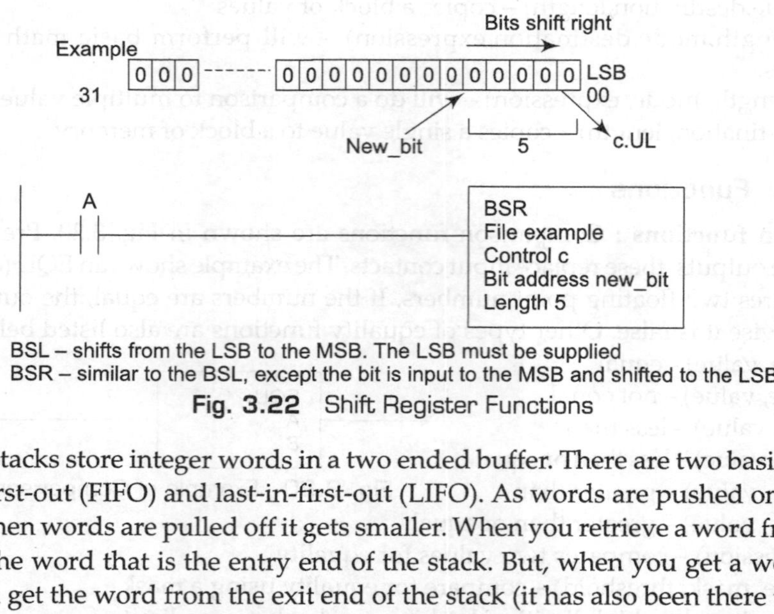

Fig. 3.22 Shift Register Functions

2. **Stacks :** Stacks store integer words in a two ended buffer. There are two basic types of stacks; first-in-first-out (FIFO) and last-in-first-out (LIFO). As words are pushed on the stack it gets larger, when words are pulled off it gets smaller. When you retrieve a word from a LIFO stack you get the word that is the entry end of the stack. But, when you get a word from a FIFO stack you get the word from the exit end of the stack (it has also been there the longest). In a FIFO stack the parts pass through an entry gate, but are stopped by the exit gate. In the LIFO stack the parts enter the stack and lower the plate, when more parts are needed the plate is raised. In this arrangement the order of the parts in the stack will be reversed.

The LIFO stack commands are shown in Fig. 3.24. As values are loaded on the stack the will be added sequentially stack[0], stack[1], stack[2], stack[3] then stack[4]. When values are unloaded they will be taken from the last loaded position, so if the stack is full the value of stack[4] will be removed first.

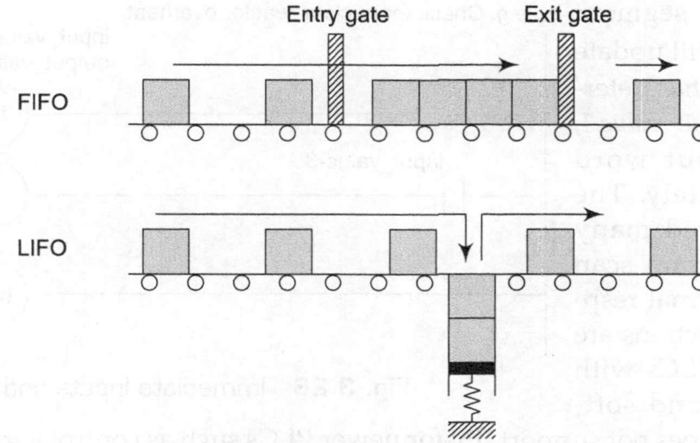

Fig. 3.23 Stack Types

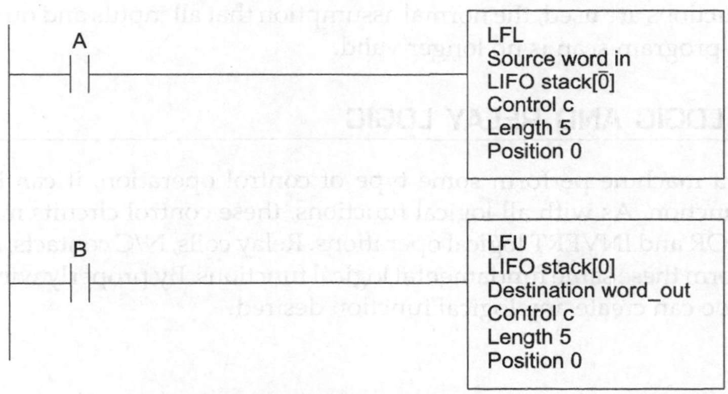

Fig. 3.24 LIFO Stack Commands

3.4.4 INPUT and OUTPUT Functions

The input scan normally records the inputs before the program scan, and the output scan normally updates the outputs after the program scan, as shown in Fig. 3.25. Immediate input and output instructions can be used to update some of the inputs or outputs during the program scan.

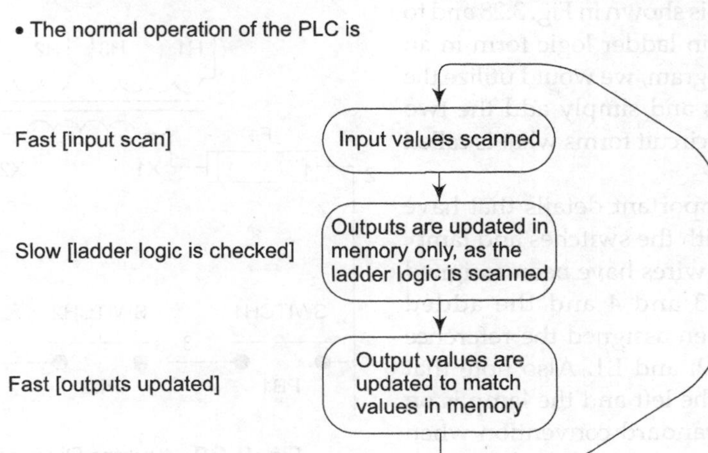

Fig. 3.25 Input Program and Output Scan

Figure 3.26 shows a segment within a program that will update the input word *input_value*, determine a new value for *output_value.1*, and update the output word *output_value* immediately. The process can be repeated many times during the program scan allowing faster than normal response times. These instructions are less useful on newer PLCs with networked hardware and software, so Allen Bradley does not support IIN for newer PLCs such as control logix, even though the IOT is supported.

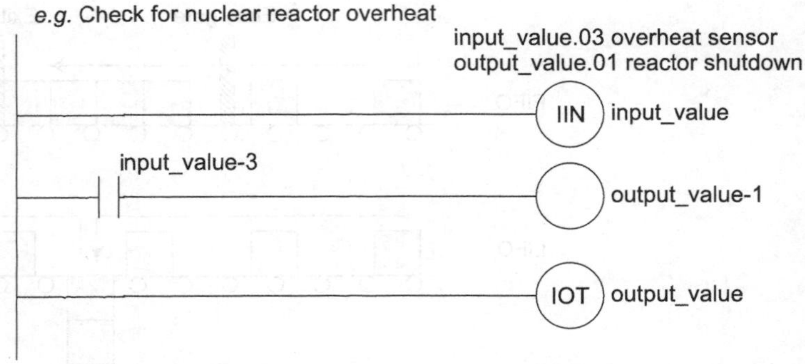

Fig. 3.26 Immediate Inputs and Outputs

When these instructions are used, the normal assumption that all inputs and outputs are updated before and after the program scan is no longer valid.

3.5 BOOLEAN LOGIC AND RELAY LOGIC

Since the relays in a machine perform some type of control operation, it can be said that they perform a logical function. As with all logical functions, these control circuits must consist of the fundamental AND, OR and INVERT logical operations. Relay coils, N/C contacts, and N/O contacts can be wired to perform these same fundamental logical functions. By properly wiring relay contacts and coils together, we can create any logical function desired.

3.5.1 AND

AND operation is generally represented by connection of two switches, a lamp and a battery. Figure 3.27 shows the actual wiring connection for two switches, a lamp and a battery in an **AND** configuration. The lamp, LAMP1, will illuminate only when SWITCH1 AND SWITCH2 are ON. The Boolean expression for this is

$$Lamp1 = (Switch1) \bullet (Switch2)$$

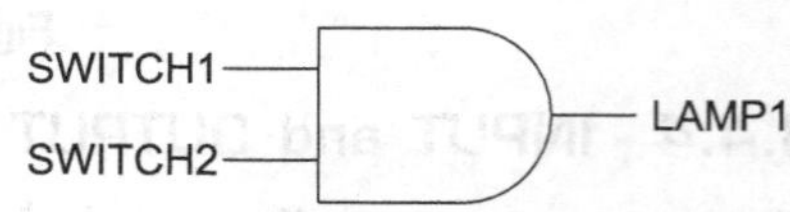

Fig. 3.27 AND Logic Circuit

The logic diagram is shown in Fig. 3.28 and to represent the circuit in ladder logic form in an electrical machine diagram, we would utilize the power from the rails and simply add the two switches. This added circuit forms what is called a **rung**

There are a few important details that have been added along with the switches and lamp. Note that the added wires have been assigned the wire numbers 3 and 4 and the added components have been assigned the reference designators PB1, PB2, and L1. Also note that the switches are on the left and the lamp is on the right. This is a standard convention when designing and drawing machine circuits.

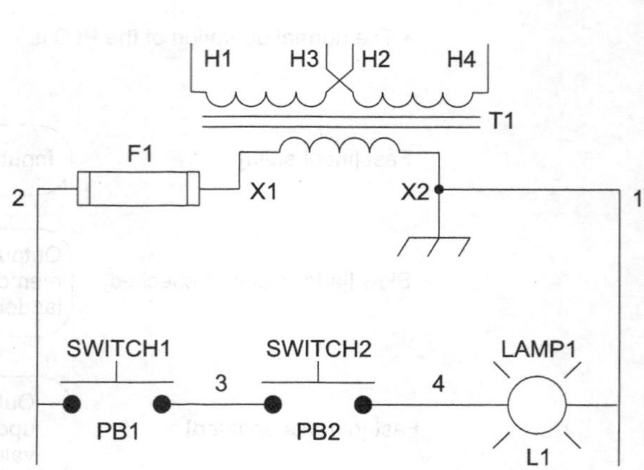

Fig. 3.28 Ladder Diagram for AND

3.5.2 OR

The same approach may be taken for the **OR** function. The circuit shown in Fig. 3.29 illustrates two switches wired as an **OR** function controlling a lamp, LAMP2. The lamp will illuminate if SWITCH1 **OR** SWITCH2 is closed; that is, depressing either of the switches will cause the lamp LAMP2 to illuminate. The Boolean expression for this circuit is

$$Lamp2 = (Switch1) + (Switch2)$$

For those more familiar with logic diagramming, the OR gate representation of the OR circuits shown in Fig. 3.29 and in Fig. 3.30. We can now add this circuit to our ladder diagram as another rung as shown in Fig. 3.31. Note that since the switches SWITCH1 and SWITCH2 are the same ones used in the top rung, they will have the same names and the same reference designators when drawn in rung 2. This means that each of these two switches have two N/O contacts on the switch assembly.

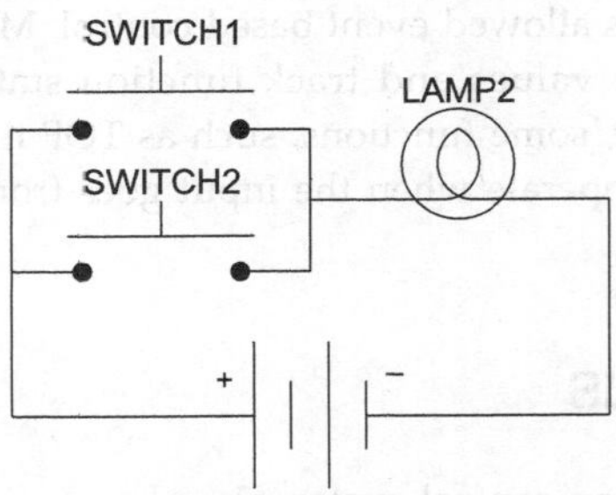

Fig. 3.29 OR Gate

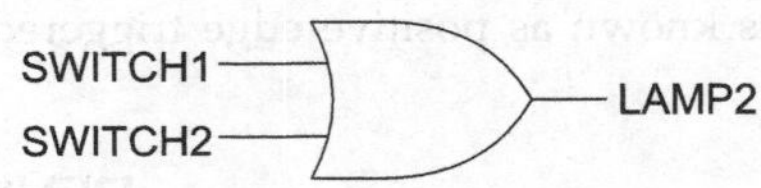

Fig. 3.30 OR Circuit

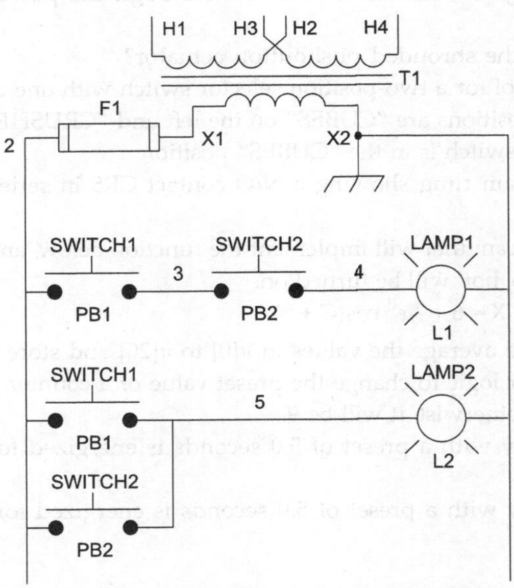

Fig. 3.31 AND Rung 2 Circuit

In a PLC, the ordering of the rungs is critical and rearranging the order could change the way the PLC program executes.

SUMMARY

Modern control systems still include relays, but these are rarely used for logic. A relay is a simple device that uses a magnetic field to control a switch, as pictured in.When a voltage is applied to the input coil, the resulting current creates a magnetic field. The magnetic field pulls a metal switch (or reed) towards it and the contacts touch, closing the switch. The contact that closes when the coil is energized is called normally open. The normally closed contacts touch when the input coil is not energized. Relays are normally drawn in schematic form using a circle to represent the input coil. The output contacts are shown with two parallel lines. Normally open contacts are shown as two lines, and will be open (non-conducting) when the input is not energized. Normally closed contacts are shown with two lines with a diagonal line through them. When the input coil is not energized the normally closed contacts will be closed (conducting). Ladder logic input contacts and output coils allow simple logical decisions. Functions extend basic ladder logic to allow other types of control. For example, the addition of timers and counters allowed event based control. Most of the functions will use PLC memory locations to get values, store values and track function status. Most function will normally become active when the input is true. But, some functions, such as TOF timers, can remain active when the input is OFF. Other functions will only operate when the input goes from false to true, this is known as positive edge triggered.

REVIEW QUESTIONS

1. What is the purpose of the control transformer in machine control systems?
2. Why are fuses necessary in controls circuits even though the power mains may already have circuit breakers?
3. What is the purpose of the shrouded pushbutton actuator?
4. Draw the electrical symbol for a two-position selector switch with one contact. The switch is named "ICE" and the selector positions are "CUBES" on the left and "CRUSHED" on the right. The contact is to be closed when the switch is in the "CUBES" position.
5. Draw an electrical diagram rung showing a N/O contact CR5 in series with a N/C contact CR11, operating a lamp L3.
6. Write a ladder logic program that will implement the function below, and if the result is greater than 100.5 then the output 'too_hot' will be turned on.

$$X = 6 + Ae^{B}\cos(C + 5)$$

7. Use an FAL instruction to average the values in *n*[0] to *n*[20] and store them in '*n*_avg'.
8. Write some simple ladder logic to change the preset value of a counter. When the input 'A' is active the preset should be 13, otherwise it will be 9.
9. If a delay-on (TON) relay with a preset of 5.0 seconds is energized for 3 seconds, explain how it reacts.
10. If a delay-off (TOF) relay with a preset of 5.0 seconds is energized for 1 second, explain how the relay reacts.

4

FILE STRUCTURE AND ADDRESSING FORMATS

LEARNING OBJECTIVES

After reading this chapter, you will be able to explain:

1. Fundamental concepts of SLC 500 BASED SYSTEMS
2. Addressing formats and elements of Output and Input, Timer data, Bit data, Counter data and various other associated files.

4.1 INTRODUCTION

The SLC 500 based systems can be configured with or without rack system. The SLC500/PLC-5 can be configured on various sizes of rack systems with different number of slots. Figure 4.1 shows the CPU and I/O configuration of SLC 5/03 on seven slot rack system.

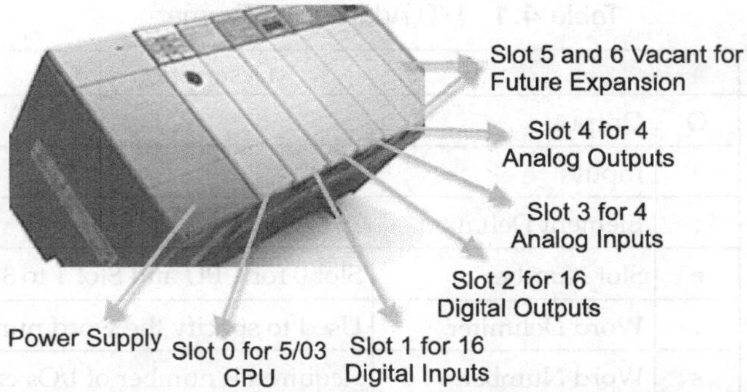

Fig. 4.1 SLC 5/03 Configuration on 7-Slot Rack System

The micrologix 1000 and micrologix 1200 based PLC systems are shown in Figs. 4.2 and 4.3 respectively. In these PLCs, there is no need of rack system where inputs and outputs are present at slot 0 for configuration.

The SLC 500 based system user memory is comprised of data files and program files as shown in Fig. 4.4. The data files 3 through 8 are the default values. Files 9 to 255 can be configured to be bit, timer, counter, control, integer, floating point, ASCII, or string files.

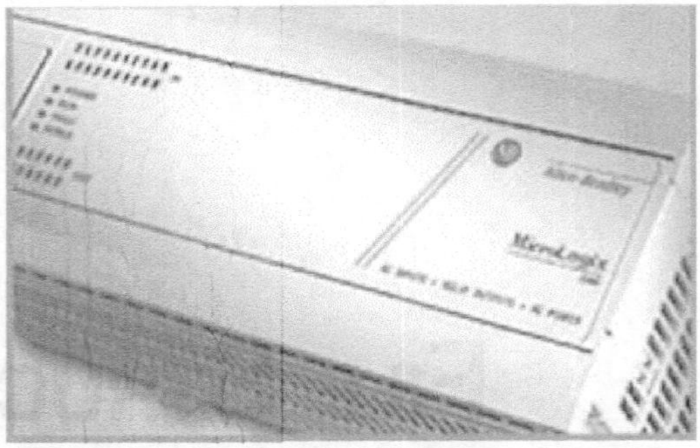

Fig. 4.2 Micrologix 1000 Based System

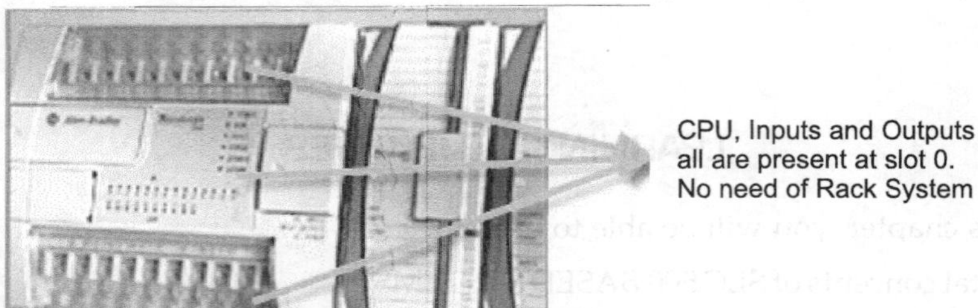

Fig. 4.3 Micrologix 1200 Based System

4.2 OUTPUT AND INPUT DATA FILES (FILES O0: AND I1:)

The Data Files 0 and 1 represent external outputs and inputs, respectively. Table 4.1 explains the addressing format for outputs and inputs. The format specifies **e** as the slot number and **s** as the word number. When referencing output and input data file words, e.s means slot and word taken together.

Table 4.1 I/O Addressing Format

Addressing Format	Description		
O:e.s/b	**O**	Output	
I:e.s/b	**I**	Input	
	:	Element Delimiter	
	e	Slot Number	Slot 0 for CPU and Slot 1 to 30 are I/O slots.
	.	Word Delimiter	Used to specify the word number if required.
	s	Word Number	Required if number of I/Os exceeds 16 in a slot.
	/	Bit Delimiter	
	b	Bit Number	Inputs: 0–15 and Outputs:0–15

Example: *SLC 5/03 I/O addressing*

1. O:2/5 means digital output 5 of slot 2.
2. O:2 means digital output word 0 of slot 2.
3. I:1/6 means digital input 6 of slot 1.

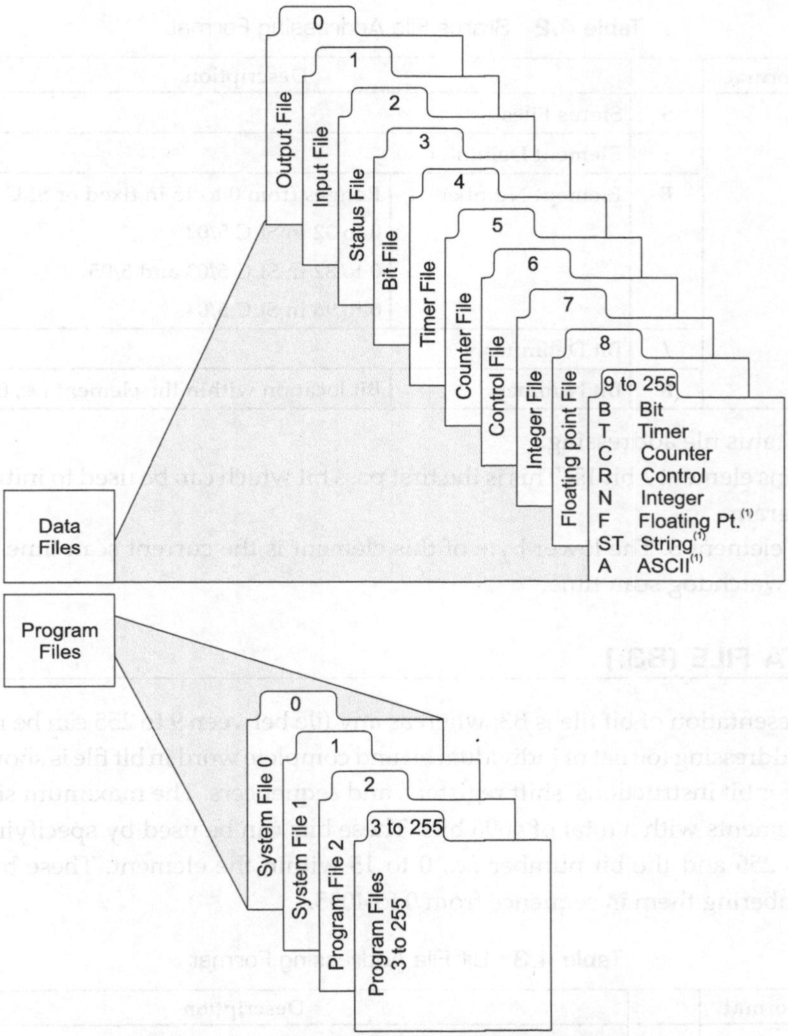

Fig. 4.4 Processor Files of SLC 500 System

4. I:1 means digital input word 0 of slot 1.
5. O:4 means analog output word 0 of slot 4.
6. I:3 means analog input word 0 of slot 3.

Example: *Micrologix 1000 and 1200 I/O addressing*

1. O:0/5 means output 5.
2. I:0/6 means input 6. Here it can be noticed that both inputs and outputs are present at the same location along with CPU.

4.3 STATUS FILE (FILE S2:)

The addressing format of individual bit and complete word in status file is shown in Table 4.2. User cannot alter the status file. To be more précised no addition or deletion is possible in contents of status file.

Table 4.2 Status File Addressing Format

Addressing Format		Description	
S:e/b	S	Status Files	
	:	Element Delimiter	
	E	Element Number	Ranges from 0 to 15 in fixed or SLC 5/01 0 to 32 in SLC 5/02 0 to 82 in SLC 5/03 and 5/05 0 to 96 in SLC 5/04.
	/	Bit Delimiter	
	b	Bit Number	Bit location within the element *i.e.,* 0 is 15

Examples of Status file addressing :

1. S:1/15 means element 1 bit 15. This is the first pass bit which can be used to initiate instruction in the program.
2. S:3 means element 3. The lower byte of this element is the current scan time and the upper half is the watchdog scan time.

4.4 BIT DATA FILE (B3:)

The default representation of bit file is B3: whereas any file between 9 to 255 can be made as bit file if required. The addressing format of individual bit and complete word in bit file is shown in Table 4.3. This file is used for bit instructions, shift registers, and sequencers. The maximum size of the file is 256 one-word elements with a total of 4096 bits. These bits can be used by specifying the element number *i.e.,* 0 to 255 and the bit number *i.e.,* 0 to 15 within the element. These bits can also be specified by numbering them in sequence from 0 to 4095.

Table 4.3 Bit File Addressing Format

Addressing Format		Description	
Bf:e/b	B	Bit File	
	f	File Number. Default is 3. Any file from 9 to 255 can be used as bit file if required.	
	:	Element Delimiter	
	e	Element Number	Ranges from 0 to 255. These are 1-word elements with 16-bits per element.
	/	Bit Delimiter	
	b	Bit Number	Bit location within the element *i.e.,* 0 is 15
Bf /b	B	Same as above	
	f	Same as above	
	/	Same as above	
	b	Bit Number	Ranges from 0–4095.

Examples :

1. B3:6/12 means bit 12 of element 6.
2. B3:254/6 means bit 6 of element 254.
3. B3/1894 means bit 1894 out of 4095.

4.5 TIMER DATA FILE (T4:)

The default representation of timer file is T4: whereas any file between 9 to 255 can be made as timer file if required. The addressing format of individual bit and complete word in timer file is explained in Table 4.4. Timer instructions use three control bits. Each data file is made of 3 word data file elements. Word 0 is the status word with various control bits, word 1 indicates the preset value, and word 2 indicates accumulator value.

Table 4.4 Timer Addressing Format

Addressing Format		Description	
Tf:e.s/b	T	Timer File	
	f	File Number. Default is 4. Any file from 9 to 255 can be used as timer file if required.	
	:	Element Delimiter	
	e	Element Number	Ranges from 0 to 255. These are 3-word elements with 16-bits per element.
	.	Word Delimiter	
	s	Word Number	Ranges from 0 to 2.
	/	Bit Delimiter	
	b	Bit Number	Ranger from 0 is 15

Examples :

1. T4:0/15 or T4:0/EN means enable bit.
2. T4:0/14 or T4:0/TT means timer timing bit.
3. T4:0/13 or T4:0/DN means done bit.
4. T4:0.1 or T4:0.PRE means preset word.
5. T4:0.2 or T4:0.ACC means accumulator word.
6. T4:0.1/6 or T4:0.PRE/6 means bit 6 of preset word.
7. T4:0.2/3 or T4:0.ACC/3 means bit 3 of accumulator word.

4.6 COUNTER DATA FILE ELEMENTS (C5:)

The default representation of counter file is C5: whereas any file between 9 to 255 can be made as counter file if required. The addressing format of individual bit and complete word in counter file is explained in Table 4.5. Each Counter address is made of a 3-word data file element. Word 0 is the status word with various control bits, word 1 indicates the preset value and word 2 is the accumulated value.

Table 4.5 Counter File Addressing Format

Addressing Format	Description		
Cf:e.s/b	**C**	Counter File	
	f	File Number. Default is 5. Any file from 9 to 255 can be used as counter file if required.	
	:	Element Delimiter	
	e	Element Number	Ranges from 0 to 255. These are 3-word elements with 16-bits per element.
	.	Word Delimiter	
	s	Word Number	Ranges from 0 to 2.
	/	Bit Delimiter	
	b	Bit Number	Ranges from 0 to 15.

Examples :

1. C4:0/15 or C4:0/CU means counter up bit.
2. C4:0/14 or C4:0/CD means counter down bit.
3. C4:0/13 or C4:0/DN means done bit.
4. C4:0/12 or C4:0/OV means overflow bit.
5. C4:0/11 or C4:0/UN means underflow bit.
6. C4:0.1 or C4:0.PRE means preset word.
7. C4:0.2 or C4:0.ACC means accumulator word.
8. C4:0.1/6 or C4:0.PRE/6 means bit 6 of preset word.
9. C4:0.2/3 or C4:0.ACC/3 means bit 3 of accumulator word.

4.7 CONTROL DATA FILE (R6:)

The default representation of control data file is R6: whereas any file between 9 to 255 can be made as control file if required. The addressing format of individual bit and complete word in control data file is explained in Table 4.6. Each Control address is made of a 3-word data file elements. It is used with bit shift, FIFO, LIFO, sequencer instructions, and ASCII instructions. Word 0 is the status word, word 1 indicates the length of stored data, and word 2 indicates the position.

Table 4.6 Control File Addressing Format

Addressing Format	Description		
Rf:e.s/b	**R**	Control File	
	f	File Number. Default is 6. Any file from 9 to 255 can be used as control file if required.	
	:	Element Delimiter	
	e	Element Number	Ranges from 0 to 255. These are 3-word elements with 16-bits per element.
	.	Word Delimiter	
	s	Word Number	Ranges from 0 to 2.
	/	Bit Delimiter	
	b	Bit Number	Ranges from 0 to 15.

Examples :

1. R6:3 means element 3 of control file 6.
2. R6:3/15 or R6:3/EN means enable bit.
3. R6:3/14 or R6:3/EU means unload enable bit.
4. R6:3/13 or R6:3/DN means done bit.
5. R6:3/12 or R6:3/EM means stack empty bit.
6. R6:3/11 or R6:3/ER means error bit.
7. R6:3/10 or R6:3/UL means unload bit.
8. R6:3/9 or R6:3/IN means inhibit bit.
9. R6:3/8 or R6:3/FD means found bit.
10. R6:3.1 or R6:3.LEN means length value.
11. R6:3.2 or R6:3.POS means position value.
12. R6:3.1/4 or R6:3.LEN/4 means bit 4 of length value.
13. R6:3.2/9 or R6:3.POS/9 means bit 9 of position value.

4.8 INTEGER DATA FILE (N7:)

The default representation of integer data file is N7: whereas any file between 9 to 255 can be made as integer file if required. The addressing format of individual bit and complete word in integer data file is explained in Table 4.7. These are 1-word elements, addressable at the element and bit level.

Table 4.7 Integer Data File Addressing Format

Addressing Format	Description		
Nf:e/b	N	Integer File	
	f	File Number. Default is 7. Any file from 9 to 255 can be used as integer file if required.	
	:	Element Delimiter	
	e	Element Number	Ranges from 0 to 255. These are 1-word elements with 16-bits per element.
	/	Bit Delimiter	
	b	Bit Number	Bit location within the element *i.e.*, 0 to 15

Examples:

1. N7:3 means element 3 of integer file 7.
2. N7:3/15 means bit 15 of element 3.
3. N12:48 means element 48 of integer file 12 which is assigned by the user.

4.9 FLOAT DATA FILE (F8:)

The default representation of floating data file is F8: whereas any file between 9 to 255 can be made as floating file if required. The addressing format of individual bit and complete word in floating data file is explained in Table 4.8. These are 2-word elements, addressable at the element and bit level.

Table 4.8 Float File Addressing Format

Addressing Format		Description	
Ff:e/b	F	Float File	
	f	File Number. Default is 8. Any file from 9 to 255 can be used as integer file if required.	
	:	Element Delimiter	
	e	Element Number	Ranges from 0 to 255. These are 2-word elements with 32 bits per element.
	/	Bit Delimiter	
	b	Bit Number	Bit location within the element *i.e.*, 0 to 15

Examples:

1. F8:3 means word 0, element 3 of float file 8.
2. F8:3/12 means word 0, bit 12 of element 3 of float file 8.
3. F15:48 means word 0, element 48 of float file 15 which is made by the user.
4. F8:3.1 means word 1, element 3 of float file 8.
5. F15:48.1 means word 1, element 48 of float file 15 which is made by the user.

SUMMARY

The SLC500/PLC-5 can be configured on various sizes of rack systems with different number of slots. In Micrologix 1000 and Micrologix 1200 based PLC systems there is no need of rack system where inputs and outputs are present at slot 0 for configuration. For output and input files. The format specifies **e** as the slot number and *s* as the word number. Bit data file is used for bit instructions, shift registers, and sequencers. The maximum size of the file is 256 one-word elements with a total of 4096 bits user cannot alter the contents of status file. The default representation of integer data file is N7. Control data file is used with bit shift, FIFO, LIFO, sequencer instructions, and ASCII instructions. Addressing formats for various files is discussed in this chapter.

REVIEW QUESTIONS

1. Explain I/O addressing format for SLC 500 and Micrologix series with the help of suitable diagram and examples.
2. Draw the diagram of the file structure of SLC 500 based systems and explain it.
3. Write short note on status data files.
4. Explain briefly the bit file structure of SLC 500 with the help of examples.
5. What do you mean by timer data file structure? Explain its addressing format with the help of examples.
6. Define counter data file structure. Explain its addressing format with the help of examples.
7. Explain the structure and operation of control data file in brief.
8. Discuss the structure of input data file. Explain its addressing format with the help of examples.
9. What is a floating data file? Discuss its addressing format with suitable examples.

5

PLC PROJECT DEVELOPMENT

LEARNING OBJECTIVES

After reading this chapter, you will be able to explain:

1. Introduction to RS logix 500 and RS linx softwares, various installation steps
2. Micrologix, SLC 5 /03 driver configurations
3. Various phases in developing a new project
4. LADSIM based simulation

5.1 INTRODUCTION

The following software's are required to develop PLC projects based on SLC 500 based system design.

1. RS logix 500 standard/starter
2. RS linx

5.1.1 RS Logix 500 Standard/Starter

This software is used to:

(*a*) Edit the program
(*b*) Save the program
(*c*) Run the program

5.1.2 RS Linx

This software contains all the required drivers for communication between PLC and PC. It is necessary software used to configure the required drivers for different Allen Bradley make PLCs in order to communicate with PC.

5.2 SOFTWARE INSTALLATION

Both the above software comes on an autorun CD. Insert the CD then installation will start automatically. If autorun does not start automatically then open the CD manually and install it. After inserting the CD a new window will appear as shown in Fig. 5.1.

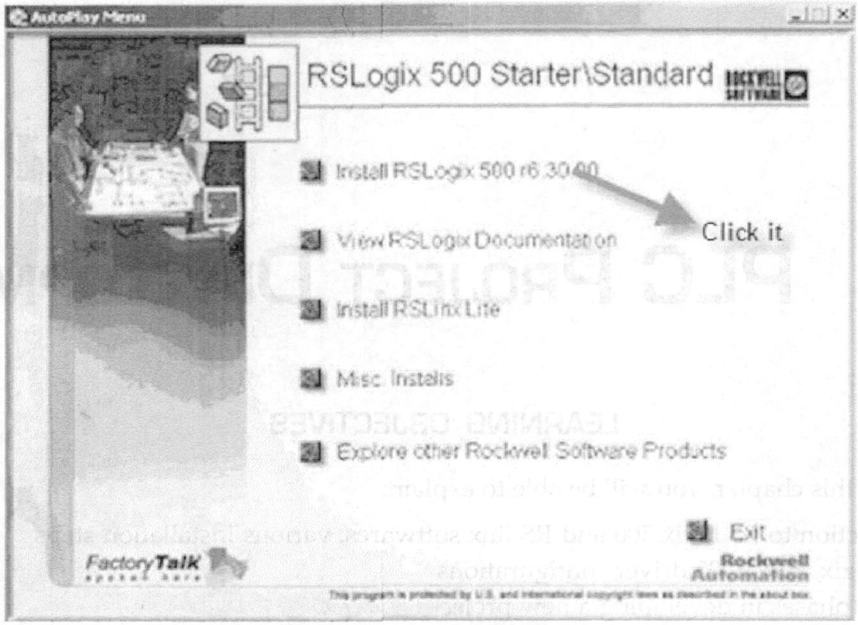

Fig. 5.1 RS Logix Install Window

In this window select install RS logix 500 r6 30.00 and double click it. Then a new window will open as shown in Fig. 5.2.

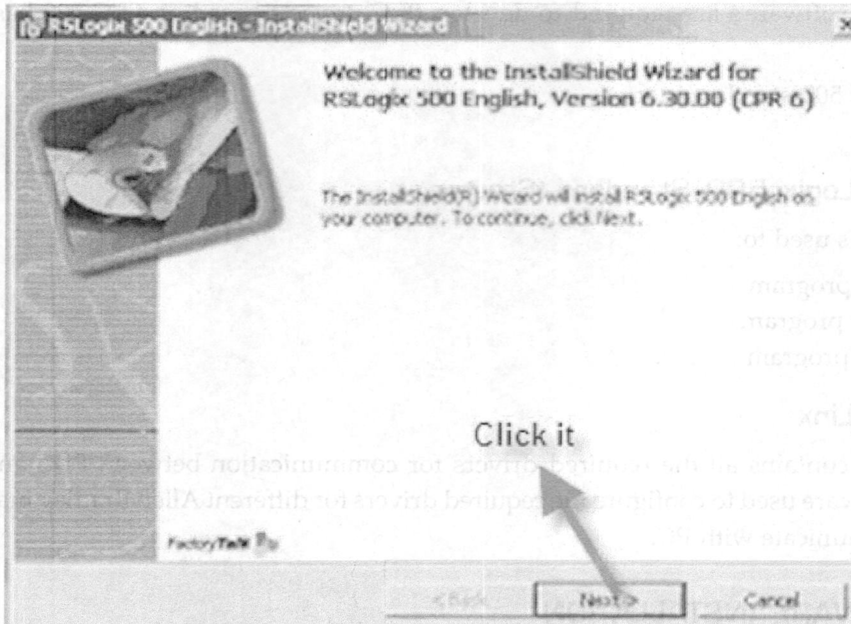

Fig. 5.2 Welcome Window

Click next in the above window to open license agreement wizard as shown in Fig. 5.3.

Fig. 5.3 License Agreement

In this window select "I accept the terms in the license agreement" and click next to open customer information window as shown in Fig. 5.4.

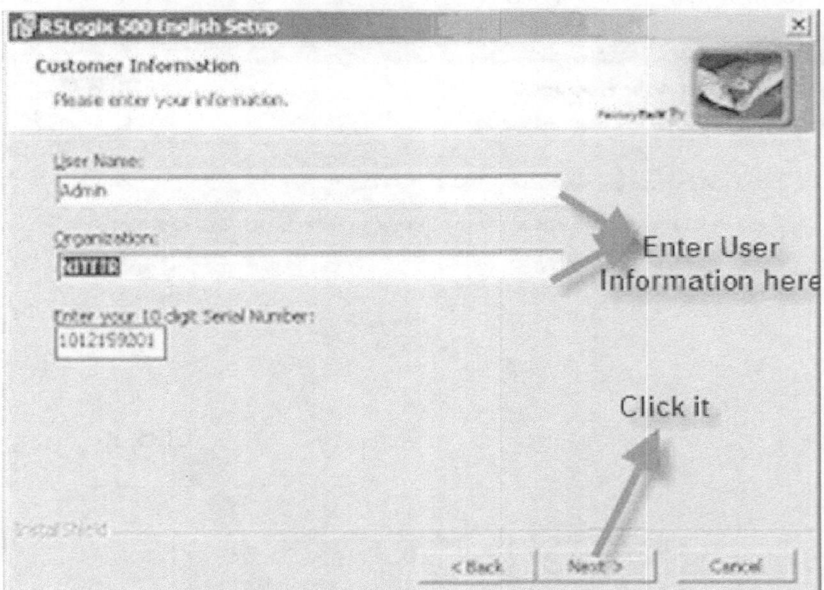

Fig. 5.4 RS Logix User Information

Enter the customer information and 10 digit serial number of the software. Click next to open a new window asking to select the "Setup type" as shown in Fig. 5.5.

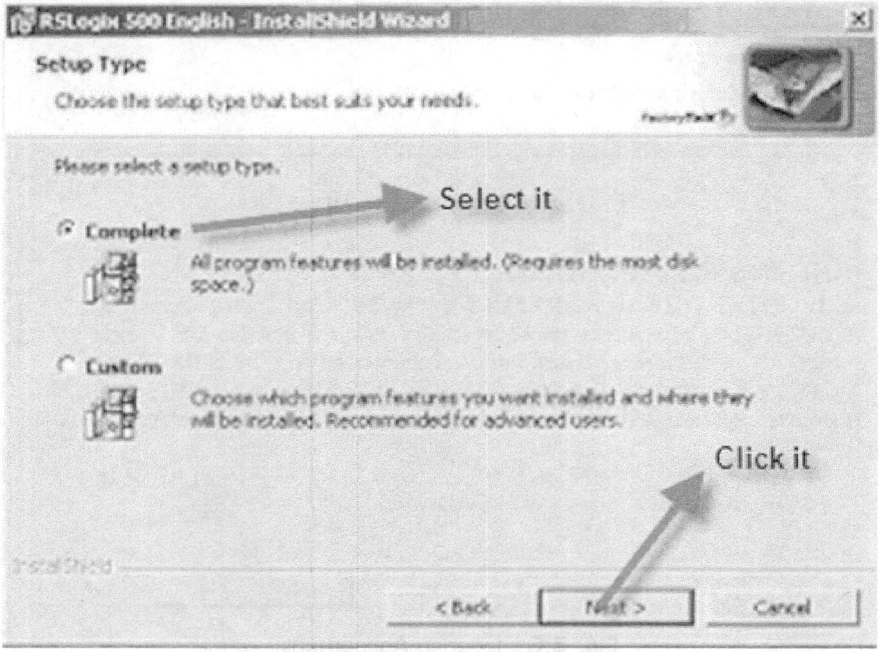

Fig. 5.5 RS Logix Setup Type Selection

Select setup type as complete in the last window and enter next then a new window will popup showing a message "ready to install program" as shown in Fig. 5.6.

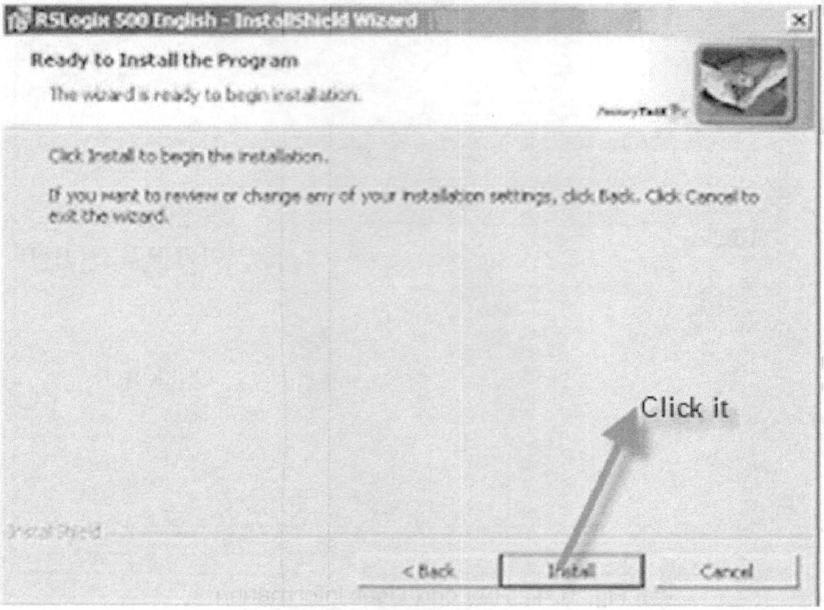

Fig. 5.6 Install Ready Window

Click install to complete the installation process. You will notice a message in new window that "install shield wizard completed" as shown in Fig. 5.7.

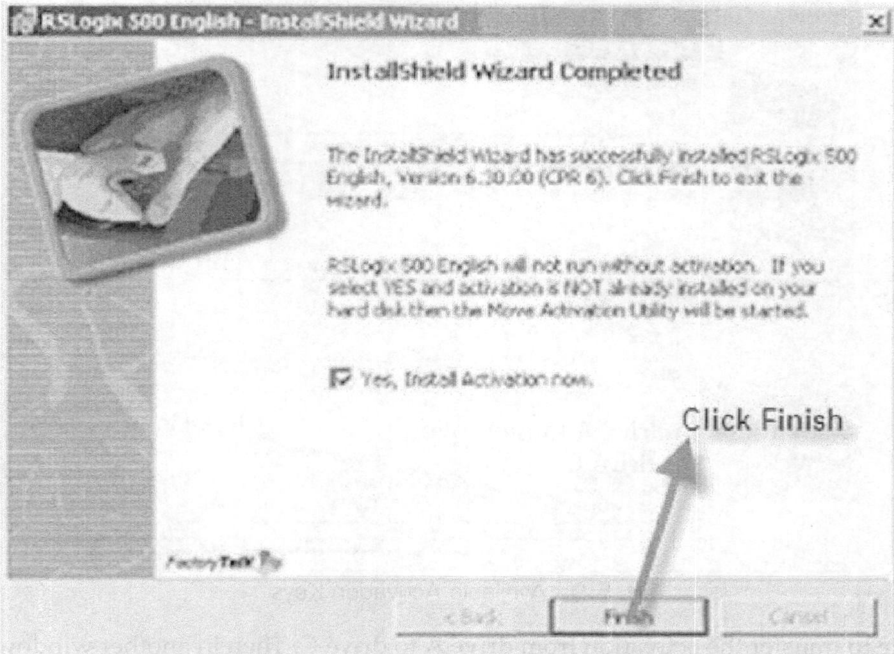

Fig. 5.7 Activation Install Window

In last window check the "install activation now" checkbox and click finish to complete the activation process as shown in Fig. 5.8.

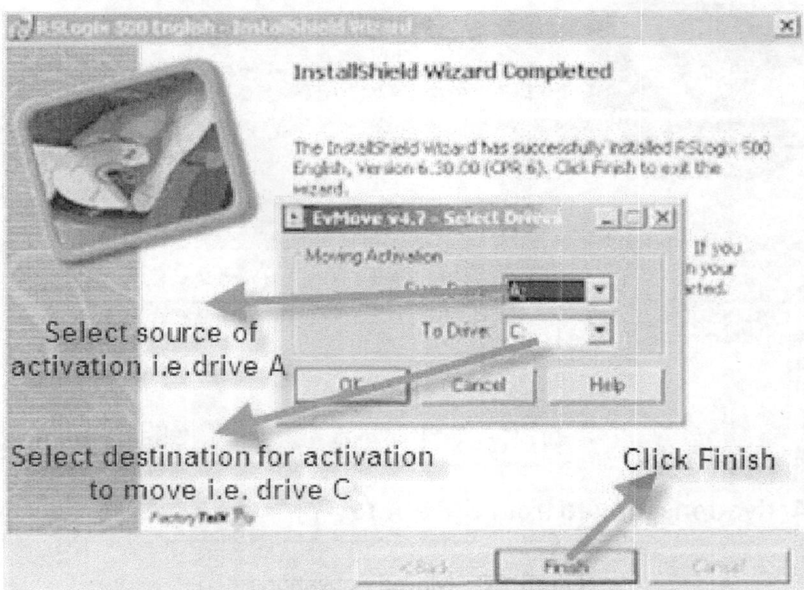

Fig. 5.8 Activation Move Window

Select source drive as A and destination drive as C and click OK. A new window will appear as shown in Fig. 5.9.

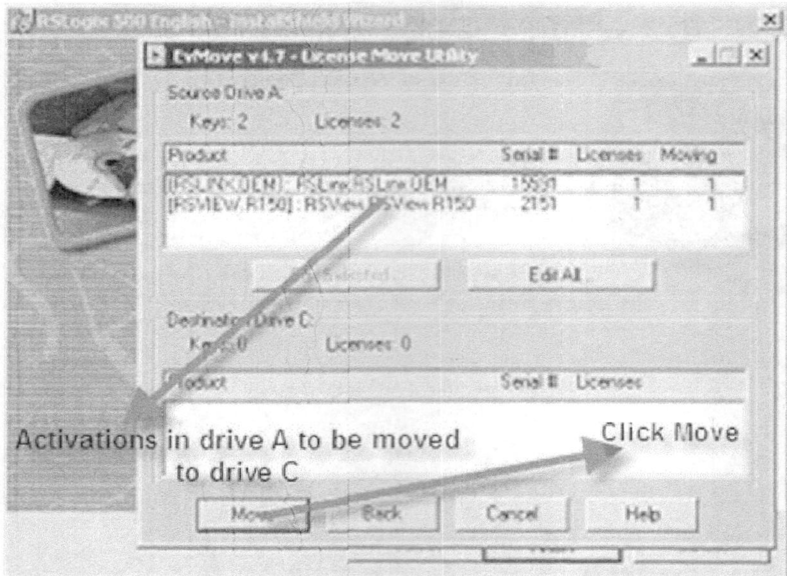

Fig. 5.9 Available Activation Keys

Click move to transfer the activation from drive A to drive C. Then in another window you will notice a message that "Activation moved successfully" as shown in Fig. 5.10.

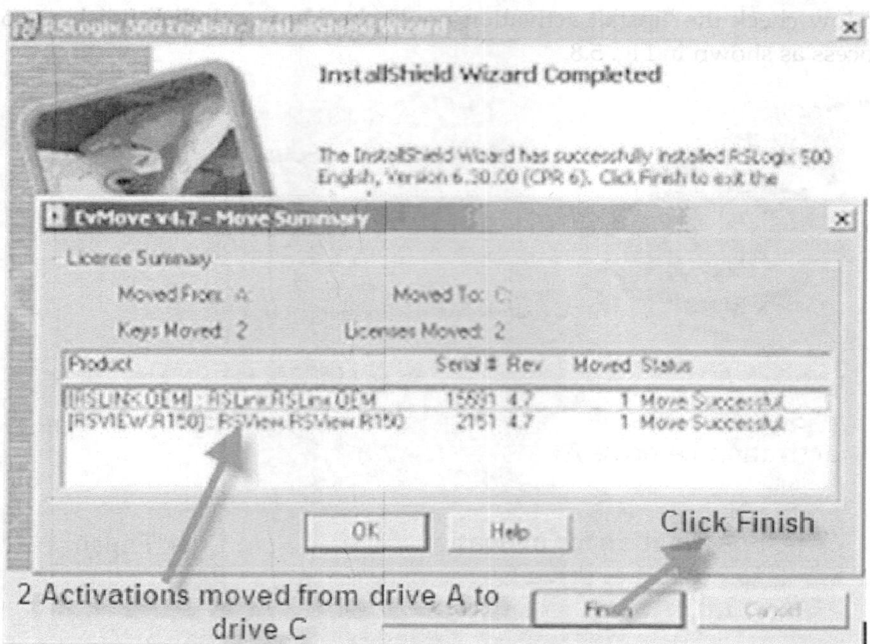

Fig. 5.10 Moved Activations

Click OK to complete the RS logix installation and activation. Now the next step is to install the RS linx software in order to configure the required drivers for installed PLCs. Once RS logix installation is complete select RS linx lite as shown in Fig. 5.11

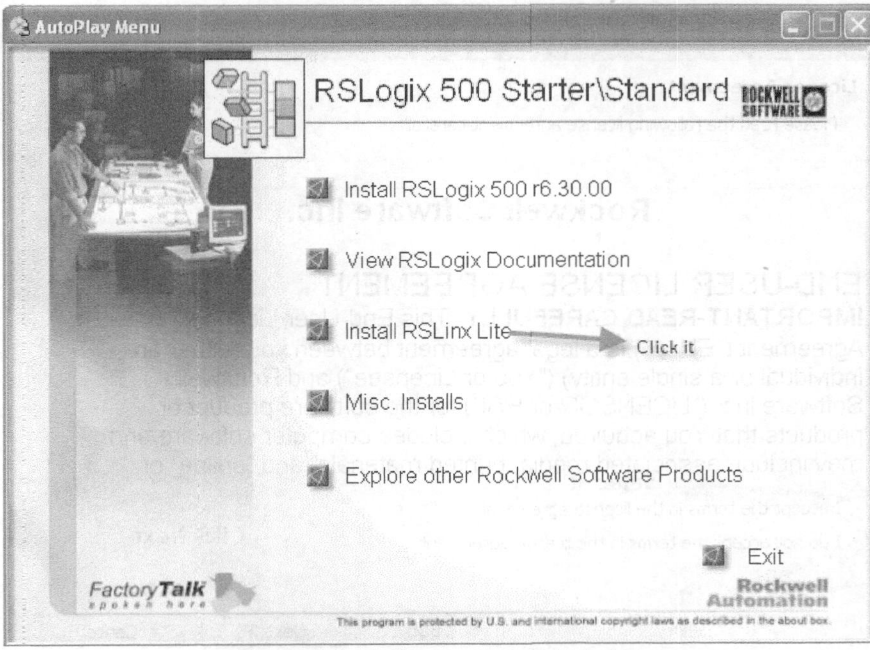

Fig. 5.11 Software Select Window

Double click install RS Linx Lite then a new window will appear as shown in Fig. 5.12.

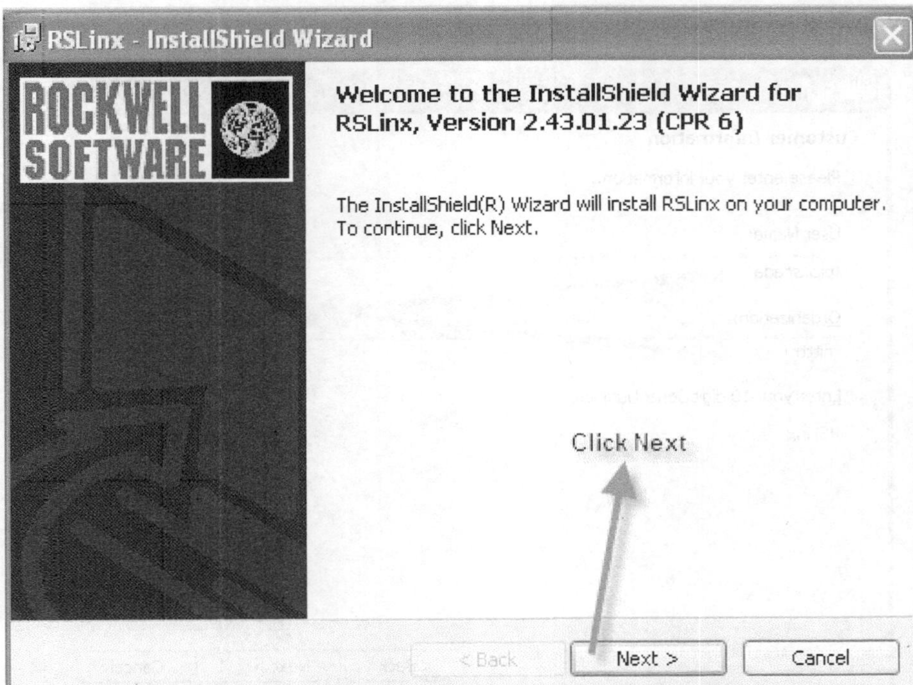

Fig. 5.12 Welcome Window

Click next to open license agreement wizard as shown in Fig. 5.13.

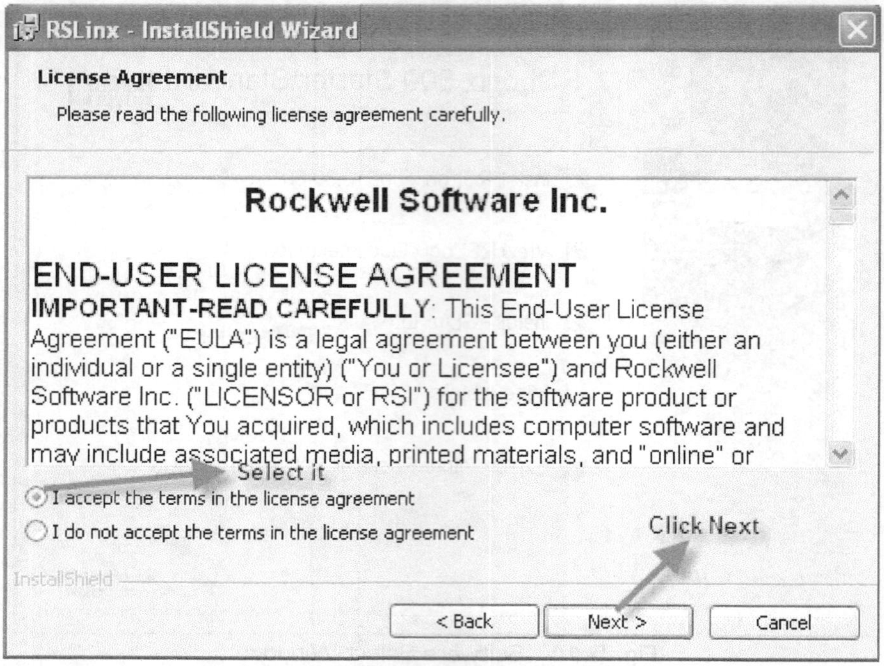

Fig. 5.13 RS Linx License Agreement

In this window select "I accept the terms in the license agreement" and click next to open customer information window as shown in Fig. 5.14

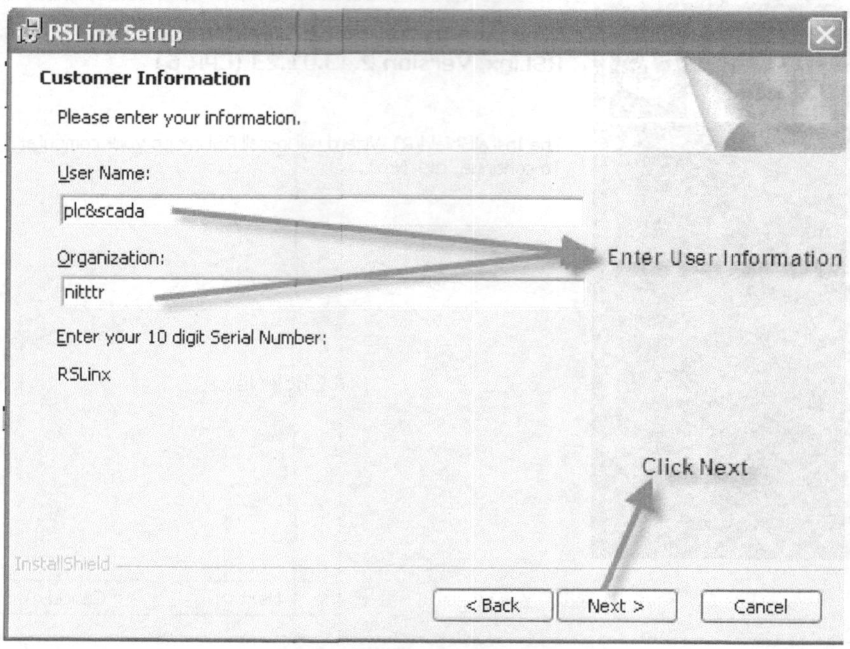

Fig. 5.14 RS Linx User Information

Click next to open a new window to select the setup type as shown in Fig. 5.15.

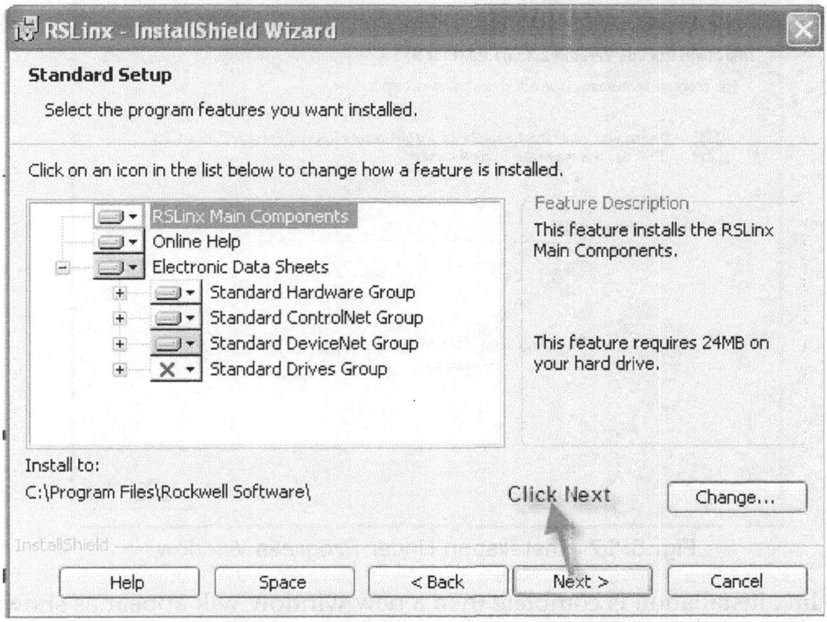

Fig. 5.15 Setup Type Selection Window

Click next, then a new window will come showing a message that "Ready to install program" as shown in Fig. 5.16.

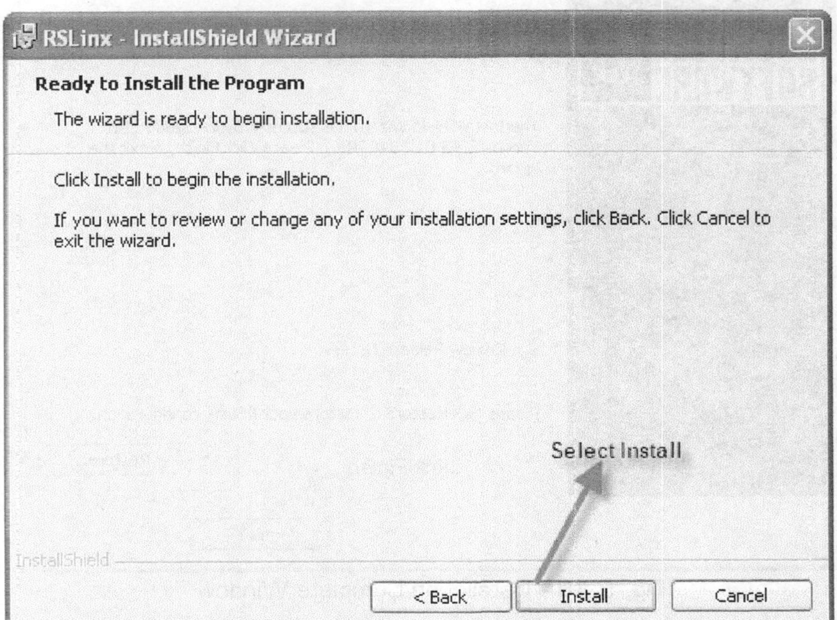

Fig. 5.16 Ready to Install Window

Then you will notice a new window with message that "Installing RS Linx" as shown in Fig. 5.17. It may take some time to install.

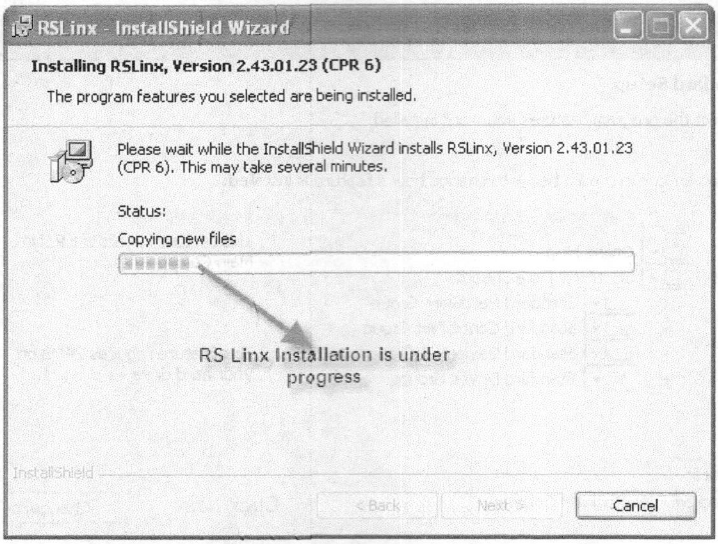

Fig. 5.17 Installation Under Progress Window

Once the RS linx installation is complete then a new window will appear as shown in Fig. 5.18 with a message that "Installation is completed successfully". Click finish to complete the installation wizard.

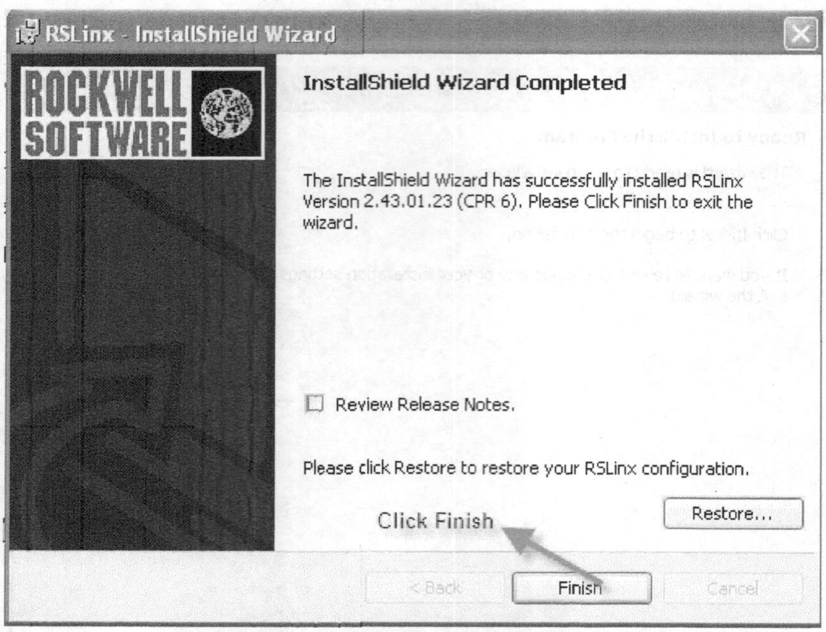

Fig. 5.18 Installation Complete Window

5.3 DRIVER CONFIGURATION

To configure the required drivers, start the RS linx software by selecting Start-Programs-Rockwell Software- RS Linx as shown in Fig. 5.19.

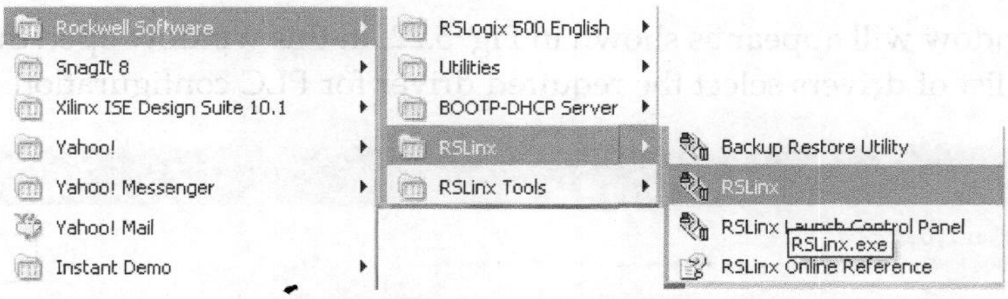

Fig. 5.19 Path for RS Linx

Then a new window will appear to configure the driver for various devices as shown in Fig. 5.20.

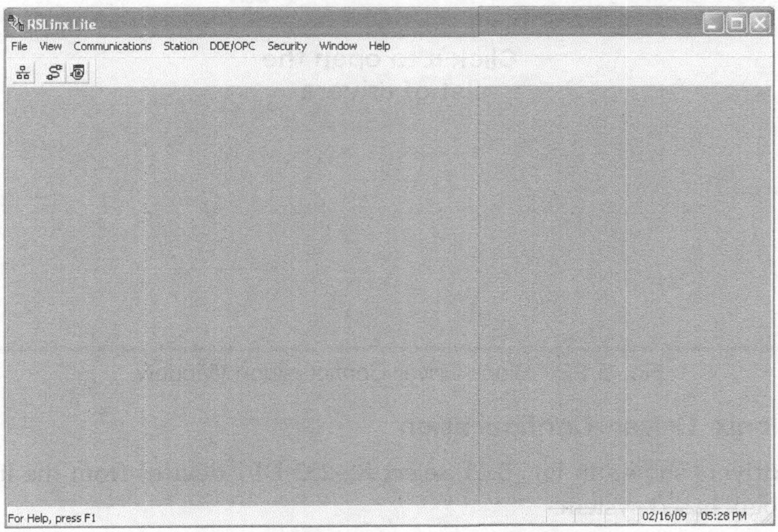

Fig. 5.20 RS Linx Lite Window

To configure a driver for any device, go to communications and double click configure drivers as shown in Fig. 5.21.

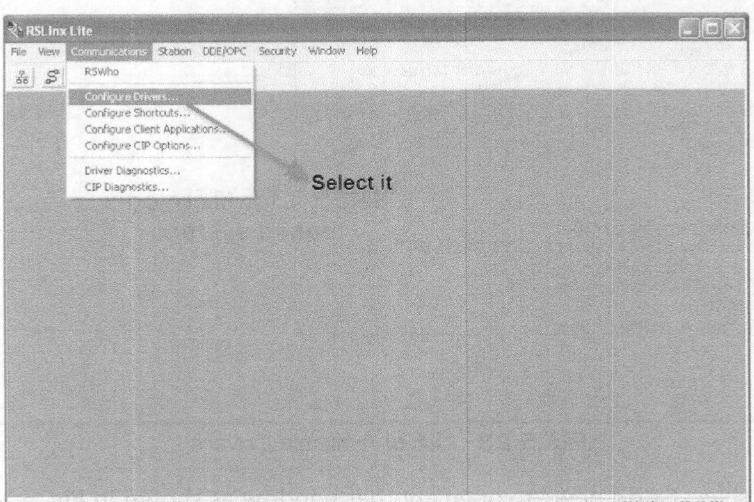

Fig. 5.21 Path to Configure Drivers

Then a new window will appear as shown in Fig. 5.22. In this window open the list of available drivers. From the list of drivers select the required driver for PLC configuration.

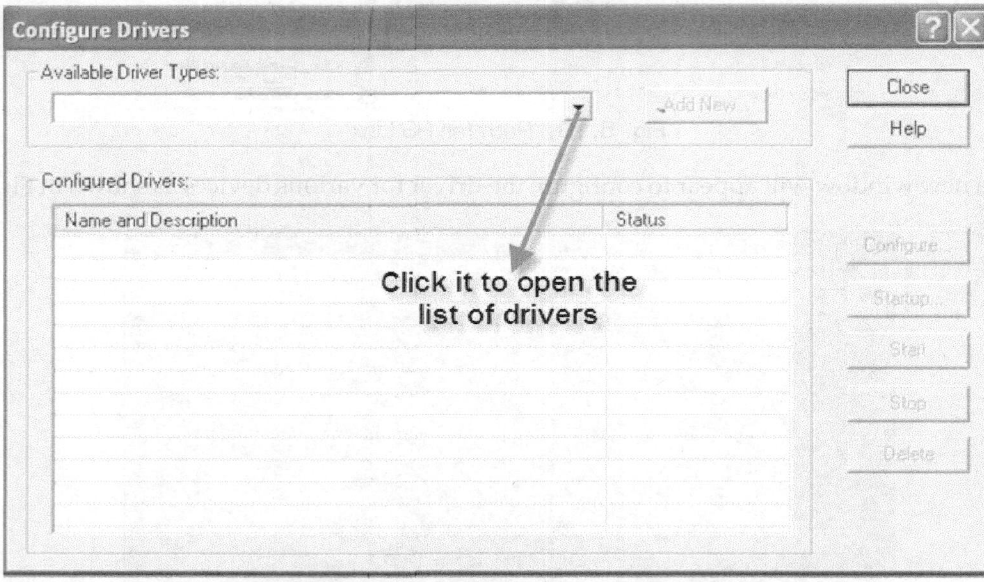

Fig. 5.22 Blank Driver Configuration Window

5.3.1 Micrologix Driver Configuration

From the list of drivers shown in Fig. 5.23, select RS-232 DF1 devices from the list to configure driver for micrologix based system.

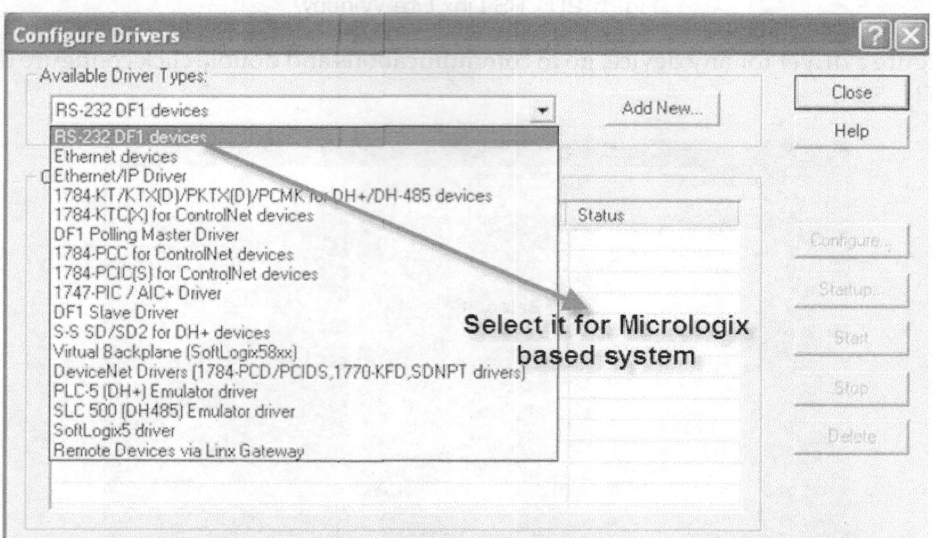

Fig. 5.23 List of Available Drivers

In last window select "Add new" to open another window as shown in Fig. 5.24. Click OK.

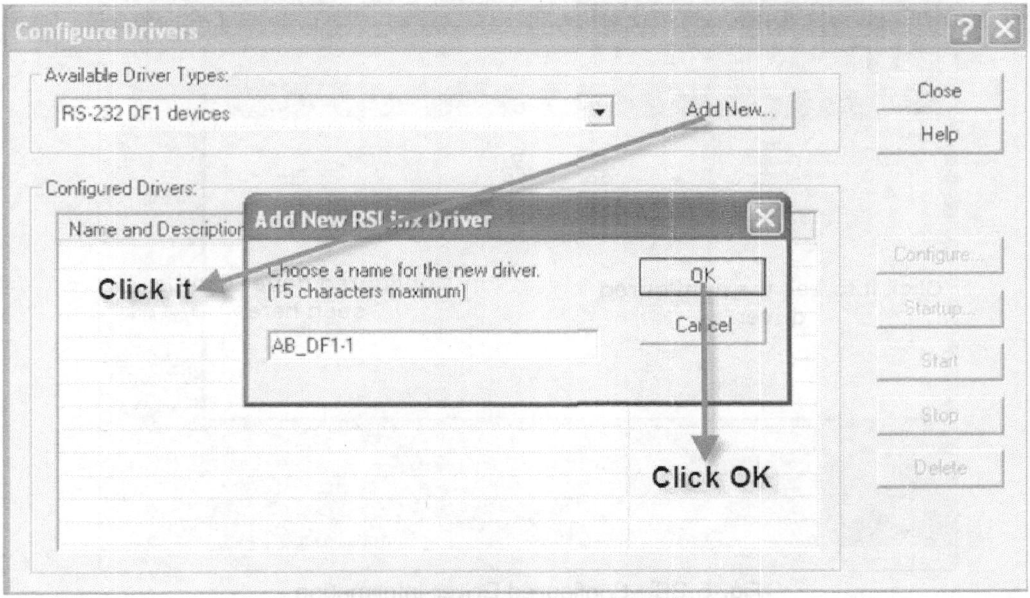

Fig. 5.24 Add New Driver

Then a new window will appear in which the driver can be configured automatically or manually by entering the parameters as shown in the figure. Now power on the PLC and select auto configure to configure the driver automatically. Then you will notice a message that driver configuration is successful. The Fig. 5.25 shows the running driver in configured drivers list.

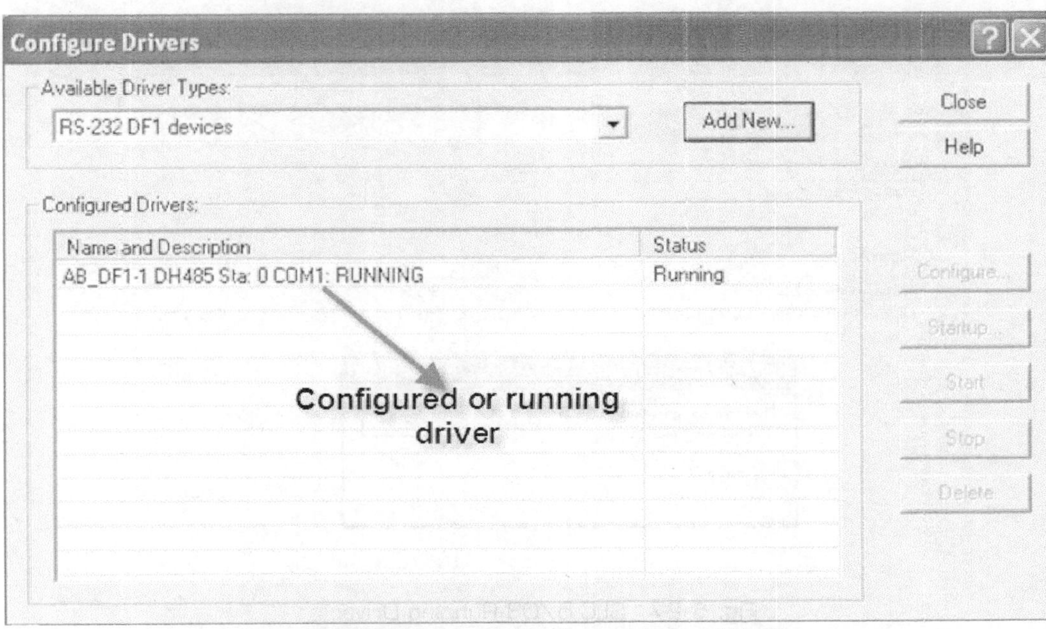

Fig. 5.25 Micrologix Configured Driver Window

Once the driver is successfully configured, it can be seen by clicking RS Who icon as shown in Fig. 5.26.

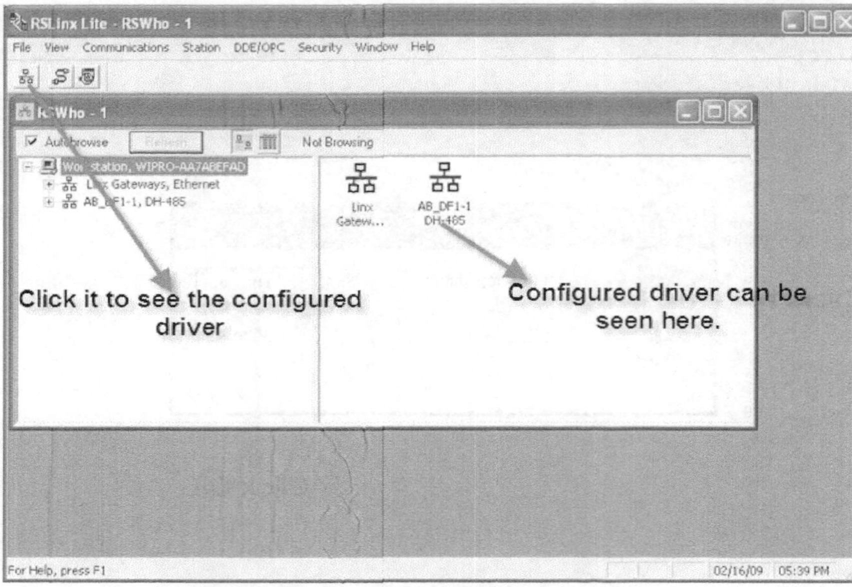

Fig. 5.26 Configured Driver Information

5.3.2 SLC 5/03 Driver Configuration

For configuring SLC 5/03 based system select 1747 PIC/AIC + Driver from the list then click add new. Click OK to finish the driver configuration in small window. Once the driver is configured correctly then it will appear in the name and description field with its status running as shown in Fig. 5.27.

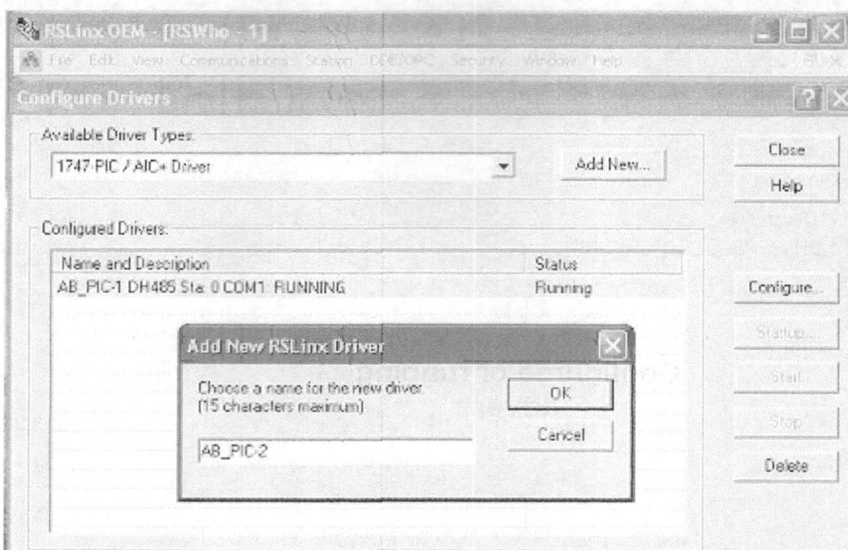

Fig. 5.27 SLC 5/03 Running Driver

Installed driver can be viewed in the RS linx window as shown in Fig. 5.28. Now we can close RS Linx driver configuration wizard.

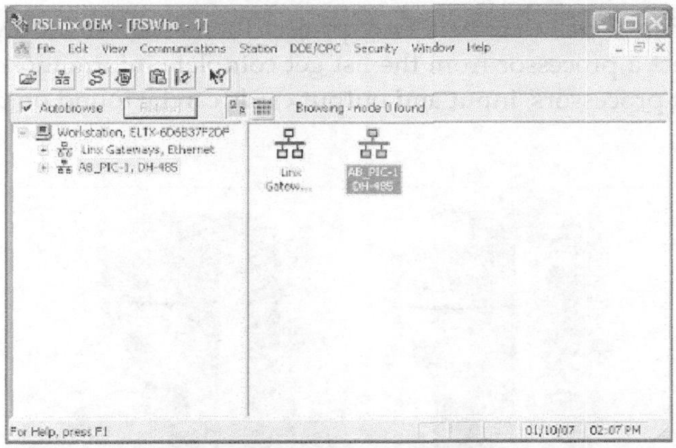

Fig. 5.28 Driver Information

5.4 PROJECT DEVELOPMENT

The three major steps in creating a new project are

1. Program editing
2. Program saving
3. Program running

1. Program Editing

To create a new project open RS logix by selecting Start-Programs-Rockwell Software- RS Logix 500 English as shown in Fig. 5.29 then a new window will open as shown in Fig. 5.30.

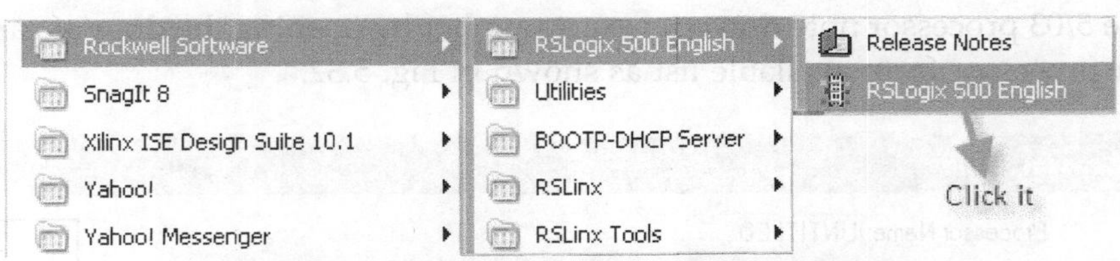

Fig. 5.29 Path to Open RS Logix 500 Window

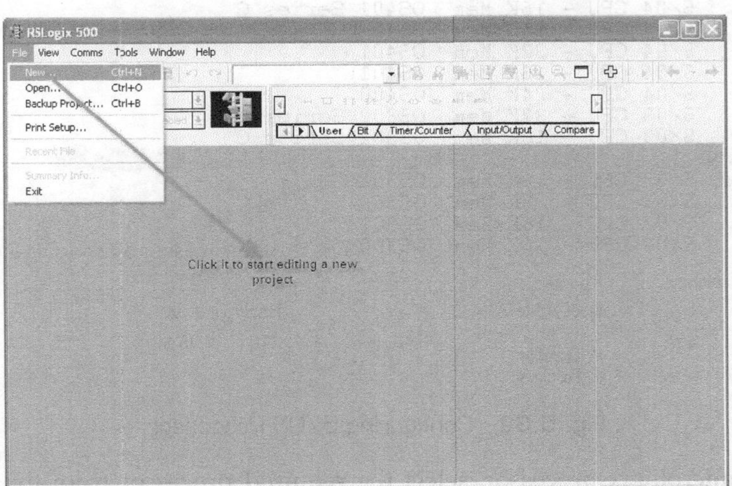

Fig. 5.30 RS Logix 500 Window

From menu select File-New to create a new project then a new window will come to select processor type. To select a processor from the list get complete information about the processor. This information about processors, input and output cards can be found on the location shown in Fig. 5.31.

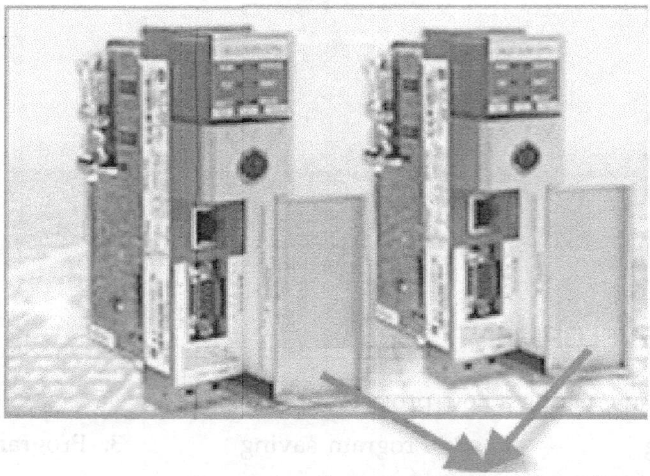

Fig. 5.31 Location for Card Information

To select the 5/03 processor note down its exact number by opening the door of card and select that particular processor form available list as shown in Fig. 5.32.

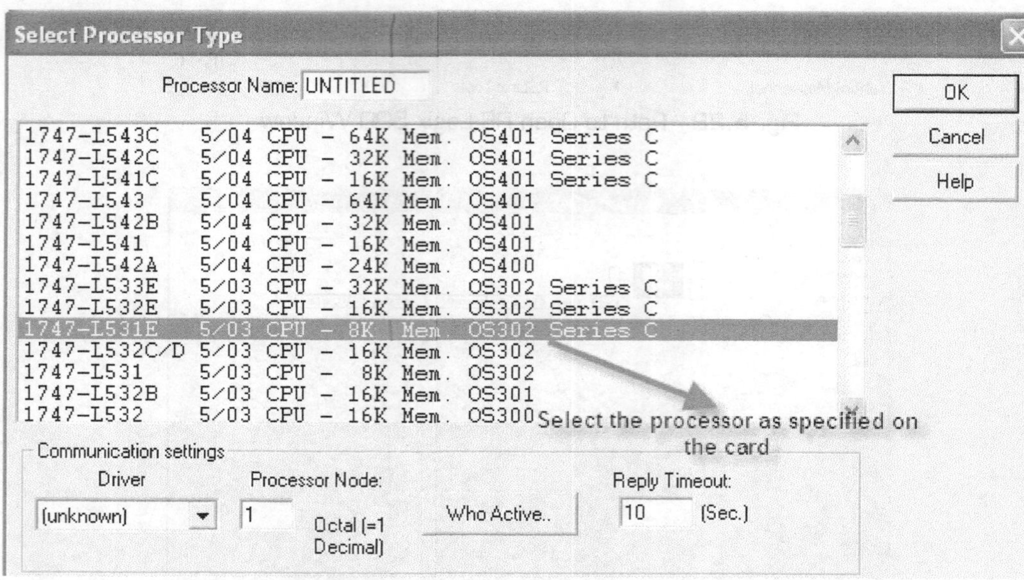

Fig. 5.32 Configuring 5/03 Processor

To configure a micrologix processor, select the required processor from the list as shown in Fig. 5.33.

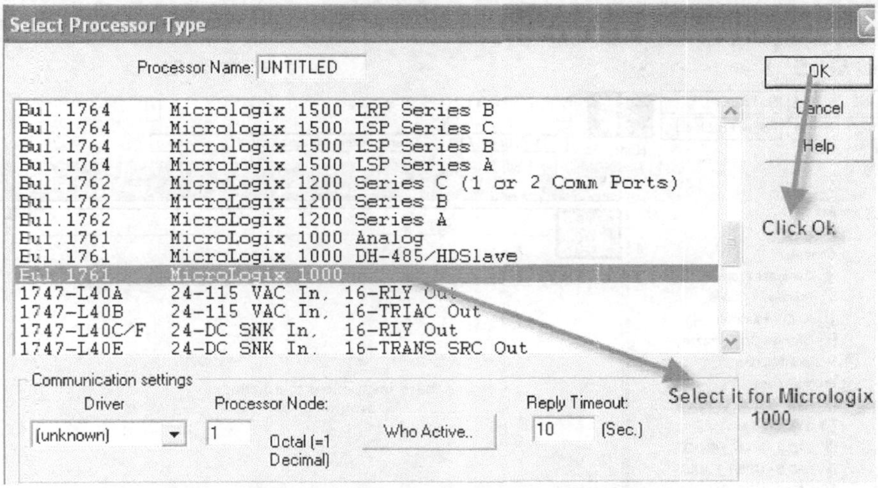

Fig. 5.33 Micrologix Processor Configuration

Once the required processor is selected then click OK to get a new program editor window. For editing a program first input and output configuration is required. In rack based systems this configuration gives information about exact slot numbers of I/O cards in the rack system. If PLC is connected with PC and powered on the select "Read I/O Config" as shown in Fig. 5.34 then it will automatically configure the inputs and outputs. In case of micrologix there is no need of I/O configuration because of the absence of rack system. All I/O and CPU are present on same location.

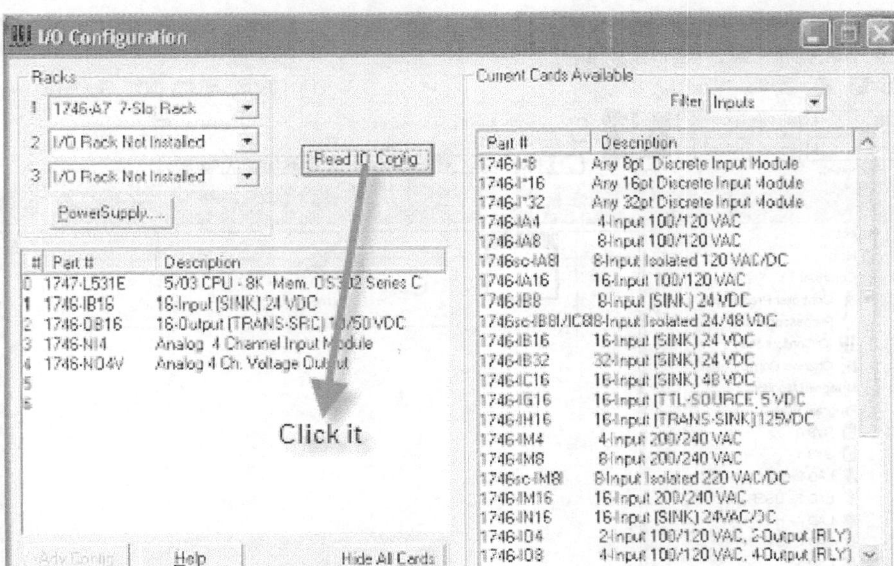

Fig. 5.34 I/O Configuration

In the above figure slot 0 shows the configured processor, slots 1 to 4 show I/O configuration and slots 5, 6 are vacant for future expansion. Click OK to get the program editor window as shown in Fig. 5.35. To edit a program for OR gate select the required instructions from top menu. To make an OR gate we need two no instructions in parallel with an output instruction.

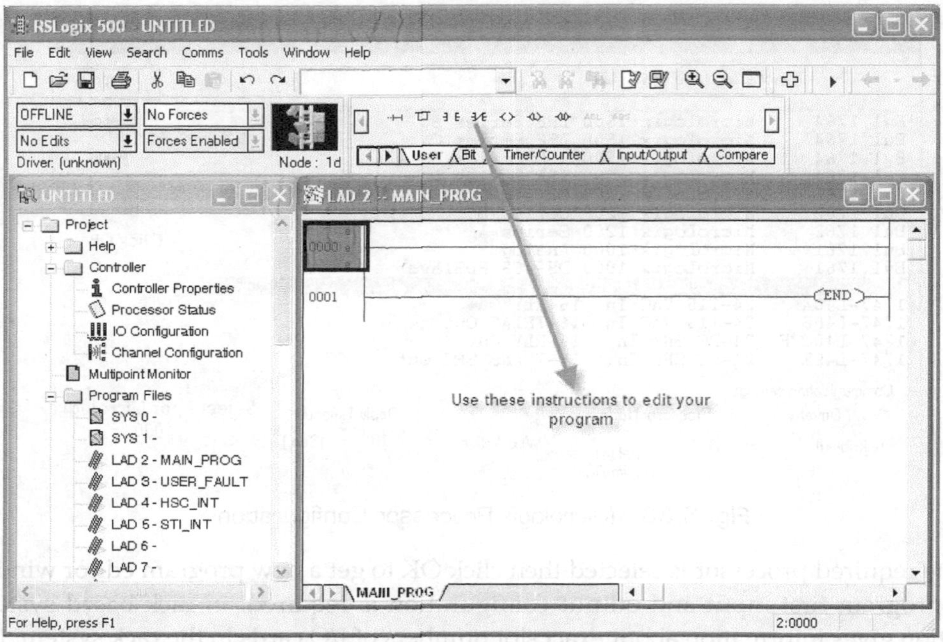

Fig. 5.35 Program Editing Window

1. Click no instruction from top list then it will appear in rung 1 of the program as shown in Fig. 5.36.

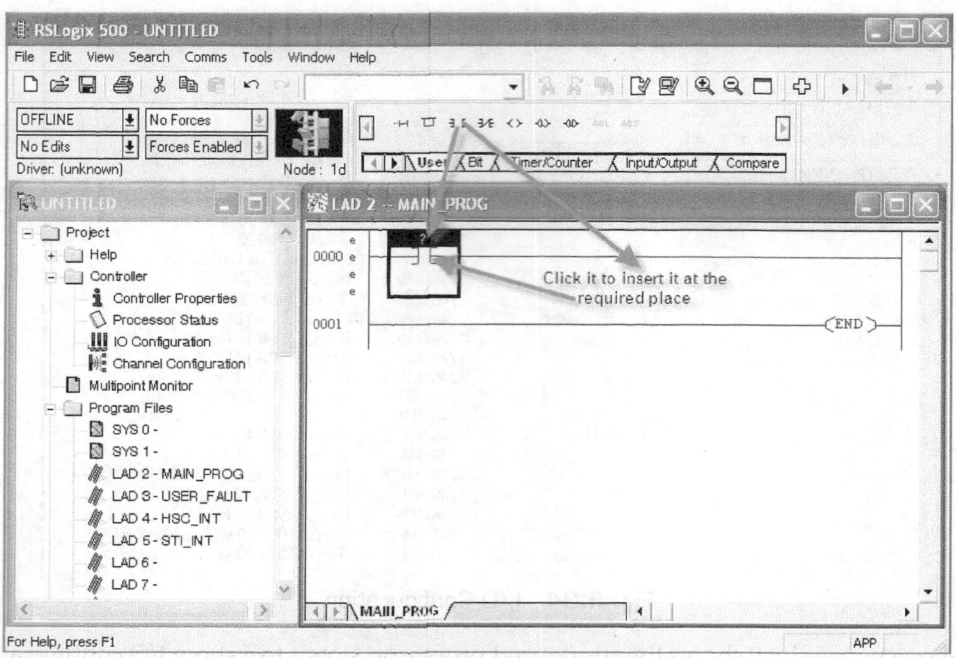

Fig. 5.36 Inserting First NO Instruction

2. Insert branch to place another no in parallel with the first one as shown in Fig. 5.37.

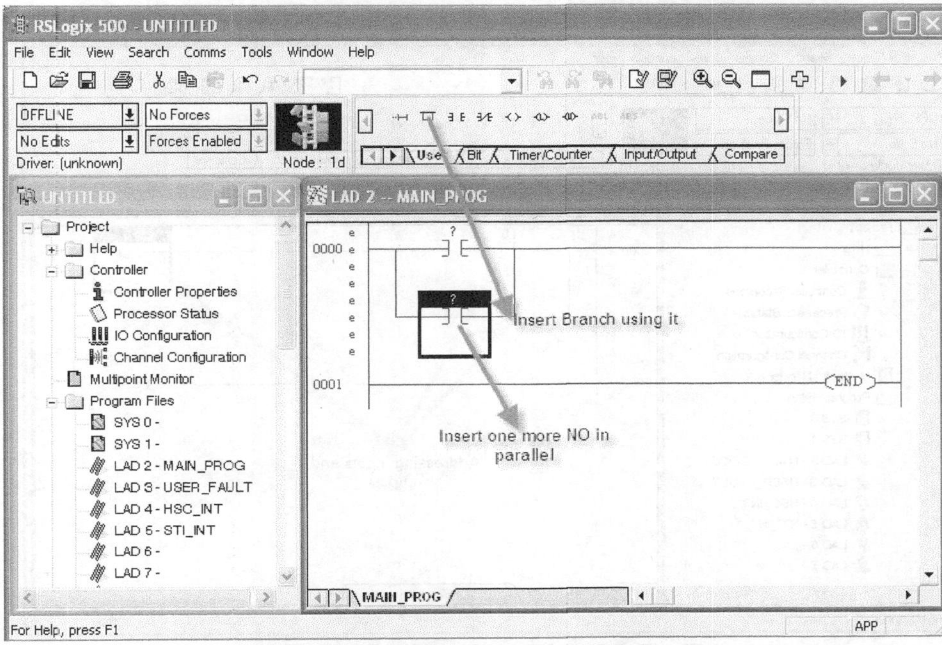

Fig. 5.37 Inserting Second NO in Branch

3. Now insert one output at the end of the rung to complete the program editing for OR gate as shown in Fig. 5.38.

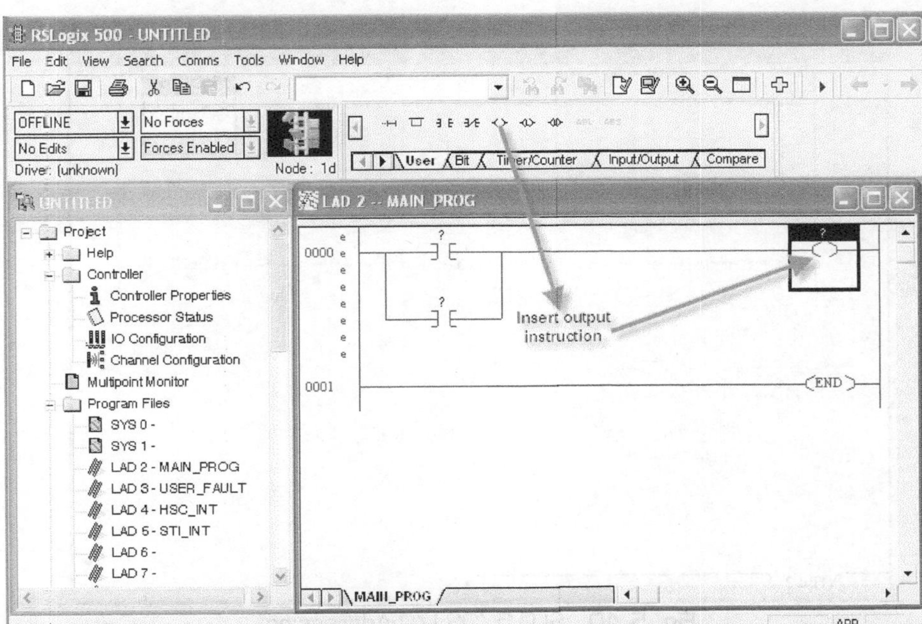

Fig. 5.38 Inserting Output Instruction

4. Enter addresses for inputs and outputs as per your configuration. For micrologix give addresses as shown in Fig. 5.39.

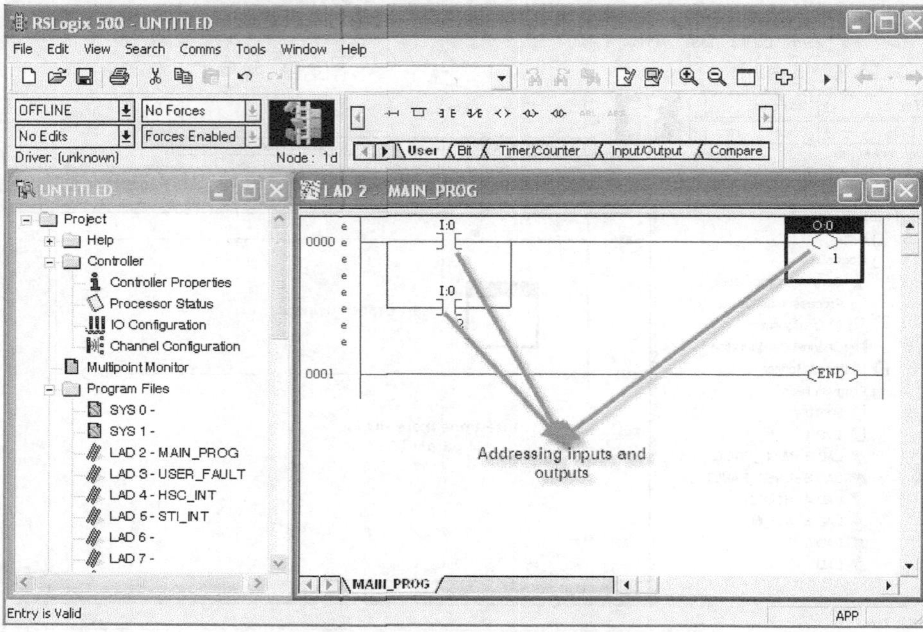

Fig. 5.39 Micrologix I/O Addressing

For SLC 5/03 give addresses as per I/O configuration given in Fig. 5.40. This I/O configuration is shown in Fig. 5.39.

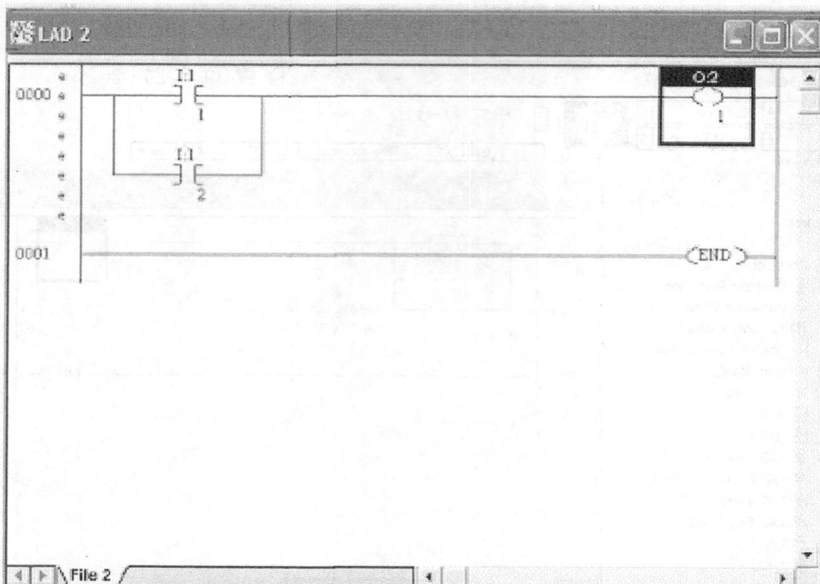

Fig. 5.40 SLC 5/03 I/O Addressing

2. Program Saving

Once the editing part is over then the next step is to save the project. For saving the project select File-save and enter the name of the file *e.g.*, demo1 and click save as shown in Fig. 5.41.

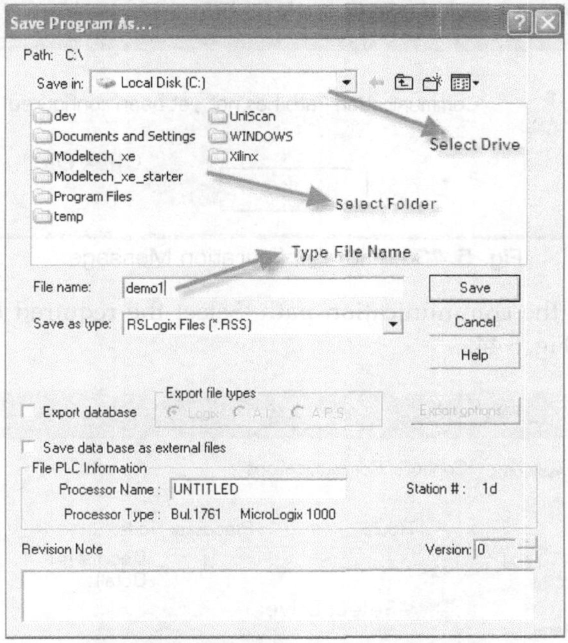

Fig. 5.41 Program Saving Window

3. Program Running

After saving the project the next step is to run it. To run the project select Comms in main menu and click download project as shown in Fig. 5.42.

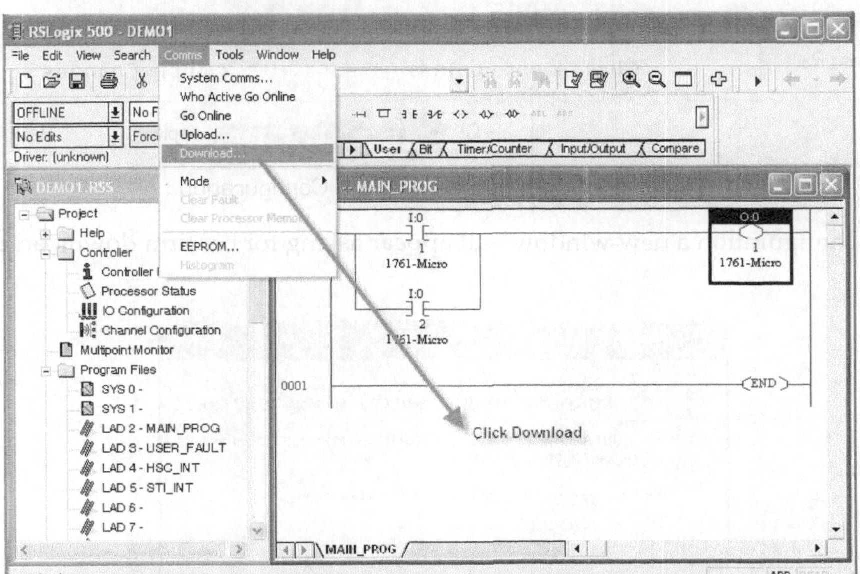

Fig. 5.42 Program Download Window

A new window will appear with a message that "Communication path has not yet been configured" as shown in Fig. 5.43.

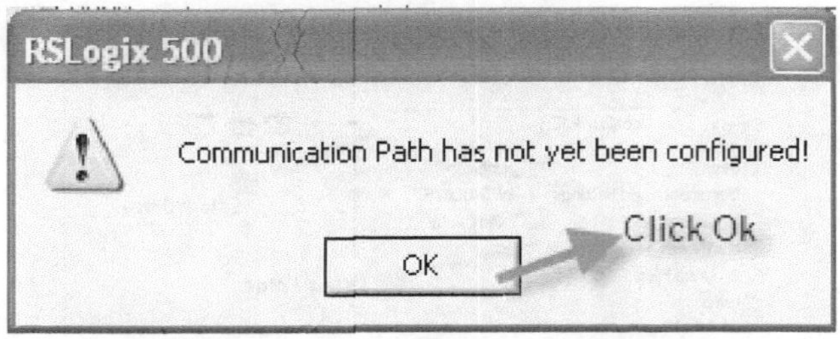

Fig. 5.43 Path Configuration Message

Click OK to configure the communication path. Select the required driver from the list *i.e.*, "AB_DF1-1" as shown in Fig. 5.44.

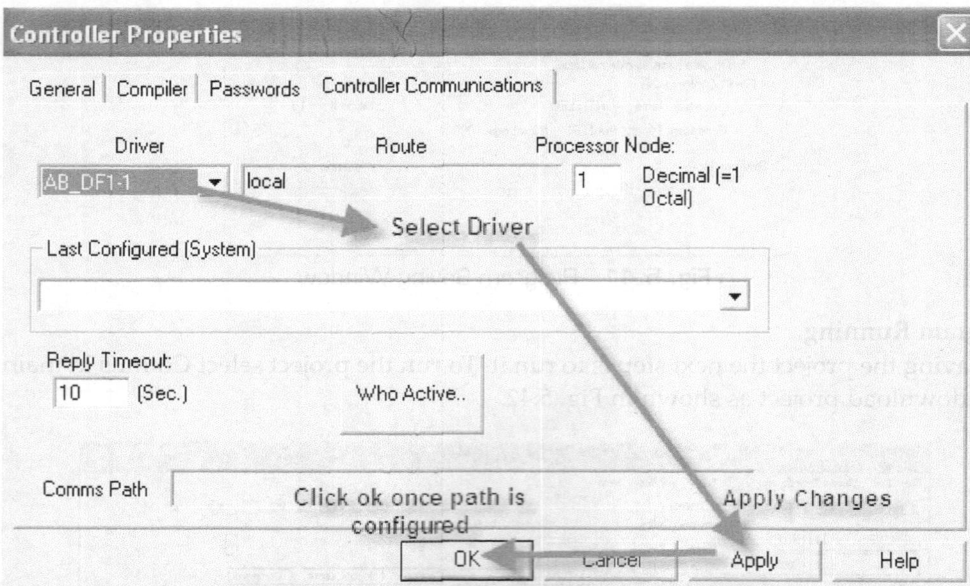

Fig. 5.44 Communication Path Configuration

After path configuration a new window will appear asking for confirm download as shown in Fig. 5.45.

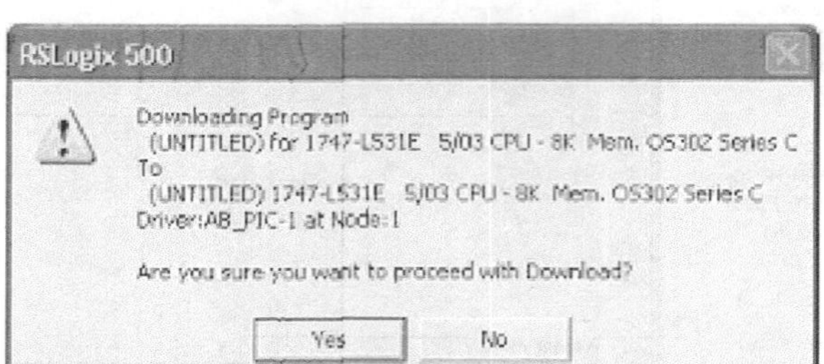

Fig. 5.45 Download Program Confirmation

Click yes to open a new window which is used to switch from run mode to program mode as shown in Fig. 5.46.

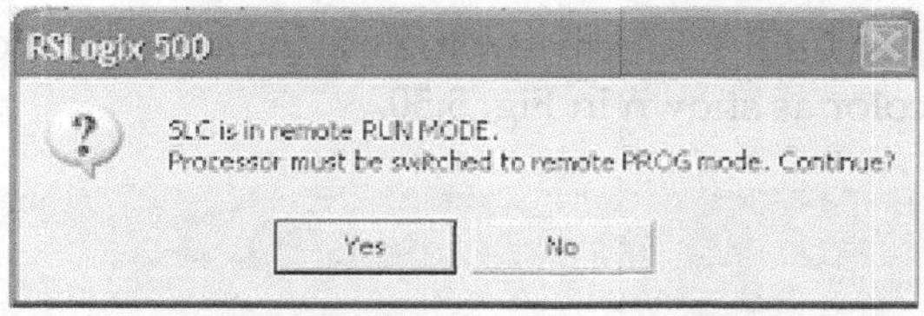

Fig. 5.46 Switch from Remote Run to Remote Program Mode

Again click yes then a new window will appear with a message that "Change back to run mode" as shown in Fig. 5.47. Click yes.

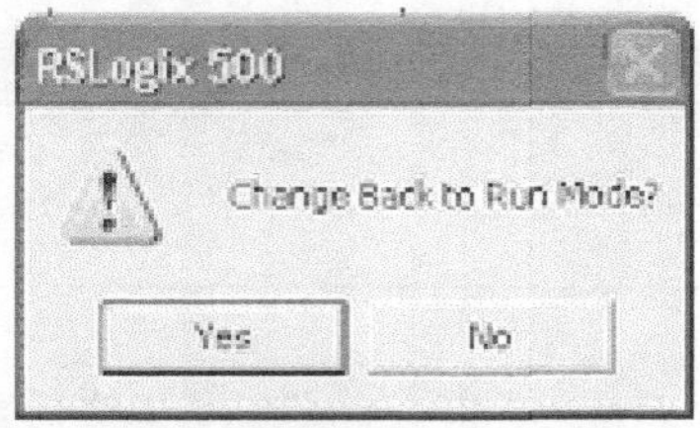

Fig. 5.47 Go Back to Run Mode

Click yes then another window will appear asking for "Do you want go online?" as shown in Fig. 5.48. Click yes.

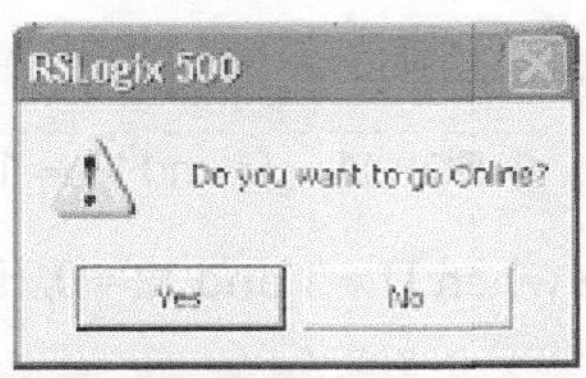

Fig. 5.48 Go Online

Then you will notice green in software lines on both sides of the rung. Here you will notice grey lines on both side of a rung. It indicates that program is in run mode as shown in Fig. 5.49. Now verify the OR gate for different inputs.

Case 1: I1 and I2 = 0 and O1 = 0 (low logic is indicated without color)

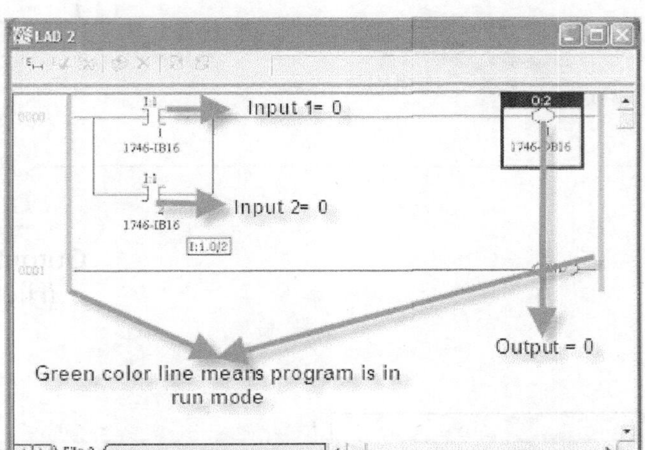

Fig. 5.49 I1 = 0 and I2 = 0

Case 2: I1 = 0 and I2 = 1, then O1 = 1, where low logic is indicated without color and high logic is indicated by green or grey color as shown in Fig. 5.50.

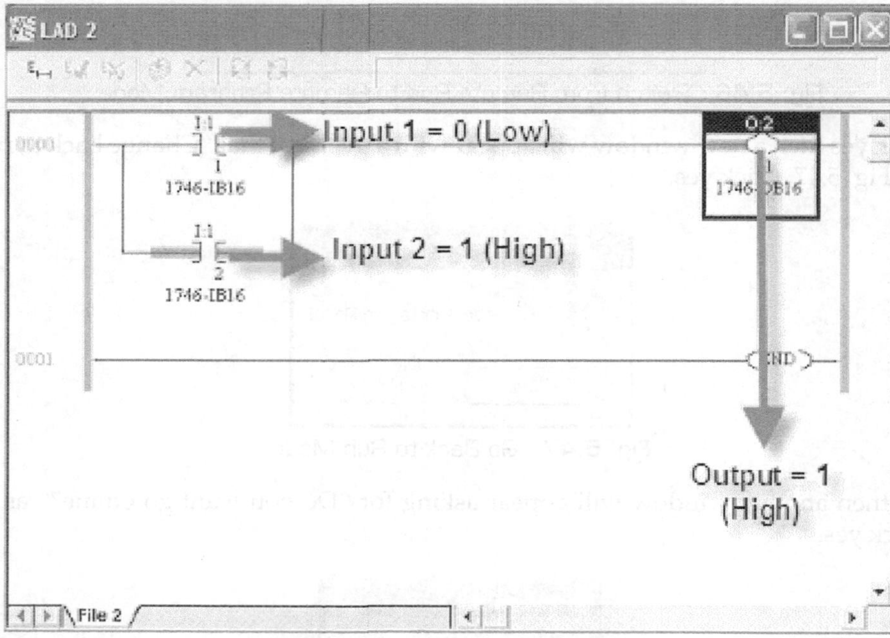

Fig. 5.50 I1 = 0 and I2 = 1

Case 3: Figure 5.51. Shows the case when I1 = 1 and I2 = 0, then O1 = 1.

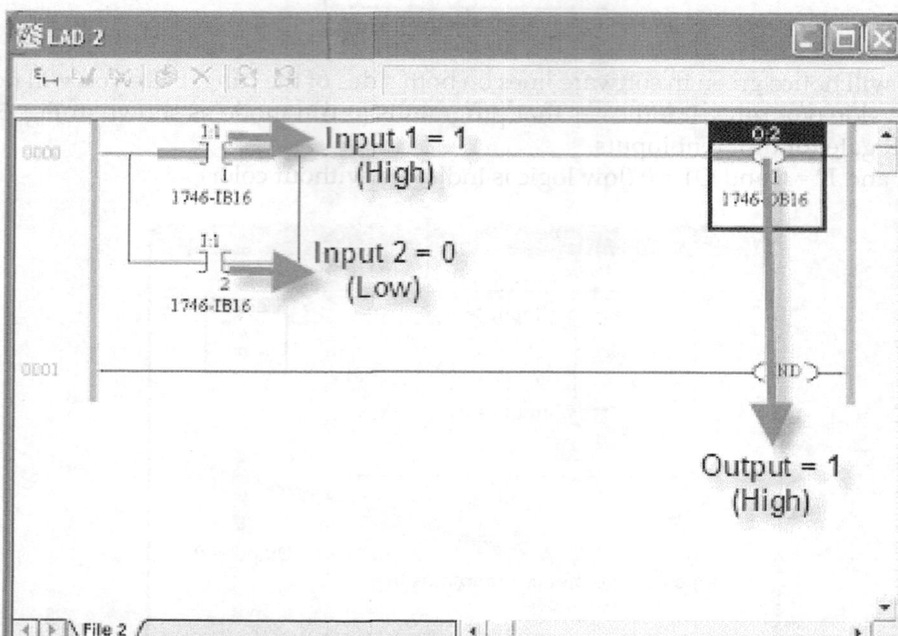

Fig. 5.51 I1 = 0 and I2 = 1

Case 4: Figure 5.52. Shows the case when I1 =1 and I2 = 1, then O1 = 0.

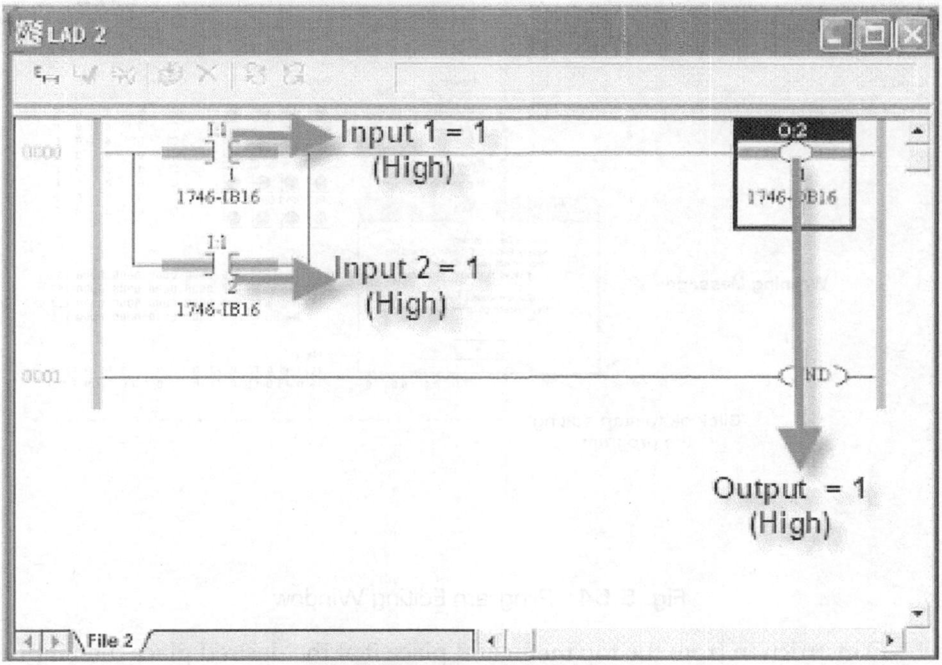

Fig. 5.52 I1 = 1 and I2 = 1

In this way we can verify our projects using different PLCs with the help of RS logix 500 programming software. Run "XORDemo" given in "Demos" folder of attached CD to view the working of above project in more detail. Before running XORDemo, install Instantdemo setup on your PC given in Instant demo folder of CD.

5.5 LADSIM BASED SIMULATION

In this section we will simulate ladder diagrams using demo version of ladsim software. The steps to simulate OR gate using ladsim are as follows:

1. Open the ladsim software by double clicking the ladsim icon shown in Fig. 5.53.

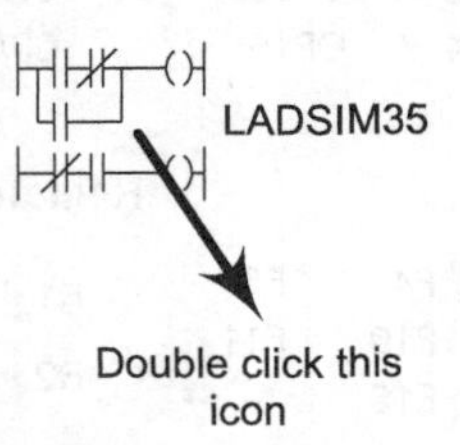

Fig. 5.53 Ladsim Icon

2. It will open a new window with a warning message that it is demo version and will stop after 20 minutes as shown in Fig. 5.54. In this version saving, printing, external I/Os and help files are disabled. Click OK to start editing the program.

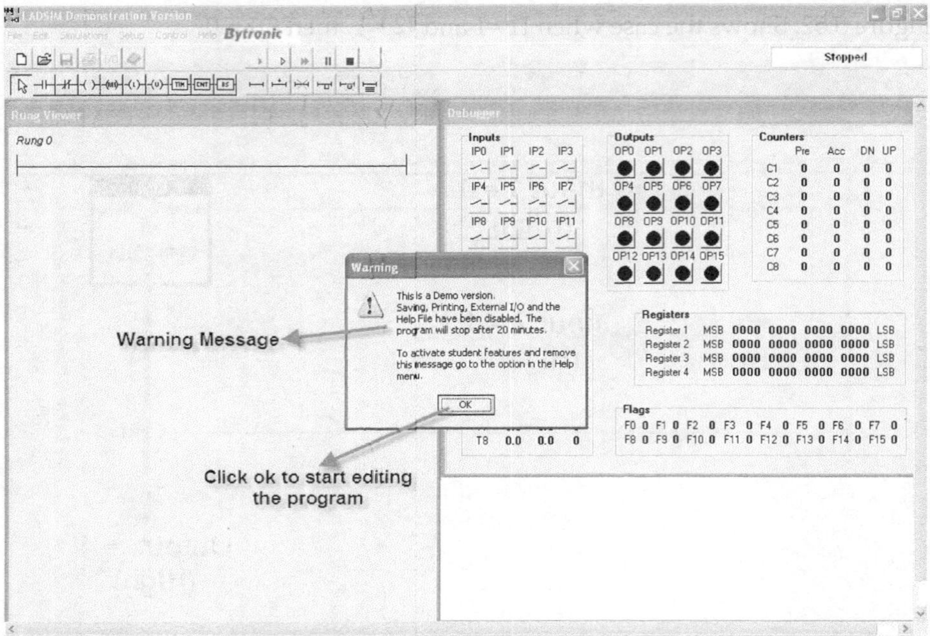

Fig. 5.54 Program Editing Window

3. Select NO instruction from the top menu and place it at the desired place then input address assignment window will appear as shown in Fig. 5.55. Give input address as IP0 and click OK.

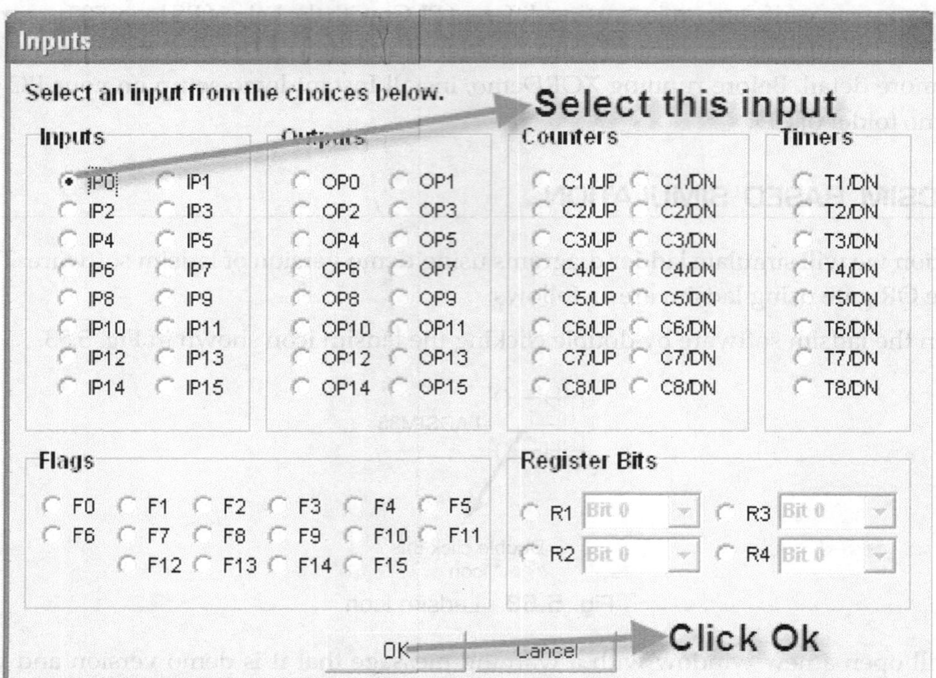

Fig. 5.55 Input Address Assignment

4. Then NO instruction with specified address can be seen in rung as shown in Fig. 5.56.

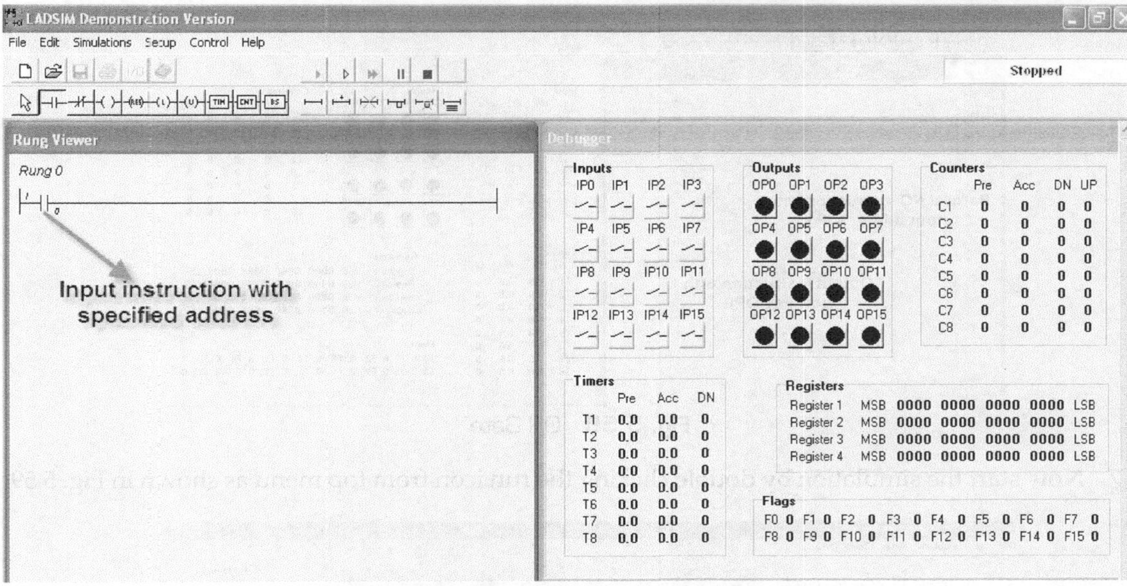

Fig. 5.56 Rung with Input Instruction

5. Insert a branch to add one more no instruction in parallel with previous one to complete the OR gate. A message will appear asking for confirmation that "Is this branch correct" click yes if branch is placed at proper place as shown in Fig. 5.57.

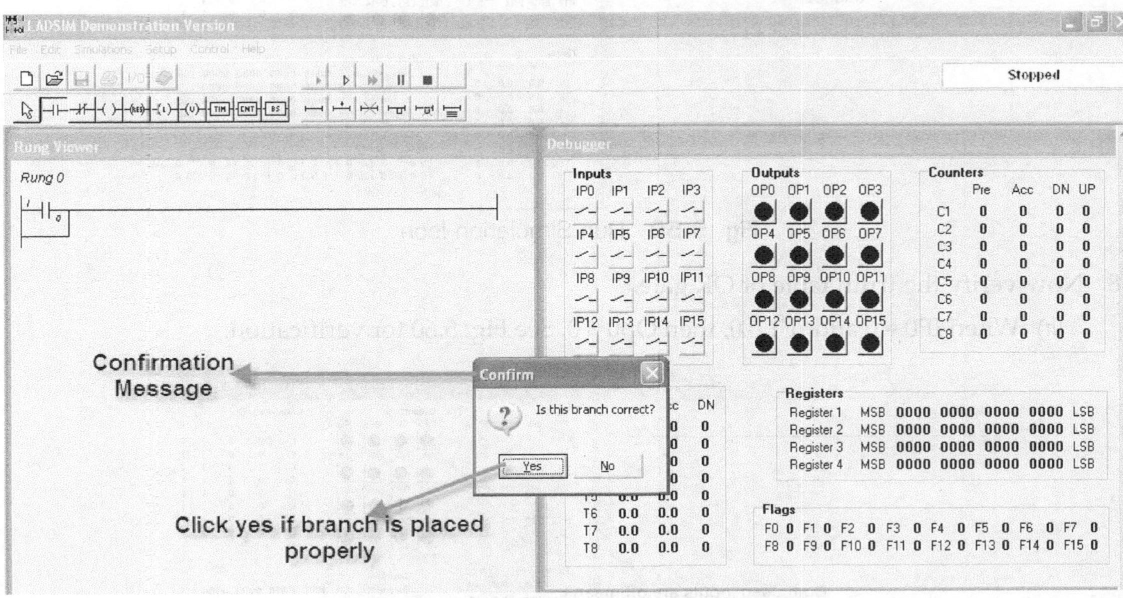

Fig. 5.57 Branch Confirmation Message

6. Insert one no instruction in created branch by giving input address as IP1 and complete the OR gate by inserting an output with address OP0 as shown in Fig. 5.58.

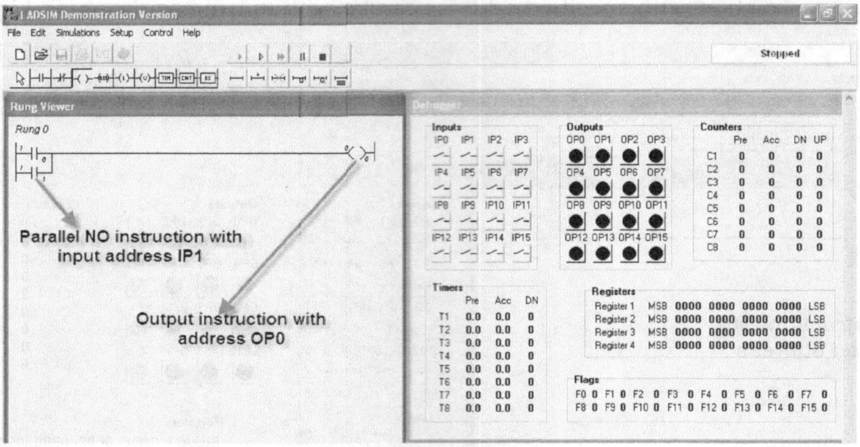

Fig. 5.58 OR Gate

7. Now start the simulation by double clicking the run icon from top menu as shown in Fig. 5.59.

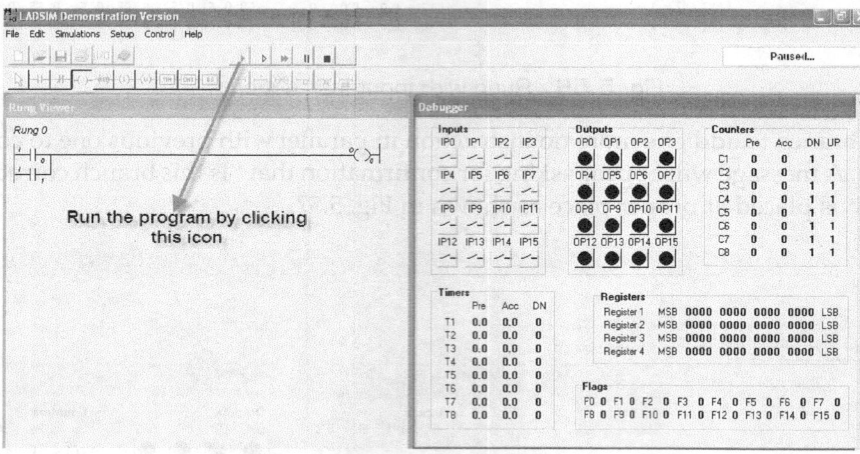

Fig. 5.59 Run Simulation Icon

8. Now verify the truth table of OR gate.
   (*a*) When IP0 = 0 and IP1 = 0, then Op0 = 0. See Fig. 5.60 for verification.

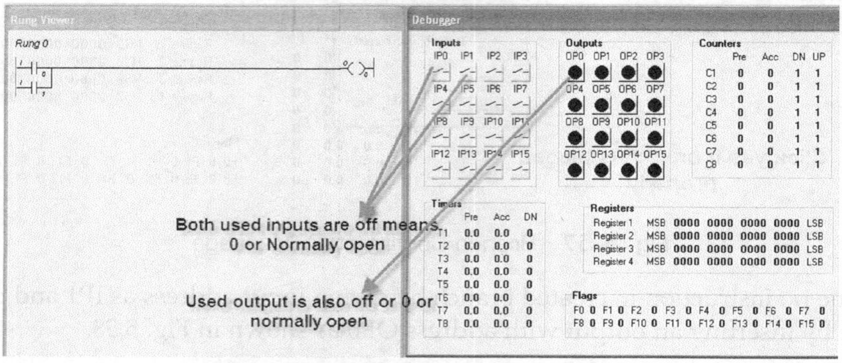

Fig. 5.60 OR Gate Output at IPO = 0 and IP1 = 0

(*b*) IP0 = 0 and IP1 = 1, then Op0 = 1. See Fig. 5.61 for verification.

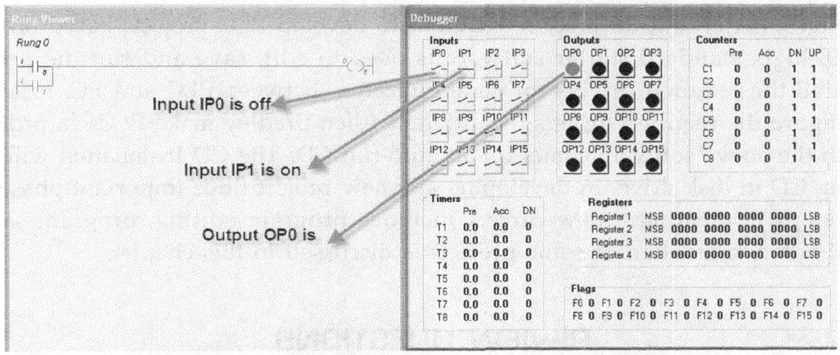

Fig. 5.61 OR Gate Output at IPO = O and IP1 = 1

(*c*) IP0 = 1 and IP1 = 0, then Op0 = 1. See Fig. 5.62 for verification.

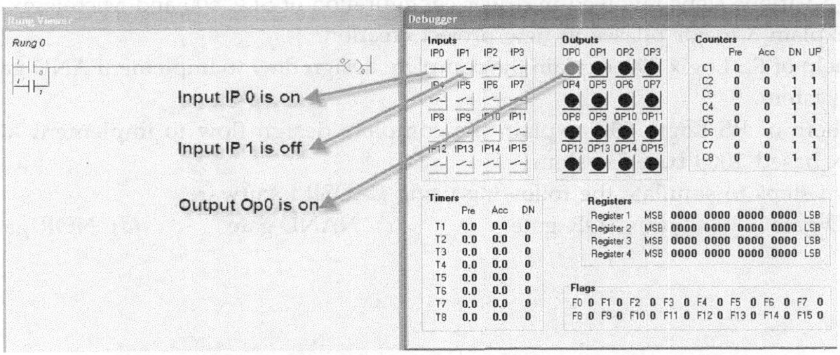

Fig. 5.62 OR Gate Output at IPO = 1 and IP1 = O

(*d*) IP0 = 1 and IP1 = 0, then Op0 = 1. See Fig. 5.63 for verification.

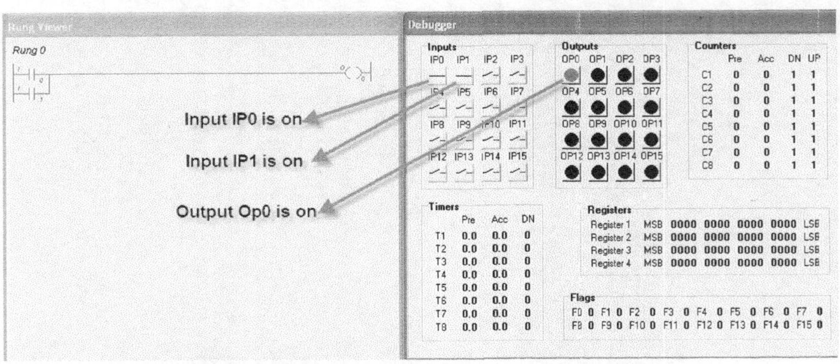

Fig. 5.63 OR Gate Output at IPO = 1 and IP1 = 1

So in this way any user program can be simulated with the help of LADSIM software. Run "XORDemoLS" given in "Demos" folder of attached CD to view the editing and simulation of above project in Ladsim software. In the remaining part of this book we will use RS logix 500 software to edit programs for various applications. The reader can try these programs on LADSIM software as self task.

SUMMARY

RS logix 500 standard/starter and RS linx are required to develop PLC projects based on SLC 500 based system design. RS logix standard/starter software is used to edit, save and run the program. RS linx software contains all the required drivers for communication between PLC and PC. It is necessary software used to configure the required drivers for different Allen Bradley make PLCs in order to communicate with PC. Both the above software comes on an auto-run CD. The CD Installation will start automatically after inserting CD in disk drive. In developing any new project three important phases are followed. The three major steps in creating a new project includes program editing, program saving, program running. LADSIM based simulation of some projects is discussed in this chapter.

REVIEW QUESTIONS

1. List major steps used in installation of RS Logix 500 and RS Linx software.
2. What do you mean by driver configuration?
3. Write down various steps involved in driver configuration of SLC 500 and Micrologix based systems.
4. List and explain 3 major phases in new project creation.
5. With the help of RS Logix 500, explain the complete design flow to implement AND gate using SLC5/03 based system.
6. With the help of RS Logix 500, explain the complete design flow to implement XOR gate using Micrologix based 1000 based system.
7. Write down steps to simulate the following using LADSIM software:

   (*a*) AND gate (*b*) XOR gate (*c*) NAND gate (*d*) NOR gate

6

INSTRUCTION SET-I

LEARNING OBJECTIVES

After reading this chapter, you will be able to explain :

Operation and significance of various bit, timer and counter instructions and reset instruction.

6.1 INTRODUCTION

In this chapter we will discuss basic instruction set. It can be divided into three categories which are as follows:

1. Bit instructions 2. Timer instructions 3. Counter instructions

6.2 BIT INSTRUCTIONS

Bit instructions operate on a single bit of data. The processor may set or reset the bit, based on logical continuity of ladder rungs during an operation. A bit can be addressed as many times as the program requires. Bit instructions are commonly used with output, input files, status file, bit file, timer file, counter file and sequencer file. Table 6.1 shows the various bit instructions along with its name and purpose.

Table 6.1 Bit Instructions

Instruction	Name	Function
XIC	Examine if closed	Used to check a bit for ON condition
XIO	Examine if open	Used to check a bit for OFF condition
OTE	Output energize	Used to turn a bit ON or OFF.
OTL	Output latch	Used to turn a bit ON and this bit retain its state
OTU	Output unlatch	Used to turn a bit OFF and this bit retain its state
OSR	One shot rise	Triggers a one-time event

6.2.1 Examine if Closed (XIC) Instruction

This instruction is equivalent to Normally Open (NO) instruction discussed in chapter 1. It is used in the ladder program to determine if a bit is ON. During execution if the XIC instruction is ON or logically 1 then the instruction is evaluated as true. During execution, if the XIC instruction is OFF or logically 0 then the instruction is evaluated as false. The XIC symbol and different states are shown in Fig. 6.1. Examples of an XIC instruction include a push button wired to an input, an output wired to a valve, a timer controlling an indicator lamp etc.

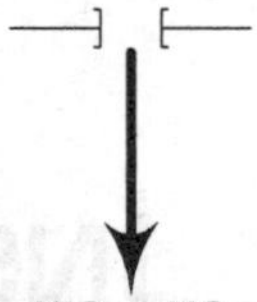

Bit State	XIC Instruction
0 or low	Instruction is evaluated as false
1 or high	Instruction is evaluated as true

Fig. 6.1 XIC Instruction Operation

6.2.2 Examine if Open (XIO) Instruction

This instruction is equivalent to Normally Closed (NC) instruction discussed in chapter 1. It is used in the ladder program to determine if a bit is OFF. During execution, if the XIC instruction is OFF or logically 0 then the instruction is evaluated as true. During execution if the XIC instruction is ON or logically 1 then the instruction is evaluated as false. The XIC symbol and different states are shown in Fig. 6.2. Examples of an XIO instruction include a sensor wired to an input, an output wired to a conveyor, a timer controlling a hooter etc.

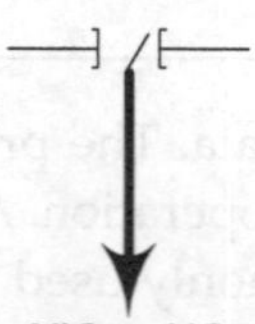

Bit State	XIC Instruction
0 or low	Instruction is evaluated as false
1 or high	Instruction is evaluated as true

Fig. 6.2 XIO Instruction Operation

6.2.3 Output Energize (OTE) Instruction

An OTE instruction is used in the ladder program to turn a bit ON when rung conditions are evaluated as true. It turns a bit OFF when rung conditions are evaluated as false. The symbol for OTE instruction is shown in Fig. 6.3. For example an output wired to a light.

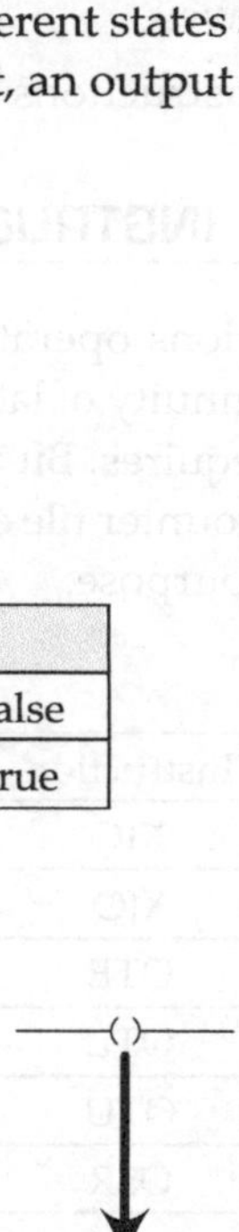

Fig. 6.3 OTE Instruction Symbol

Example: *Demonstration of XIC, XIO and OTE instructions.*

The ladder diagram given below in Fig. 6.4 shows the implementation and verification of XOR gate using RS Logix 500 software to demonstrate the working of XIC, XIO and OTE instructions. The inputs and output shown with green or grey color means these are logically 1 or high.

Case 1. When input 1 = 0 and input 2 = 0, then output 1 = 0

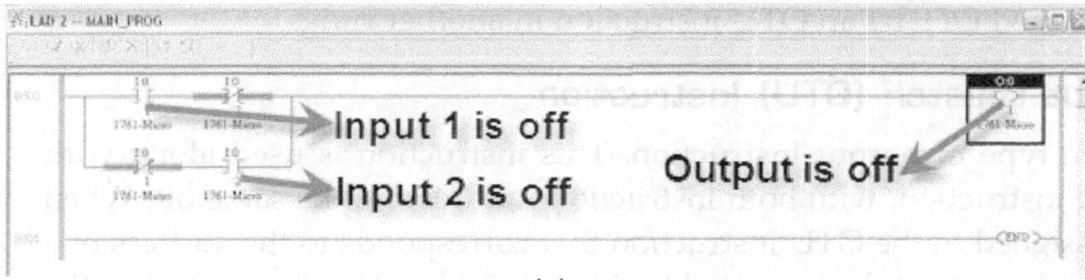

(a)

Case 2. When input 1 = 0 and input 2 = 1, then output 1 = 1

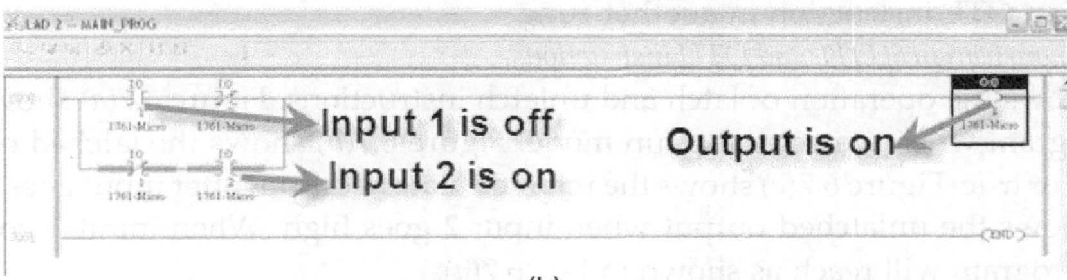

(b)

Case 3. When input 1 = 1 and input 2 = 0, then output 1 = 1

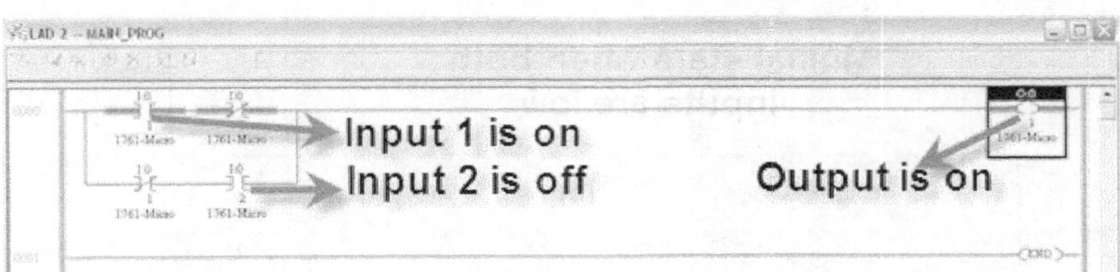

(c)

Case 4. When input 1 = 1 and input 2 = 1, then output 1 = 0

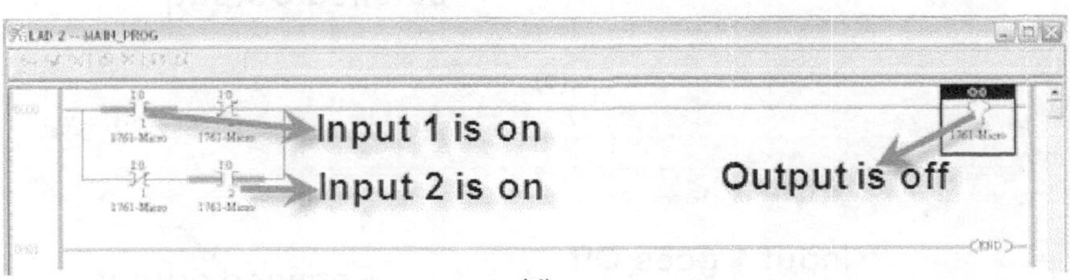

(d)

Fig. 6.4 Ladder Diagram XOR Gate

6.2.4 Output Latch (OTL) Instruction

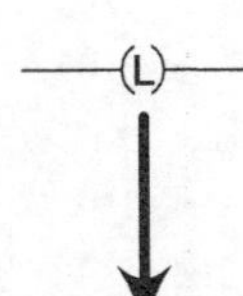

OTL or Output Latch Instruction
Fig. 6.5

This is retentive type of output instruction. An OTL can only turn on a bit, while OTU can only turn off a bit. This instruction is used along with unlatch or OTU instruction, with both instructions addressing the same bit. When an address is assigned to the OTL instruction that corresponds to the address of a physical output, the output device wired to this terminal is energized when the bit is set. The symbol for OTL instruction is shown in Fig. 6.5.

When rung conditions become true to false then the bit remains set and the corresponding output device remains energized. Thereafter, the bit remains ON, regardless of the rung condition, until the bit is turned OFF by an OTU instruction in another rung.

6.2.5 Output Unlatch (OTU) Instruction

This is retentive type of output instruction. This instruction is used along with unlatch or OTU instruction, with both instructions addressing the same bit. When an address is assigned to the OTU instruction that corresponds to the address of a physical output, the output device wired to this terminal is de-energized when the bit is set. The symbol for OTU instruction is shown in Fig. 6.6.

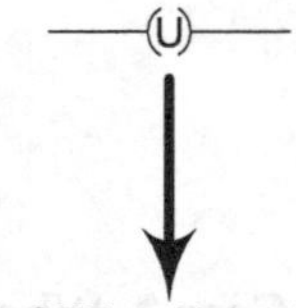

OTU or Unlatch Instruction

Fig. 6.6

Thereafter, the bit remains OFF, regardless of the rung condition, until it is turned on typically by an OTL instruction in another rung.

Example: *Demonstration of OTL and OTU instructions.*

Figure 6.7 shows the operation of latch and unlatch instructions. Figure 6.7(*a*) shows the initial state of the program when we enter the run mode. Figure 6.7(*b*) shows the latched output when input goes high or true. Figure 6.7(*c*) shows the retained state of output after input goes low or false. Figure 6.7(*d*) shows the unlatched output when input 2 goes high. When input 2 goes low then initial state of program will reach as shown in Fig. 6.7(*a*).

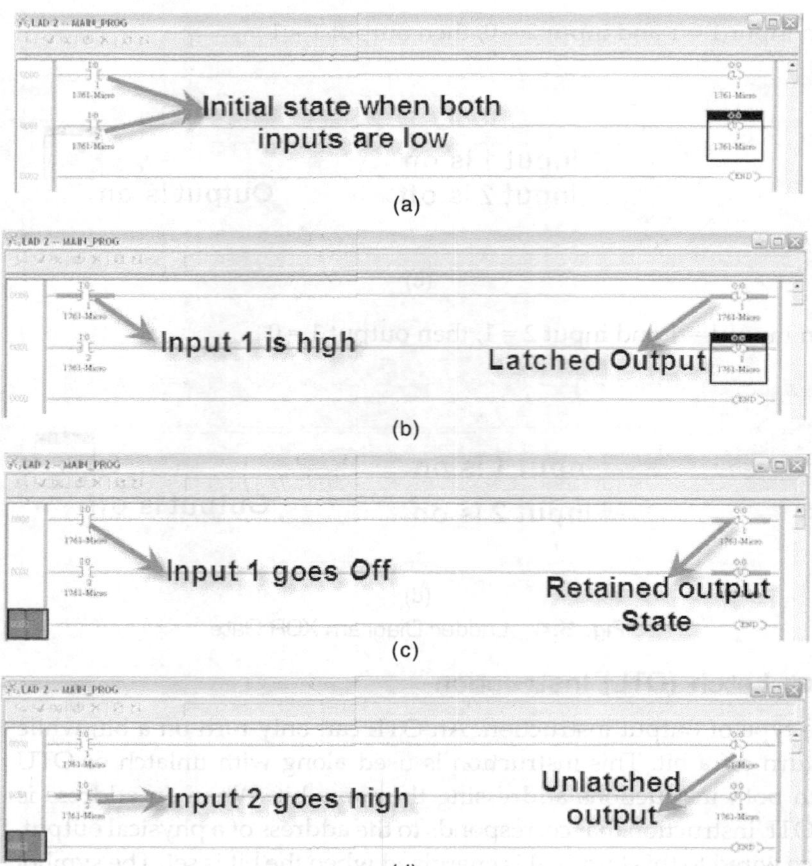

Fig. 6.7 Latch and Unlatch Operation

6.2.6 One-Shot Rising (OSR)

The OSR instruction is a retentive input instruction that triggers an event to occur one time. The symbol for OSR instruction is shown in Fig. 6.8.

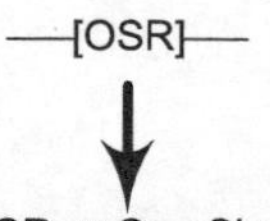

OSR or One Shot Rise Instruction

Fig. 6.8

When the rung conditions preceding the OSR instruction go from false-to-true, the OSR instruction will be true for one scan. After one scan is complete, the OSR instruction becomes false, even if the rung conditions preceding it remain true. The OSR instruction will only become true again if the rung conditions preceding it changes from false-to-true. The SLC 500 and SLC 5/01 processors allow to use one OSR instruction per output in a rung; the OSR cannot be within a branch. The SLC 5/02 and higher processors allow using one OSR instruction per output in a rung; putting the OSR within a branch is permitted.

To program this instruction use a bit address from either the bit or integer data file. The bit address used for OSR instruction must be unique, don't use it elsewhere in the program. Do not use an input or output address to program the address parameter of the OSR instruction.

Example: *Demonstration of an OSR instruction.*

Figure 6.9 shows the operation of OSR instruction. Figure 6.9(*a*) shows the initial state of the program when we enter the run mode. Figure 6.9(*b*) shows the latched output when input goes high or true. Figure 6.9(*c*) shows the retained state of output after input goes low or false. Figure 6.9(*d*) shows the unlatched output when input 2 goes high. When input 2 goes low then initial state of program will reach as shown in Fig. 6.9(*a*).

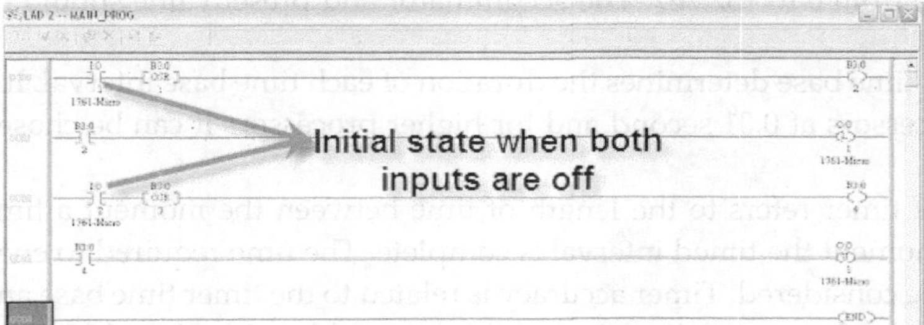

(a)

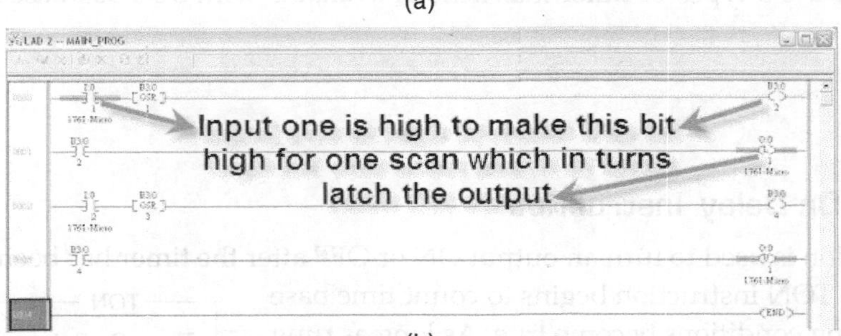

(b)

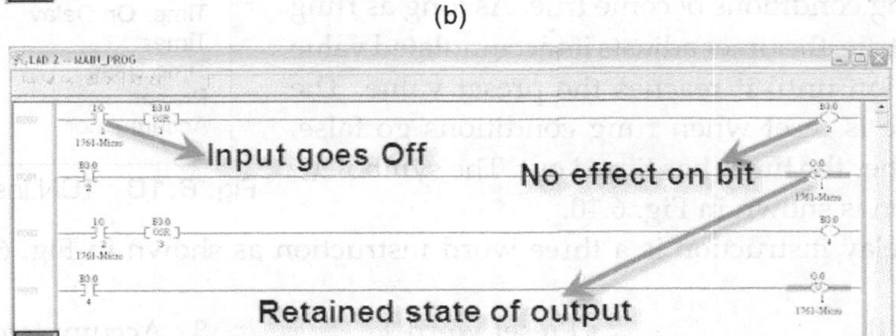

(c)

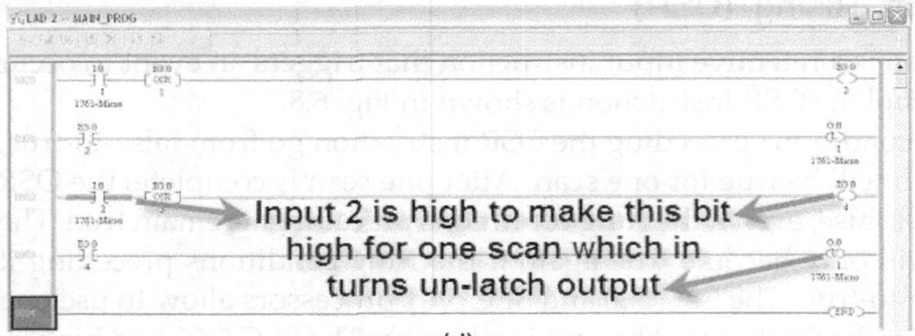

(d)

Fig. 6.9 OSR Instruction Operation

6.3 TIMER INSTRUCTIONS

To program a timer instruction, programming of some parameters is required which are as follows:

(*i*) **Accumulator Value :** This is the time elapsed since the timer was last reset. When enabled, the timer updates its accumulator value continually.

(*ii*) **Preset Value :** It is the value at which the timer must reach before the controller sets the done bit. When the accumulator value becomes equal to or greater than the preset value, the done (DN) bit is set. This bit is used to control an output device. Preset and accumulator values for timers range from 0 to +32,767. The accumulator and preset value should not be a negative number.

Time base: The time base determines the duration of each time base interval. It is fixed for fixed and SLC 5/01 processors at 0.01 second and for higher processors it can be chosen from different available options.

The accuracy of timer refers to the length of time between the moment a timer instruction is enabled and the moment the timed interval is complete. The time required to energize the output device must also be considered. Timer accuracy is related to the timer time base and the processors scan length. There are 3 types of timer instructions available with SLC 500 based systems which are as follows:

1. TON timer
2. TOF timer
3. RTO timer

6.3.1 Timer On-Delay Instruction

The TON instruction is used to turn an output ON or OFF after the timer has been ON for a preset time interval. The TON instruction begins to count time base intervals when rung conditions become true. As long as rung conditions remain true, the timer adjusts its accumulated value after each evaluation until it reaches the preset value. The accumulated value is reset when rung conditions go false, regardless of whether the timer has timed out. The symbol for timer on instruction is shown in Fig. 6.10.

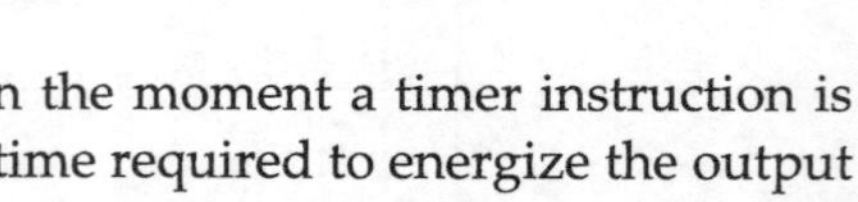

Fig. 6.10 TON Instruction Symbol

The timer on delay instruction is a three word instruction as shown in Fig. 6.11. These three words are:

1. Status word 0
2. Preset word 1
3. Accumulator word 2

Word Number	Bit Numbers															
	15	14	13	12	11	10	9	8	7	6	5	4	3	2	1	0
0	EN	TT	DN						Internal Use							
1	Preset Value (PRE)															
2	Accumulator Value (ACC)															

Fig. 6.11 Three Words of TON Instruction

The word 0 of timer instruction consists of three status bits which are as follows and explained in Table 6.2.

(*a*) Enable bit (*b*) Timer timing bit (*c*) Done bit

Table 6.2 Status of Control Bits of TON Timer

Bit Description	Set	Reset
Timer Done Bit DN (Bit 13)	When accumulator value is equal or greater than preset value	When rung conditions go false
Timer Timing Bit TT(Bit 14)	When rung conditions are true and accumulator value is less than the preset value	When rung condition goes false or the done bit is set
Timer Enable Bit EN (Bit 15)	When rung conditions are true	When rung conditions are false

Example: *Demonstration of TON instruction operation.*

Figure 6.12 shows the operation of TON instruction. It explains the conditions for setting and resetting of three control bits. Figure 6.12(*a*) shows the initial state of the program when we enter the run mode. Figure 6.12(*b*) shows the condition when timer is enabled and doing its operation. Figure 6.12(*c*) shows the condition when timer is enabled and completes its operation. When input 1 goes low then initial state of program will reach as shown in Fig. 6.12(*a*).

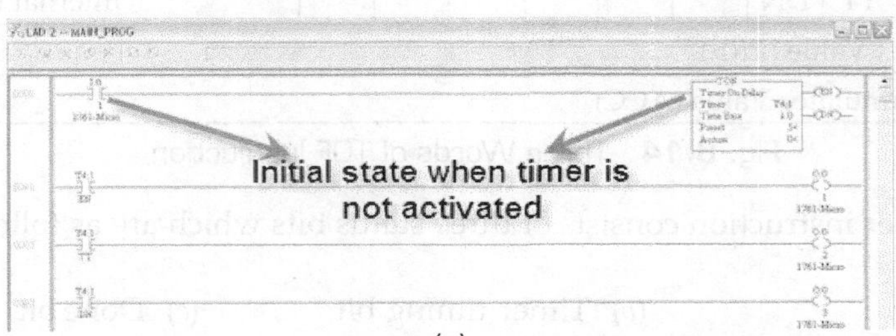

(a)

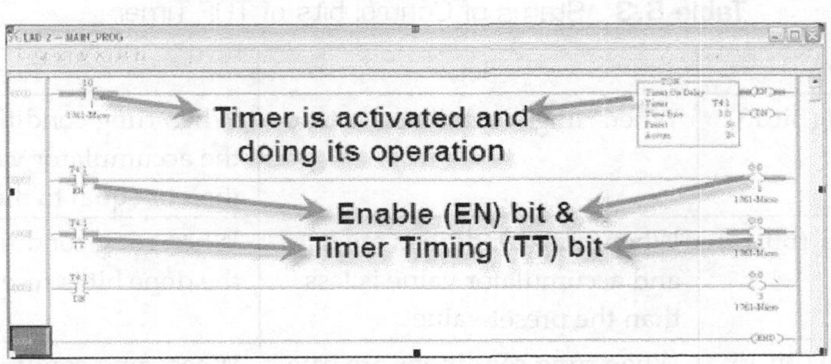

(b)

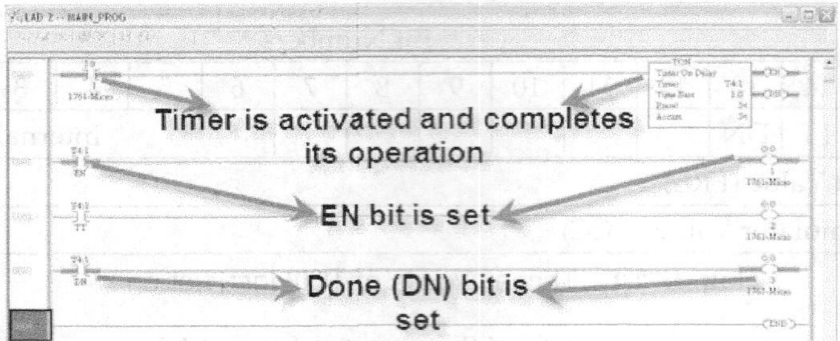

(c)

Fig. 6.12 TON Instruction Operation

6.3.2 Timer Off-Delay Instruction

The TOF instruction is used to turn an output ON or OFF after its rung has been OFF for a preset time interval. The TOF instruction begins to count time base intervals when the rung makes a true-to-false transition. As long as rung conditions remain false, the timer increments its accumulated value based on the time base for each scan until it reaches the preset value. The accumulated value is reset when rung conditions go true regardless of whether the timer has timed out. The symbol for timer off instruction is shown in Fig. 6.13.

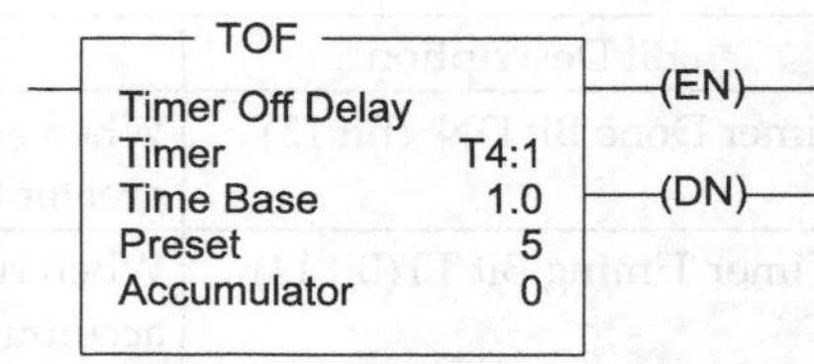

Fig. 6.13 TOF Instruction Symbol

The timer off delay instruction is also a three word instruction as shown in Fig. 6.14. These three words are:

1. Status word 0, 2. Preset word 1, 3. Accumulator word 2

Word Number	Bit Numbers															
	15	14	13	12	11	10	9	8	7	6	5	4	3	2	1	0
0	EN	TT	DN							Internal Use						
1	Preset Value (PRE)															
2	Accumulator Value (ACC)															

Fig. 6.14 Three Words of TOF Instruction

The word 0 of timer instruction consists of three status bits which are as follows and explained in Table 6.3.

(*a*) Enable bit (*b*) Timer timing bit (*c*) Done bit

Table 6.3 Status of Control bits of TOF Timer

Bit Description	Set	Reset
Timer Done Bit DN (Bit 13)	When rung conditions are true	When rung conditions go false and the accumulator value is greater than or equal to the preset value
Timer Timing Bit TT (Bit 14)	When rung conditions are false and accumulator value is less than the preset value	When rung condition goes true or the done bit is reset
Timer Enable Bit EN (Bit 15)	When rung conditions are true	When rung conditions are false

Example: *Demonstration of TOF instruction operation*

Figure 6.15 shows the operation of TOF instruction. It explains the conditions for setting and resetting of three control bits. Figure 6.15(*a*) shows the initial state of the program when we enter the run mode. Figure 6.15(*b*) shows the condition when input 1 is high then timer has not started its operation. Figure 6.15(*c*) shows the condition when input 1 makes true to false transition and timer starts its operation. When input 1 goes high then initial state of program will reach as shown in Fig. 6.15(*a*).

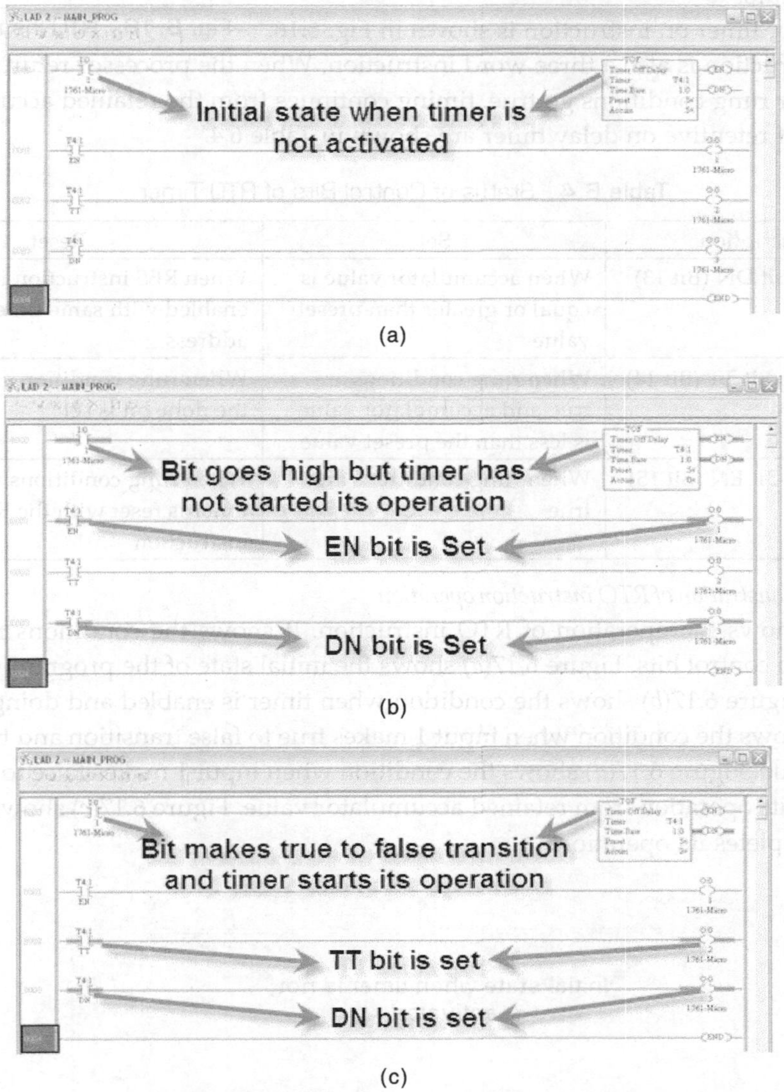

Fig. 6.15 TOF Instruction Operation

6.3.3 Retentive Timer (RTO)

The RTO instruction is used to turn an output ON or OFF after its timer has been ON for a preset time interval. The RTO instruction is a retentive type of instruction that begins to count time base intervals when rung conditions become true. The RTO instruction retains its accumulated value when any of the following occurs:

(*i*) Rung conditions become false.

(*ii*) The processor changes its operation from the run or test mode to the program mode.

(*iii*) The processor loses power (provided that battery backup is maintained).

(*iv*) A fault occurs.

The symbol for timer on instruction is shown in Fig. 6.16.

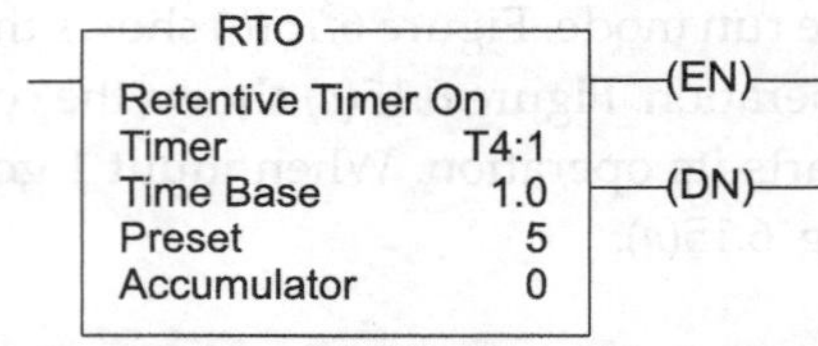

Fig. 6.16 RTO Instruction Symbol

The RTO instruction is also a three word instruction. When the processor returns to the run or test mode and/or rung conditions go true, timing continues from the retained accumulator value. The status bits of retentive on delay timer are shown in Table 6.4.

Table 6.4 Status of Control Bits of RTO Timer

Bit Description	Set	Reset
Timer Done Bit DN (Bit 13)	When accumulator value is equal or greater than preset value	When RES instruction is enabled with same timer address
Timer Timing Bit TT (Bit 14)	When rung conditions are true and accumulator value is less than the preset value	When rung condition goes false or the done bit is set
Timer Enable Bit EN (Bit 15)	When rung conditions are true	When rung conditions are false or timer is reset with the RES instruction

Example: *Demonstration of RTO instruction operation*

Figure 6.17 shows the operation of RTO instruction. It shows the conditions for setting and resetting of three control bits. Figure 6.17(*a*) shows the initial state of the program when we enter the run mode. Figure 6.17(*b*) shows the condition when timer is enabled and doing its operation. Figure 6.17(*c*) shows the condition when input 1 makes true to false transition and timer retains its accumulator value. Figure 6.17(*d*) shows the condition when input 1 makes false to true transition and timer starts its operation from retained accumulator value. Figure 6.17(*e*) shows the condition when timer completes its operation.

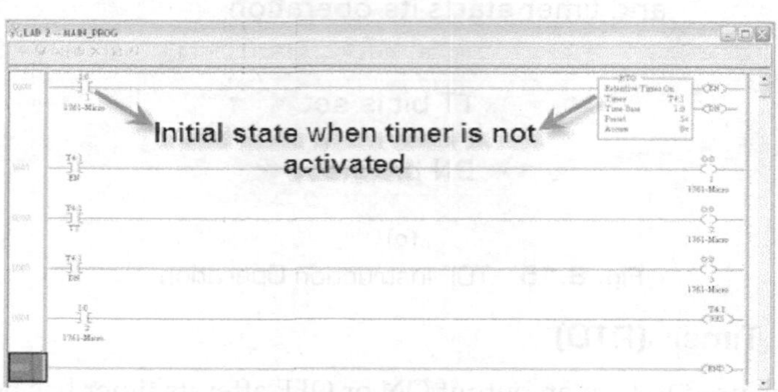

(a)

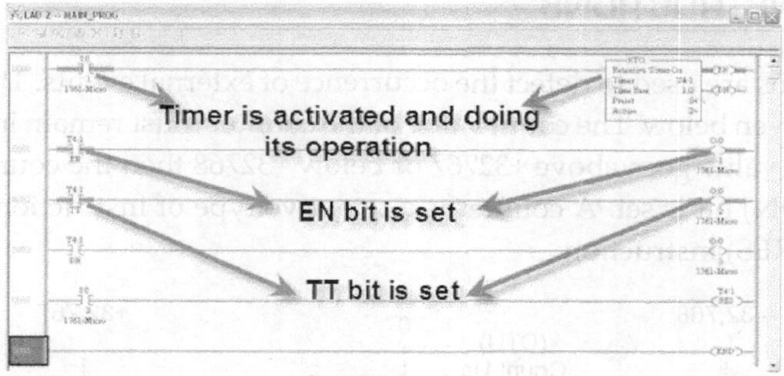

(b)

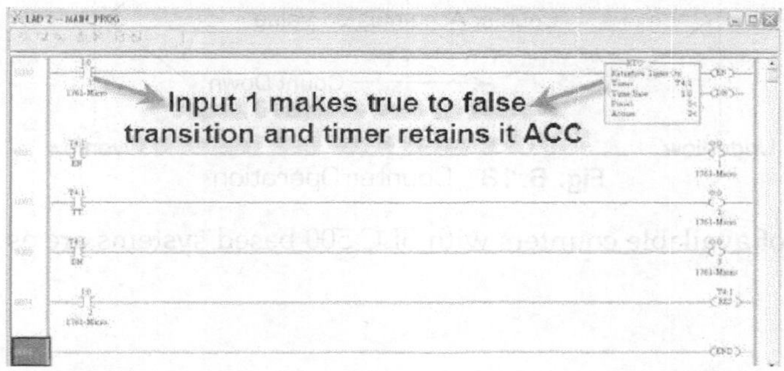

(c)

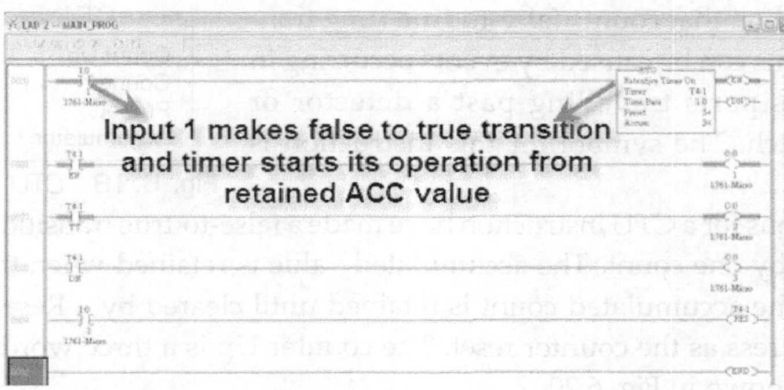

(d)

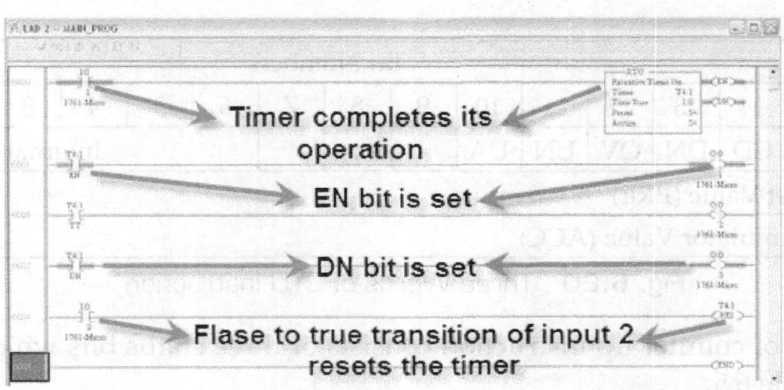

(e)

Fig. 6.17 RTO Instruction Operation

6.4 COUNTER INSTRUCTIONS

The counter instructions are used to detect the occurrence of external events. The counter operation is shown in Fig. 6.18 given below. The count value of the counter must remain in the range of –32768 to +32767. If the count value goes above +32767 or below –32768 then the counter status Overflow (OV) or Underflow (UN) bit is set. A counter is a retentive type of instruction and can be reset to zero using the Reset (RES) instruction.

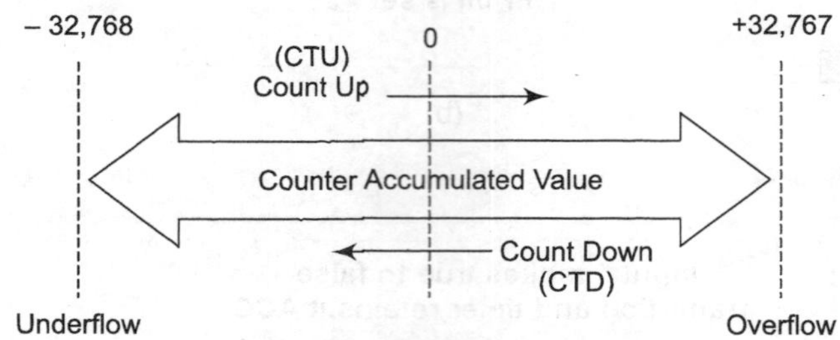

Fig. 6.18 Counter Operation

The various types of available counters with SLC 500 based systems are as follows:

1. Up counter
2. Down counter

6.4.1 Count-Up (CTU)

The CTU is an instruction that counts false-to-true rung transitions. Rung transitions can be caused by events occurring in the program such as parts travelling past a detector or actuating a limit switch. The symbol for this instruction is shown in Fig. 6.19.

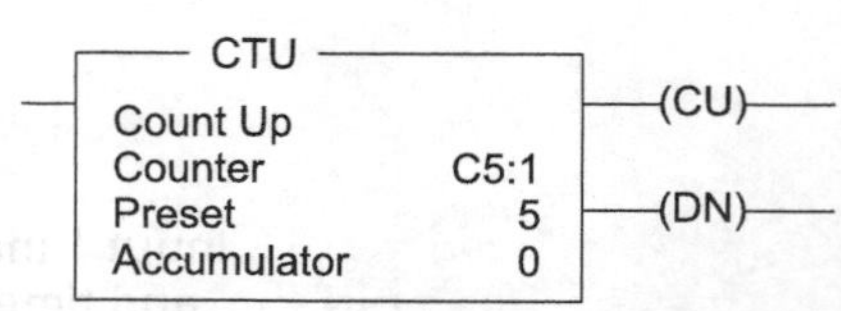

Fig. 6.19 CTU Instruction Symbol

When rung conditions for a CTU instruction have made a false-to-true transition, the accumulated value is incremented by one count. The accumulated value is retained when the rung conditions again become false. The accumulated count is retained until cleared by a Reset (RES) instruction that has the same address as the counter reset. The counter Up is a three word instruction which are given below and shown in Fig. 6.20.

1. Control word 0
2. Preset word 1
3. Accumulator word 2

Word Number	Bit Numbers															
	15	14	13	12	11	10	9	8	7	6	5	4	3	2	1	0
0	CU	CD	DN	OV	UN	UA				Internal Use						
1	Preset Value (PRE)															
2	Accumulator Value (ACC)															

Fig. 6.20 Three Words of CTU Instruction

The control word of counter up instruction consists of three status bits which are given below and explained in Table 6.5.

(*i*) Overflow bit (*ii*) Done bit (*iii*) Count Up enable bit

Table 6.5 Status of Control Bits of CTU Counter

Bit Description	Set	Reset
Overflow Bit OV (Bit 12)	When accumulator value wraps around to –32,768,	When RES instruction is enabled with same counter address or the counter is decremented less than or equal to +32,767 with a CTD instruction
Timer Done Bit DN (Bit 13)	When accumulator value is equal or greater than the preset value	When accumulator value becomes less than the preset value
Counter Up Enable Bit CU (Bit 15)	When rung conditions are true	When rung conditions are False or counter is reset with the RES instruction

The accumulated value is retained after the CTU instruction goes false, or when power is removed from and then restored to the controller. Also, the ON or OFF status of counter done, overflow, and underflow bits is retentive. The accumulated value and control bits are reset when the appropriate RES instruction is enabled.

Example: *Demonstration of CTU instruction operation*

Figure 6.21 shows the operation of CTU instruction. Figure 6.21(*a*) shows the initial state of the program when we enter the run mode. Figure 6.21(*b*) shows when input 1 makes true to false transition and counter is doing its operation. Figure 6.21(*c*) shows when counter completes its operation. Figure 6.21(*d*) shows when counter is reset with a RES instruction.

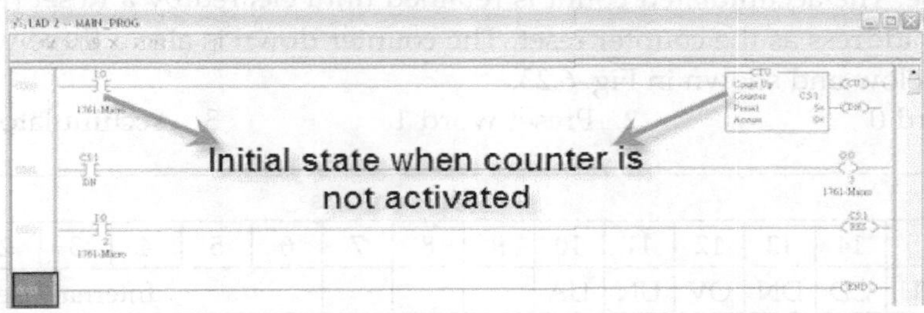

(a)

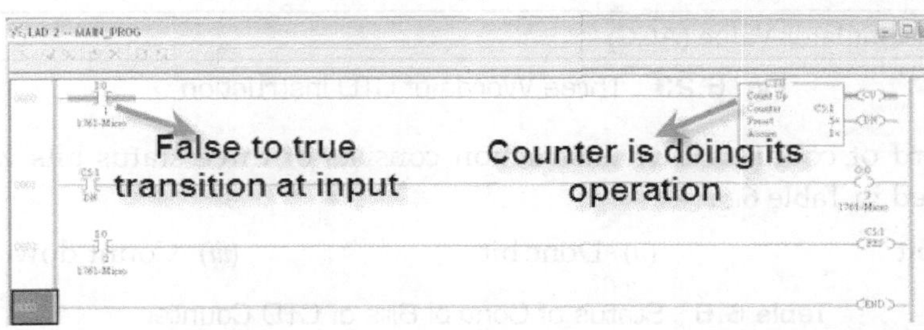

(b)

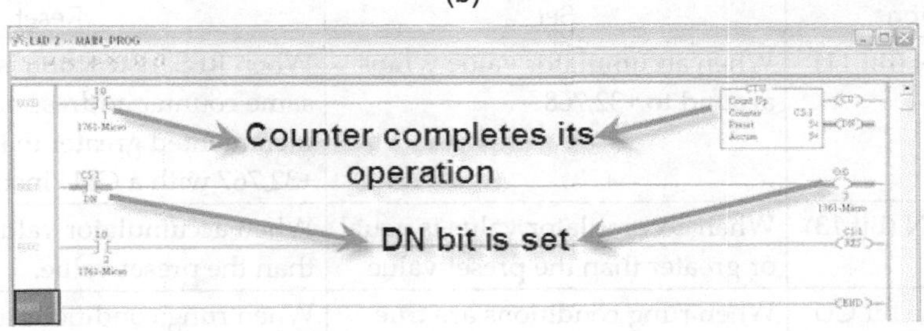

(c)

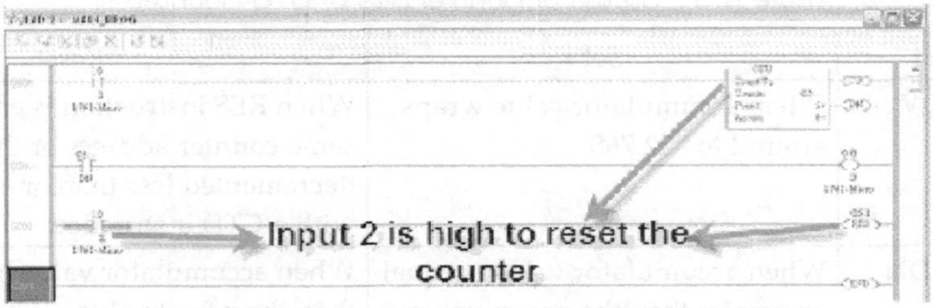

(d)

Fig. 6.21 CTU Instruction Operation

6.4.2 Count-Down (CTD)

The CTD is an instruction that counts false-to-true rung transitions. Rung transitions can be caused by events occurring in the program such as parts travelling past a detector or actuating a limit switch. The symbol for this instruction is shown in Fig. 6.22.

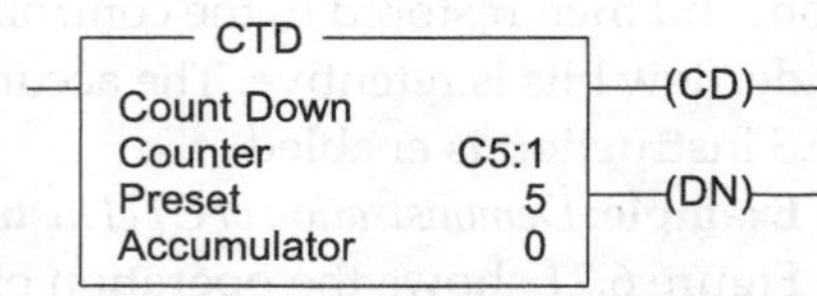

Fig. 6.22 CTD Instruction Symbol

When rung conditions for a CTD instruction have made a false-to-true transition, the accumulator value is decremented by one count. The accumulator counts are retained when the rung conditions again become false. The accumulated count is retained until cleared by a Reset (RES) instruction that has the same address as the counter reset. The counter down is also a three word instruction which are given below and shown in Fig. 6.23.

1. Control word 0
2. Preset word 1
3. Accumulator word 2

Word Number	Bit Numbers															
	15	14	13	12	11	10	9	8	7	6	5	4	3	2	1	0
0	CU	CD	DN	OV	UN	UA				Internal Use						
1	Preset Value (PRE)															
2	Accumulator Value (ACC)															

Fig. 6.23 Three Words of CTD Instruction

The control word of counter down instruction consists of three status bits which are given below and explained in Table 6.6.

(*i*) Underflow bit (*ii*) Done bit (*iii*) Count down enable bit

Table 6.6 Status of Control Bits of CTD Counter

Bit Description	Set	Reset
Underflow Bit UN (Bit 11)	When accumulator value wraps around to +32,768	When RES instruction is enabled with same counter address or the counter is incremented greater than or equal to +32,767 with a CTU instruction
Timer Done Bit DN (Bit 13)	When accumulator value is equal or greater than the preset value	When accumulator value becomes less than the preset value.
Counter Up Enable Bit CU (Bit 15)	When rung conditions are true	When rung conditions are false or counter is reset with the RES instruction

The accumulated value is retained after the CTD instruction goes false, or when power is removed from and then restored to the controller. The accumulated value and control bits are reset when the appropriate RES instruction is executed.

Example: *Demonstration of CTD instruction operation.*

Figure 6.24 shows the operation of CTD instruction. Figure 6.24(*a*) shows the initial state of the program when we enter the run mode. Figure 6.24(*b*) shows when input 1 makes true to false transition and accumulator of counter is decrementing by 1. Figure 6.21(*c*) shows when accumulator of counter goes below 0 then it reset the done bit.

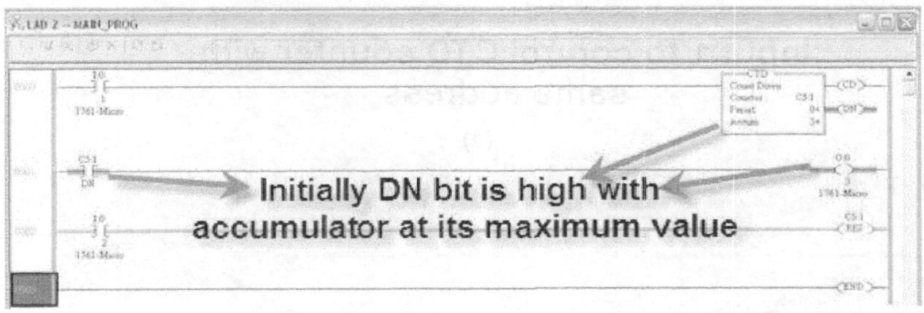

(a)

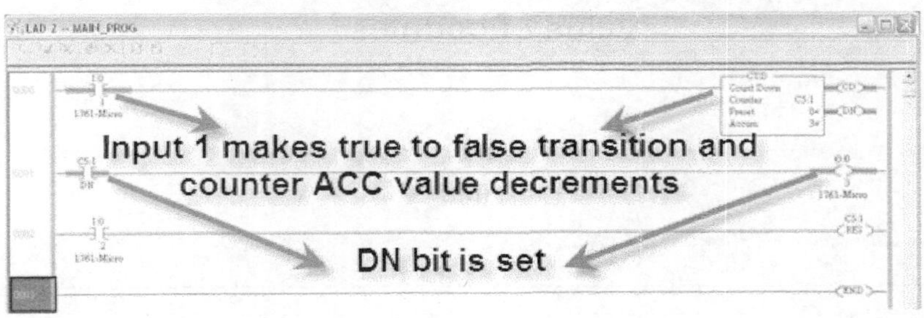

(b)

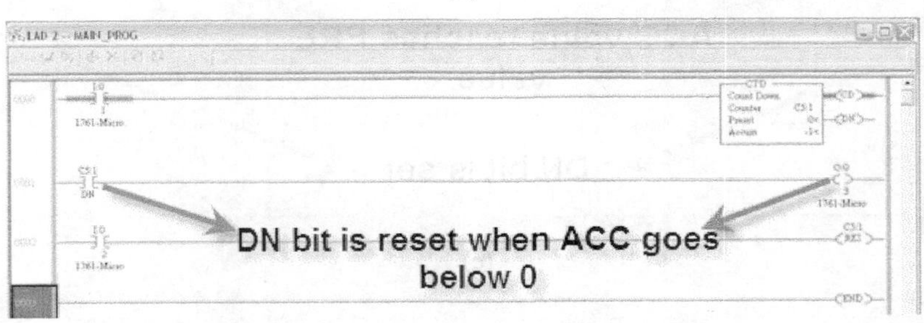

(c)

Fig. 6.24 CTD Instruction Operation

6.4.3 Up-Down Counter

The up counter and down counter can be combined to make an up-down counter. Both up and down counters must have the same address in the program where up-down counter is required.

Example: *Demonstration of CTU instruction operation*

Figure 6.25 shows the working of Up-down counter by using CTD and CTU instructions with same address. Figure 6.25(*a*) shows the initial state of the program when we enter the run mode. Figure 6.25(*b*) shows when input 1 makes true to false transition to increment accumulator. Figure 6.25(*c*) shows when accumulator of counter goes equal to preset value then it set the done bit.

Figure 6.25(*d*) shows when input 2 makes true to false transition to decrement accumulator and reset the done bit.

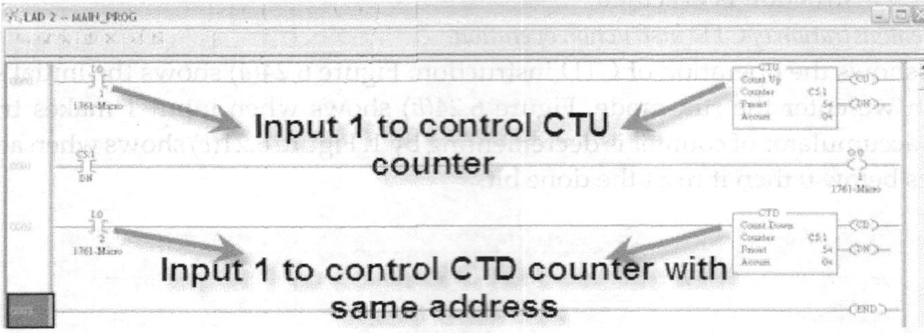

(a)

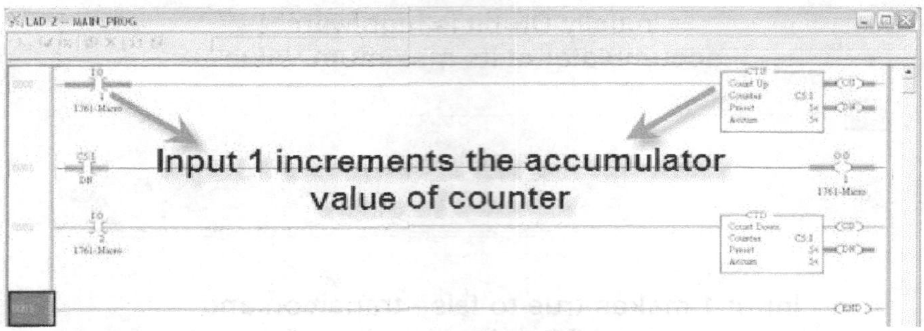

(b)

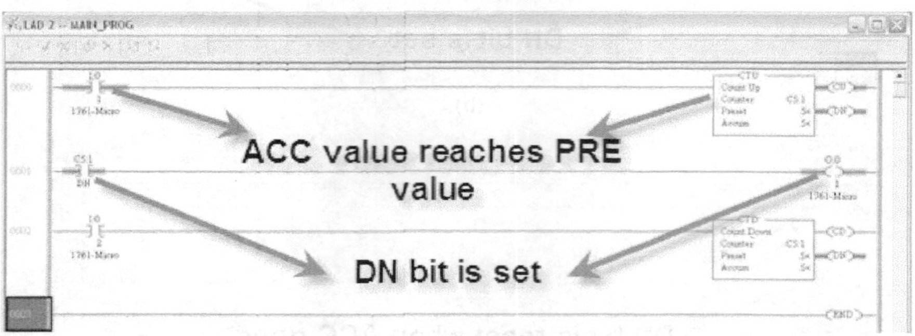

(c)

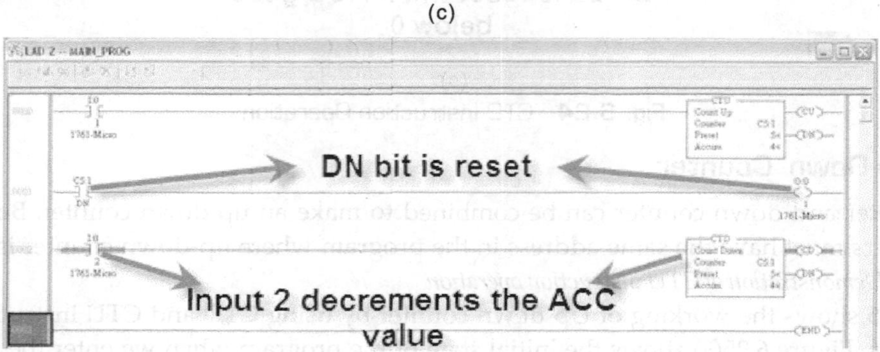

(d)

Fig. 6.25 Counter Up-Down Operation

6.5 RESET (RES)

An RES instruction is used to reset a timer or counter instruction. When the RES instruction is enabled, it resets the Timer On Delay (TON), Retentive Timer (RTO), Count Up (CTU), or Count Down (CTD) instruction having the same address as the RES instruction. The symbol for a reset instruction is shown in Fig. 6.26

Fig. 6.26 Reset Instruction

SUMMARY

Instruction set of RS Logix 500 can be broadly categorized into bit instructions, timer instructions and counter instructions. Bit instructions operate on a single bit of data. The processor may set or reset the bit, based on logical continuity of ladder rungs during an operation. Bit instructions are commonly used with output files, input files, status file, bit file, timer file, counter file and sequencer file. Bit instructions explained in this chapter discussed in this chapter includes XIC, XIO, OTE, OTL, OUT, OSR. The counter instructions are used to detect the occurrence of external events. A counter is a retentive type of instruction and can be reset to zero using the reset (RES) instruction. To program a timer instruction, programming of preset value and time base is essential. An RES instruction is used to reset a timer or counter instruction. When the RES instruction is enabled, it resets the Timer On Delay (TON), Retentive Timer (RTO), Count Up (CTU), or Count Down (CTD) instruction having the same address as the RES instruction.

REVIEW QUESTIONS

1. Name 3 main categories of basic instruction set.
2. Define bit instructions. Name various types of bit instructions available with SLC 500 based systems.
3. Compare and explain the working of XIC and XIO instructions.
4. Explain OTL and OUT instruction with the help of a suitable example.
5. Explain the working of an OSR instruction.
6. Compare OTL, OUT and OSR instructions.
7. Why do we need timer instructions? Explain various parameters associated with timer instructions.
8. What is a TON instruction? Explain its working with the help of a suitable example.
9. What is a TOF instruction? Explain its working with the help of a suitable example.
10. What is a RTO instruction? Explain its working with the help of a suitable example.
11. Why do we need counters? Explain the working of counters with help of a diagram.
12. What is a CTU instruction? Explain its working with the help of a suitable example.
13. What is a CTD instruction? Explain its working with the help of a suitable example.
14. Make up-down counter using CTU and CTD instructions and explain its working.

7

INSTRUCTION SET-II

LEARNING OBJECTIVES

After reading this chapter, you will be able to explain the operation, control field and symbols of :

1. Data handling instructions
2. Comparison instructions
3. Sequencer instructions

7.1 INTRODUCTION

In this chapter we will discuss the operation , symbols and control words additional instruction set listed below:

1. Data handling instructions
2. Comparison instructions
3. Sequencer instructions

7.2 DATA HANDLING INSTRUCTIONS

The various data handling instructions along with their functions are shown in Table 7.1.

Table 7.1 Data Handling Instructions

Instruction	Name	Function
COP	Copy file	It copies data from the source file to the destination file.
FLL	Fill file	It loads a source value into each position in the destination file.
MOV	Move	It moves the source value to the destination.
MVM	Masked move	It moves data from a source location to selected portion of the destination.
AND	And	It performs a bitwise AND operation.
OR	Or	It performs a bitwise OR operation.
XOR	Exclusive or	It performs a bitwise XOR operation.
NOT	Not	It performs a bitwise NOT operation.

7.2.1 Copy Instruction

This COP instruction copies blocks of data from one location into another. The destination file type determines the number of words that an instruction transfers. For example, if the destination file type is a counter and the source file type is an integer, three integer words are transferred for each element in the counter-type file. The symbol for COP instruction is shown in Fig. 7.1 and Fig 7.2 shows the operation of this instruction.

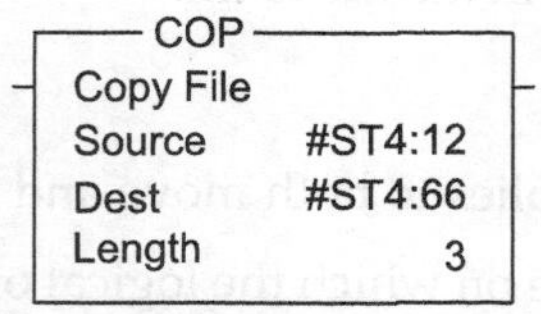

Fig. 7.1 COP Instruction Symbol

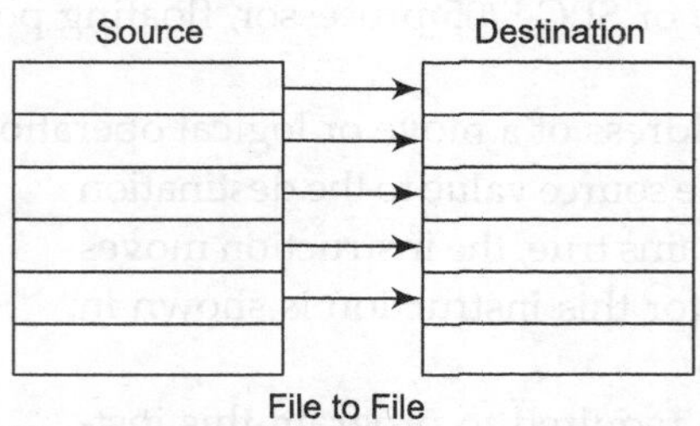

Fig. 7.2 COP Instruction Operation

To program this instruction enter the following parameters:

1. Source is the address of the file to copy. The file indicator (#) must be used in the address.
2. Destination is the starting address where the instruction stores the copy. Again the file indicator (#) must be used in the address.
3. Length is the number of *elements* in the file to copy.

All elements are copied from the source file into the destination file each time the instruction is executed. Elements are copied in ascending order.

7.2.2 Fill Instruction

This FLL instruction loads elements of a file with either a program constant or value from an element address. The symbol for FLL instruction is shown in Fig. 7.3. The instruction fills the words of a file with a source value. It uses no status bits. The following Fig. 7.4 shows how file instruction data is manipulated.

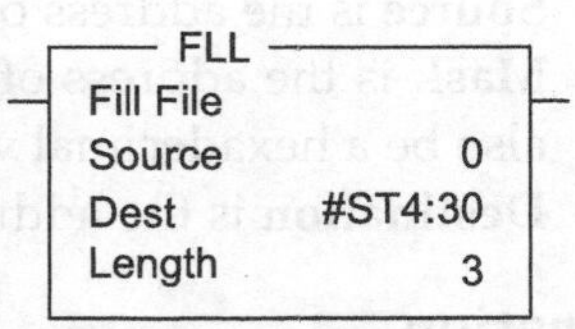

Fig. 7.3 FLL Instruction Symbol

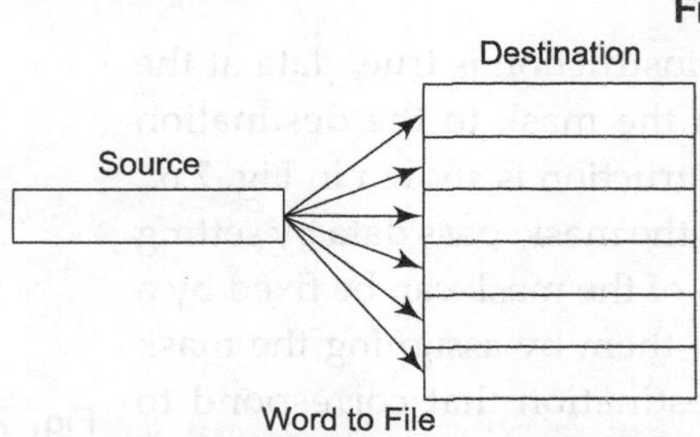

Fig. 7.4 FLL Instruction Operation

Enter the following parameters when programming this instruction:

1. Source is the program constant or element address. The file indicator (#) is not required for an element address. For an SLC 5/03, SLC 5/04, or SLC 5/05 processor, floating point and string values are supported.

2. Destination is the destination starting address of the file to fill. The file indicator (#) is required in the address. Again for an SLC 5/03, SLC 5/04, or SLC 5/05 processor, floating point and string values are supported.
3. Length is the number of *elements* in the file to fill.

7.2.3 Move Instruction

The following general information applies to both move and logical instructions.

1. Source is the address of the value on which the logical or move operation is to be performed. The source can be a word address or a program constant.
2. For an SLC 5/03, SLC 5/04, or SLC 5/05 processor, floating point and string values are supported.
3. Destination is the result address of a move or logical operation. It must be a word address.

The MOV instruction moves the source value to the destination location. As long as the rung remains true, the instruction moves the data each scan. The symbol for this instruction is shown in Fig. 7.5.

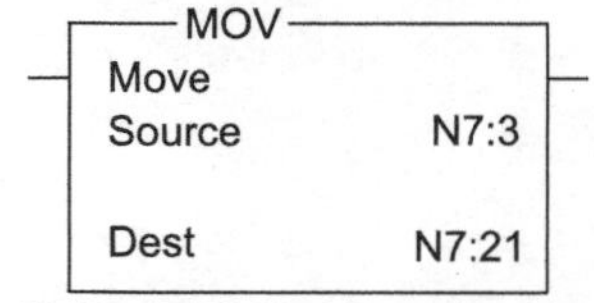

Fig. 7.5 MOV Instruction Symbol

The following parameters are required to program this instruction:

1. **Source** is the address or constant of the data you want to move.
2. **Destination** is the address where the instruction moves the data.

7.2.4 Masked Move Instruction

The MVM instruction is a word instruction that moves data from a source location to a destination, and allows portions of the destination data to be masked by a separate word. As long as the rung remains true, the instruction moves the data each scan. The following parameters are required to program this instruction:

1. **Source** is the address of the data you want to move.
2. **Mask** is the address of the mask through which the instruction moves data; the mask can also be a hexadecimal value.
3. **Destination** is the address where the instruction moves the data.

Operation

When the rung containing this instruction is true, data at the source address passes through the mask to the destination address. The symbol for this instruction is shown in Fig. 7.6.

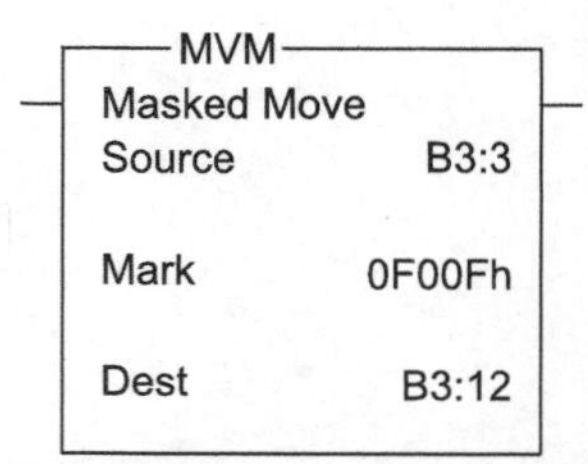

Fig. 7.6 MVM Instruction Symbol

Mask data by resetting bits in the mask; pass data by setting bits in the mask to one. The bits of the mask can be fixed by a constant value, or you can vary them by assigning the mask a direct address. Bits in the destination that correspond to zeros in the mask are not altered. The operation is shown in Fig. 7.7.

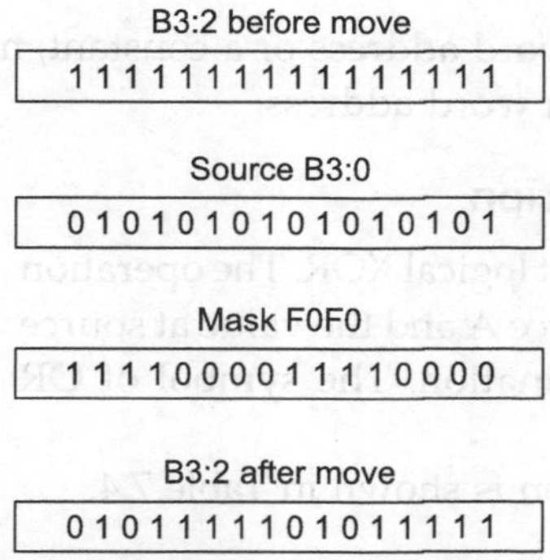

Fig. 7.7 MVM Instruction Operation

7.2.5 AND Instruction

This instruction performs a bit-by-bit logical AND operation. The operation is performed using the value at source A and the value at source B. The result is stored in the destination. The symbol of AND instruction is shown in Fig. 7.8.

The truth table for this instruction is shown in Table 7.2.

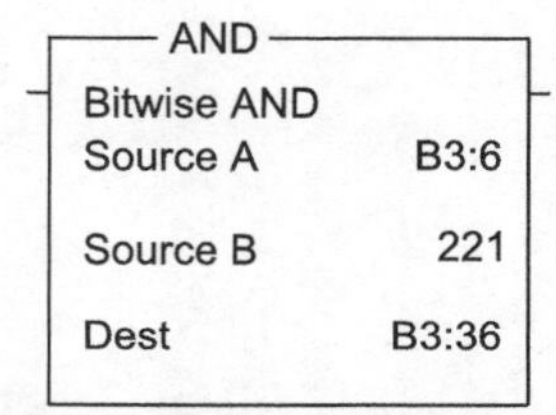

Fig. 7.8 AND Instruction Symbol

Table 7.2 AND Instruction Truth Table

A	B	Dest
0	0	0
1	0	0
0	1	0
1	1	1

Source A and B can either be a word address or a constant; however, both sources cannot be a constant. The destination must be a word address.

7.2.6 OR Instruction

This instruction performs a bit-by-bit logical OR. The operation is performed using the value at source A and the value at source B. The result is stored in the destination. The symbol of OR instruction is shown in Fig. 7.9.

The truth table for this instruction is shown in Table 7.3.

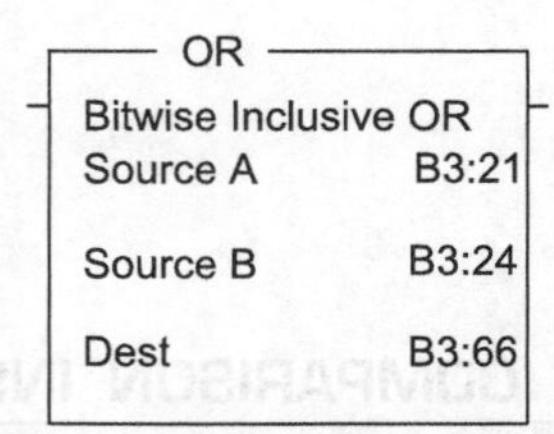

Fig. 7.9 OR Instruction Symbol

Table 7.3 OR Instruction Truth Table

A	B	Dest
0	0	0
1	0	1
0	1	1
1	1	1

Source A and B can either be a word address or a constant; however, both sources cannot be a constant. The destination must be a word address.

7.2.7 Exclusive OR Instruction

This instruction performs a bit-by-bit logical XOR. The operation is performed using the value at source A and the value at source B. The result is stored in the destination. The symbol of OR instruction is shown in Fig. 7.10.

The truth table for this instruction is shown in Table 7.4.

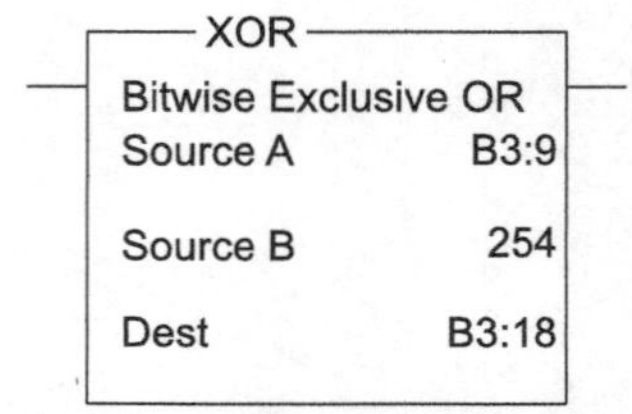

Fig. 7.10 XOR Instruction Symbol

Table 7.4 XOR Instruction Truth Table

A	B	Dest
0	0	0
1	0	1
0	1	1
1	1	0

Source A and B can either be a word address or a constant; however, both sources cannot be a constant. The destination must be a word address.

7.2.8 NOT Instruction

This instruction performs a bit-by-bit logical NOT. The operation is performed using the value at source A. The result *i.e.* one's complement of A is stored in the destination. The symbol of OR instruction is shown in Fig. 7.11 and truth table is shown in Table 7.5. The source and destination must be word addresses.

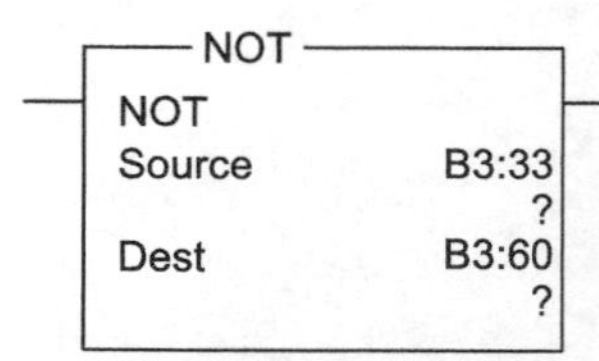

Fig. 7.11 NOT Instruction Symbol

Table 7.5 NOT Instruction Truth Table

A	Dest
0	1
1	0

7.3 COMPARISON INSTRUCTIONS

These instructions are used to compare two values with each other. For example, an LES instruction is presented with two values. If the first value is less than the second, then the comparison instruction is true. The various comparison instructions along with their function are shown in Table 7.6.

Table 7.6 Comparison Instructions

Instruction	Name	Function
EQU	Equal	Test whether one value is equal to a second value.
NEQ	Not equal	Test whether one value is not equal to a second value.

LES	Less than	Test whether one value is less than a second value.
LEQ	Less than or equal	Test whether one value is less than or equal to a second value.
GRT	Greater than	Test whether one value is greater than a second value.
GEQ	Greater than or equal	Test whether one value is greater than or equal to a second value.
MEQ	Masked comparison for equal	Test portion of two values to see whether they are equal
LIM	Limit test	Test whether the one value is within the limit range of two other values.

7.3.1 Equal (EQU)

The EQU instruction is used to test whether two values are equal. If source A and source B are equal, the instruction is logically true. If these values are not equal, the instruction is logically false. Source A must be an address. Source B can either be a program constant or an address. The symbol for this instruction is shown in Fig. 7.12.

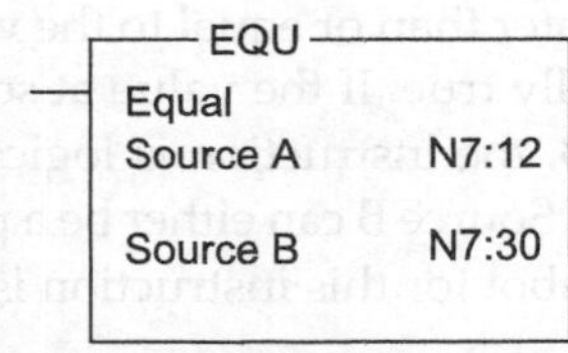

Fig. 7.12 EQU Instruction Symbol

7.3.2 Not Equal (NEQ)

The NEQ instruction is used to test whether two values are not equal. If source A and source B are not equal, the instruction is logically true. If the two values are equal, the instruction is logically false. Source A must be an address. Source B can be either a program constant or an address. The symbol for this instruction is shown in Fig. 7.13.

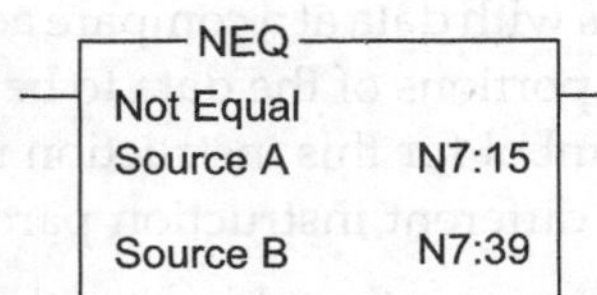

Fig. 7.13 NEQ Instruction Symbol

7.3.3 Less Than (LES)

Use the LES instruction to test whether one value of source A is less than another value of source B. If source A is less than the value at source B, the instruction is logically true. If the value at source A is greater than or equal to the value at source B, the instruction is logically false. Source A must be an address. Source B can either be a program constant or an address. The symbol for this instruction is shown in Fig. 7.14.

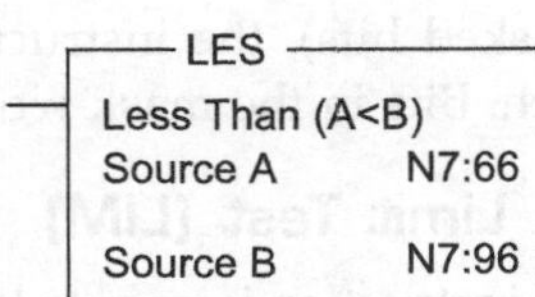

Fig. 7.14 LES Instruction Symbol

7.3.4 Less Than or Equal (LEQ)

Use the LEQ instruction to test whether one value of source A is less than or equal to another value of source B. If the value at source A is less than or equal to the value at source B, the instruction is logically true. If the value at source A is greater than the value at source B, the instruction is logically false. Source A

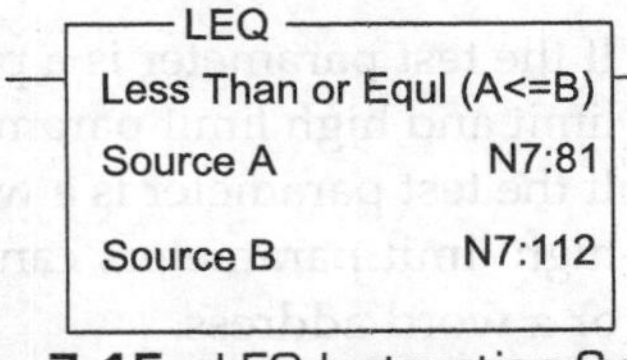

Fig. 7.15 LEQ Instruction Symbol

must be an address. Source B can either be a program constant or an address. The symbol for this instruction is shown in Fig. 7.15.

7.3.5 Greater Than (GRT)

Use the GRT instruction to test whether one value of source A is greater than another value of source B. If the value at source A is greater than the value at source B, the instruction is logically true. If the value at source A is less than or equal to the value at source B, the instruction is logically false. Source A must be an address. Source B can either be a program constant or an address. The symbol for this instruction is shown in Fig. 7.16.

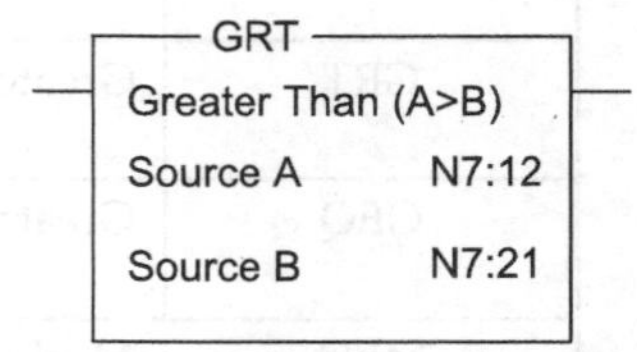

Fig. 7.16 GRT Instruction Symbol

7.3.6 Greater Than or Equal (GEQ)

The GEQ instruction is used to test whether one value (source A) is greater than or equal to another (source B). If the value at source A is greater than or equal to the value at source B, the instruction is logically true. If the value at source A is less than the value at source B, the instruction is logically false. Source A must be an address. Source B can either be a program constant or an address. The symbol for this instruction is shown in Fig. 7.17.

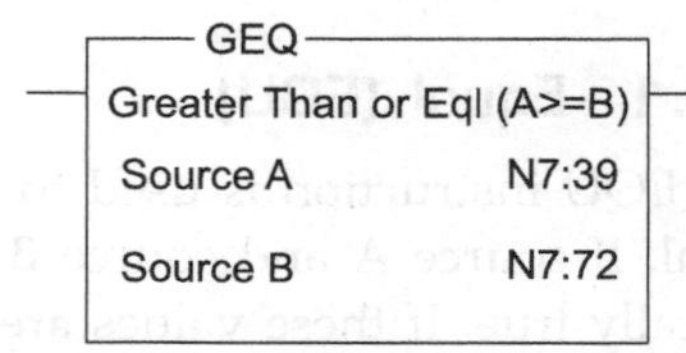

Fig. 7.17 GEQ Instruction Symbol

7.3.7 Masked Comparison for Equal (MEQ)

The MEQ instruction is used to compare data at a source address with data at a compare address. Use of this instruction allows portions of the data to be masked by a separate word. The symbol for this instruction is shown in Fig. 7.18.

The different instruction parameters are:

1. Source is the address of the value you want to compare.
2. Mask is the address of the mask through which the instruction moves data. The mask can also be a hexadecimal value.
3. Compare is an integer value or the address of the reference.

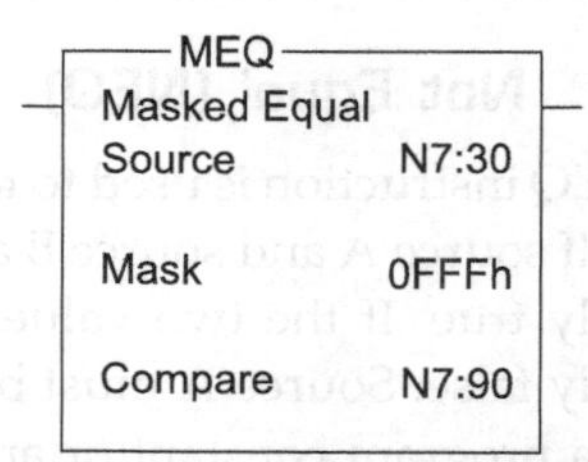

Fig. 7.18 MEQ Instruction Symbol

If the 16 bits of data at the source address are equal to the 16 bits of data at the compare address (less masked bits), the instruction is true. The instruction becomes false as soon as it detects a mismatch. Bits in the mask word mask data when reset; they pass data when set.

7.3.8 Limit Test (LIM)

The LIM instruction is used to test for values within or outside a specified range, depending on how you set the limits. The symbol for this instruction is shown in Fig. 7.19.

The low limit, test, and high limit values can be word addresses or constants, restricted to the following combinations:

1. If the test parameter is a program constant, both the low limit and high limit parameters must be word addresses.
2. If the test parameter is a word address; the low limit and high limit parameters can be either a program constant or a word address.

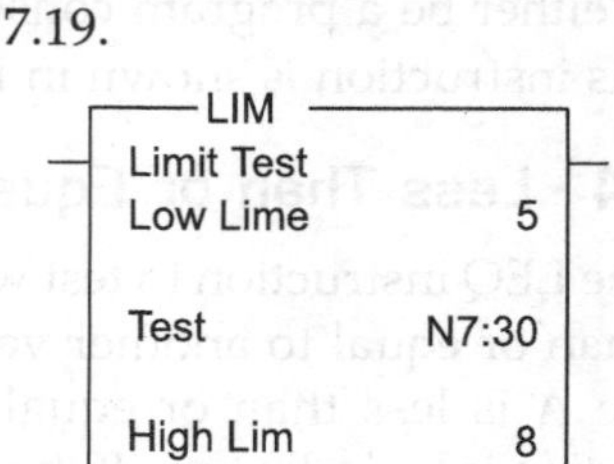

Fig. 7.19 LIM Instruction Symbol

Example: *If low limit less than high limit*

If the low limit has a value equal to or less than the high limit, the instruction is true when the test value is between the limits or is equal to either limit. If the test value is outside the limits, the instruction is false, as shown in Fig. 7.20(*a*).

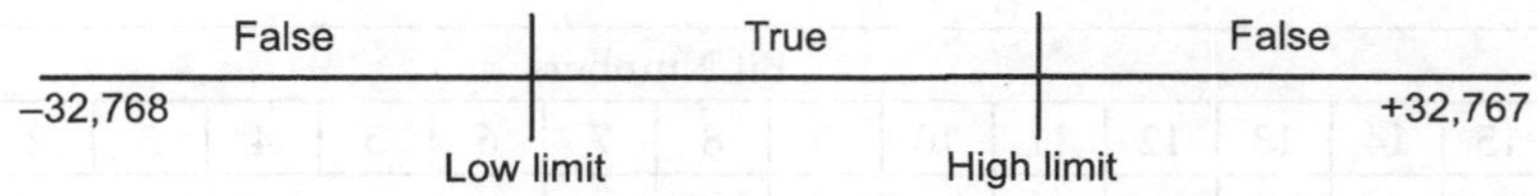

Low Limit	High Limit	Instruction is True	Instruction is False
7	8	When test value is 5 through 8	When test value is –32,768 through 4 and 9 through +32,767

(a)

Example: *If low limit greater than high limit*

If the low limit has a value greater than the high limit, the instruction is false when the Test value is between the limits. If the test value is equal to either limit or outside the limits, the instruction is true, as shown in Fig. 7.20(*b*).

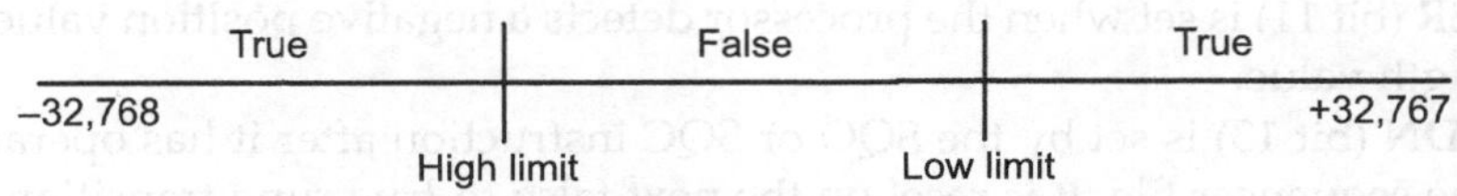

Low Limit	High Limit	Instruction is True	Instruction is False
8	5	When test value is –32,768 through 5 and 8 through +32,767	6 and 7

(b)

Fig. 7.20 LIM Instruction Operation

7.4 SEQUENCER INSTRUCTIONS

The various sequencer instructions along with their function are shown in Table 7.7 which is given below.

Table 7.7 Sequencer Instructions

Instruction	Name	Function
SQO and SQC	Sequencer output and Sequencer compare	It is used to control sequential machine operation by transferring 16-bit data through a mask to image addresses.
SQL	Sequencer load	It captures referenced conditions by manually stepping the machine through its operating sequences.

These instructions transfer 16-bit data to word addresses for the control of sequential machine operations. For programming the SQO and SQC instructions following parameters are used:

1. **File** is the address of the sequencer file. Use the file indicator (#) for this address.
2. **Mask** (SQO, SQC) is a hexadecimal code or the address of the mask word or file through which the instruction moves data. Set mask bits to pass data and reset mask bits to mask data.
3. **Source** is the address of the input word or file for a SQC from which the instruction obtains data for comparison to its sequencer file.

4. **Destination** is the address of the output word or file for a SQO to which the instruction moves data from its sequencer file.
5. **Control** (SQO, SQC) is the control structure that stores the status byte of the instruction, the length of the sequencer file, and the instantaneous position in the file as shown in Fig. 7.21.

Word Number	Bit Numbers															
	15	14	13	12	11	10	9	8	7	6	5	4	3	2	1	0
0	EN		DN		ER			FD		Not used						
1	Length of sequencer file															
2	Position															

Fig. 7.21 Sequencer Control Words

Status bits of the control structure include:

(*a*) **Found Bit FD** (bit 08) SQC only. When the status of all non-masked bits in the source address match those of the corresponding reference word, the FD bit is set. This bit is assessed each time the SQC instruction is evaluated while the rung is true.

(*b*) **Error Bit ER** (bit 11) is set when the processor detects a negative position value, or a negative or zero length value.

(*c*) **Done Bit DN** (bit 13) is set by the SQO or SQC instruction after it has operated on the last word in the sequencer file. It is reset on the next false-to-true rung transition after the rung goes false.

(*d*) **Enable EN** (bit 15) is set by a false-to-true rung transition and indicates the SQO or SQC instruction is enabled.

(*e*) **Length** is the number of steps of the sequencer file starting at position 1. The maximum number you can enter is 255 words. Position 0 is the startup position. The instruction resets or wraps to position 1 at each cycle completion.

(*f*) **Position** is the word location or step in the sequencer file from or to which the instruction moves data. A position value that points past the end of the programmed file causes a runtime major error to occur.

The reset (RES) instruction is used to reset a sequencer. All control bits (except FD) will be reset to zero. The Position will also be set to zero.

7.4.1 Sequencer Output Instruction (SQO)

This output instruction steps through the sequencer file whose bits have been set to control various output devices. The false-to-true transition instruction previous to sequencer increments the sequencer to the next step. Data stored there is transferred through a mask to the destination address specified in the instruction. Current data is written to the corresponding destination word every scan that the rung remains true. The done bit is set when the last word of the sequencer file is transferred. On the next false-to-true rung transition, the instruction resets the position to step one.

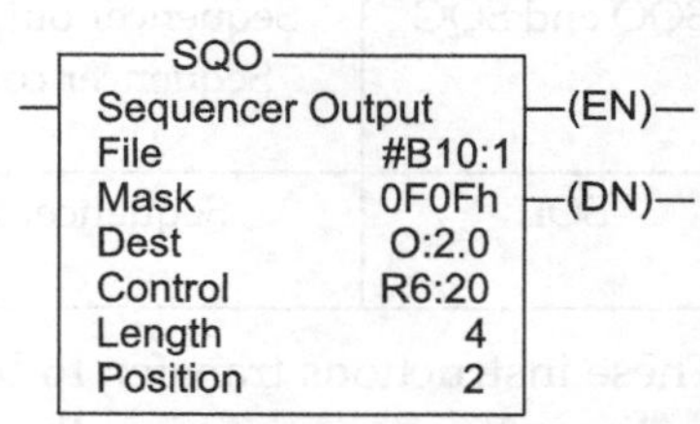

Fig. 7.22 SQO Instruction Symbol

The bits mask data when reset and pass data when set. The instruction will not change the value in the destination word unless mask bits are set. The mask can be fixed or variable. A hexadecimal code is used as fixed mask and element address or a file address is used for changing the mask with each step. The symbol of SQO instruction is shown in Fig. 7.22 and its working is shown in Fig. 7.23.

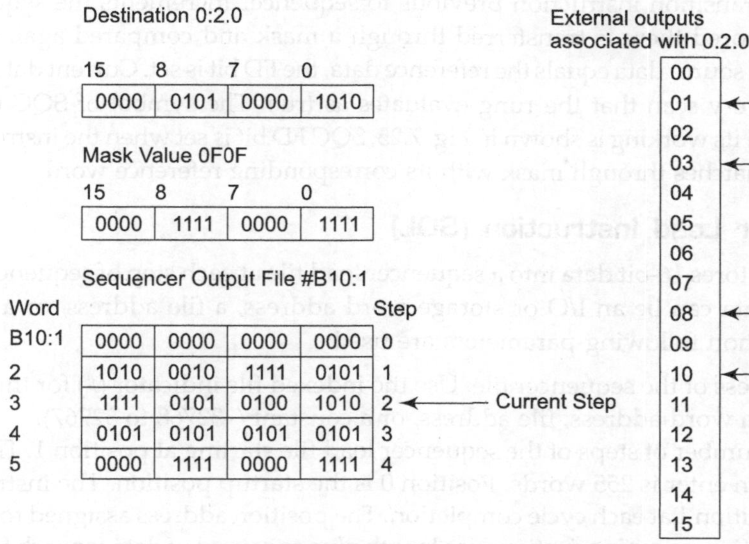

Fig. 7.23 SQO Instruction Operation

7.4.2 Sequencer Compare Instruction (SQC)

When the status of all non-masked bits in the source word matches with the all the bits of corresponding reference word then the instruction sets the found bit (FD) in the control word. Otherwise, the found bit (FD) is cleared. The bits mask data when reset and pass data when set. The instruction will not change the value in the destination word unless you set the mask bits. The mask can be fixed or variable. A hexadecimal code is used as fixed mask and element address or a file address is used for changing the mask with each step.

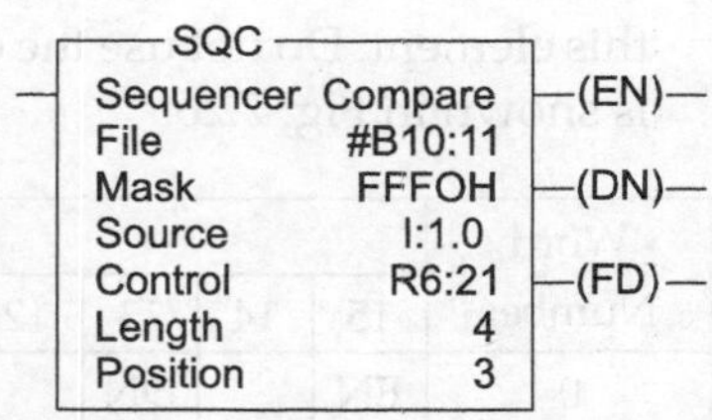

Fig. 7.24 SQC Instruction Symbol

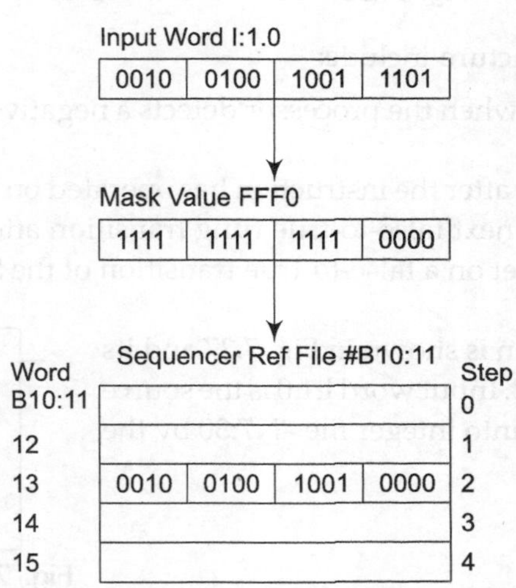

Fig. 7.25 SQC Instruction Operation

The false-to-true transition instruction previous to sequencer increments the sequencer file to the next step. Data stored there is transferred through a mask and compared against the source data for equality. If the source data equals the reference data, the FD bit is set. Current data is compared against the source every scan that the rung evaluates as true. The symbol of SQC instruction is shown in Fig. 7.24 and its working is shown in Fig. 7.25. SQC FD bit is set when the instruction detects that an input word matches through mask with its corresponding reference word.

7.4.3 Sequencer Load Instruction (SQL)

The SQL instruction stores 16-bit data into a sequencer load file at each step of sequencer operation. The source of this data can be an I/O or storage word address, a file address, or a constant. To program this instruction following parameters are used:

(*a*) **File** is the address of the sequencer file. Use the indexed file indicator (#) for this address.
(*b*) **Source** can be a word address, file address, or a constant (–32768 to 32767).
(*c*) **Length** is the number of steps of the sequencer load file starting at position 1. The maximum number you can enter is 255 words. Position 0 is the startup position. The instruction resets or wraps to position 1 at each cycle completion. The position address assigned for a sequencer file is step zero. Sequencer instructions use length plus one word of data for each file referenced in the instruction. This applies to the source if addressed as a file.
(*d*) **Position** is the word location or step in the sequencer file to which data is moved.
(*e*) **Control** is a control file address. The status bits, length value, and position value are stored in this element. Do not use the control file address for any other instruction. The control element is shown in Fig. 7.26.

Word Number	Bit Numbers															
	15	14	13	12	11	10	9	8	7	6	5	4	3	2	1	0
0	EN		DN		ER				Not used							
1	Length of sequencer file															
2	Position															

Fig. 7.26 SQL Control Words

Status bits of the control structure include:

(*a*) **Error Bit ER** (bit 11) is set when the processor detects a negative position value, or a negative or zero length value.
(*b*) **Done Bit DN** (bit 13) is set after the instruction has operated on the last word in the sequencer load file. It is reset on the next false-to-true rung transition after the rung goes false.
(*c*) **Enable Bit EN** (bit 15) is set on a false-to-true transition of the SQL rung and reset on a true-to-false transition.

The symbol for this instruction is shown in Fig. 7.27 and its operation is explained in Fig. 7.28. Input word I:1.0 is the source and data in this word is loaded into integer file #N7:30 by the sequencer load instruction.

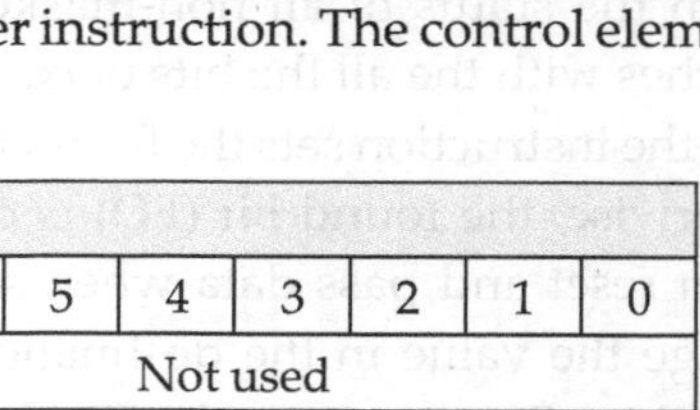

Fig. 7.27 SQL Instruction Symbol

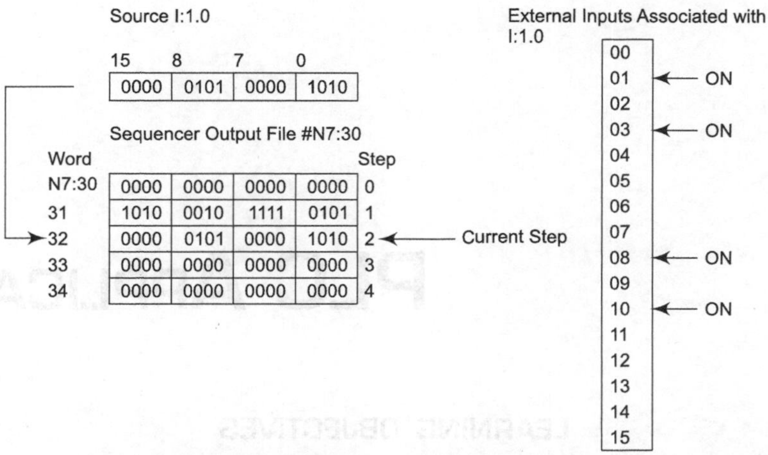

Fig. 7.28 SQL Instruction Operation

When rung conditions change from false-to-true, the SQL enable bit (EN) is set. The control element R6:4 increments to the next position in the sequencer file, and loads the contents of source I:1.0 into this location. The SQL instruction continues to load the current data into this location each scan that the rung remains true. When the rung goes false, the enable bit (EN) is reset. The instruction loads data into a new file element at each false-to-true transition of the rung. When step 4 is completed, the done bit (DN) is set. Operation cycles to position 1 at the next false-to-true transition of the rung after position 4.

SUMMARY

Data handling, sequencing and comparison instructions are discussed in this chapter. COP instruction copies blocks of data from one location into another. The destination file type determines the number of words that an instruction transfers. Instruction loads elements of a file with either a program constant or value from an element address. The EQU instruction is used to test whether two values are equal. Use the GRT instruction to test whether one value of source A is greater than another value of source B. The NEQ instruction is used to test whether two values are not equal. Sequencer Output Instruction (SQO) steps through the sequencer file whose bits have been set to control various output devices. Sequencer Load Instruction (SQL) stores 16-bit data into a sequencer load file at each step of sequencer operation.

REVIEW QUESTIONS

1. Name various data handling instructions and explain their functions.
2. What do you understand by COP instruction? Explain its working with the help of a suitable diagram.
3. Define and explain the working of FLL instruction.
4. List various move instructions and explain their working.
5. Compare MOV and MVM instructions.
6. What are logical instructions? Explain their working in detail.
7. Discuss why do we need comparison instructions?
8. Write down various types of comparison instructions and explain their working with suitable examples.
9. Name various parameters associated with sequencers and explain them.
10. List out various status bits of sequencers and explain them.
11. Explain the working of following sequencer instructions:
    (*a*) SQO (*b*) SQC (*c*) SQL

8

PLC Applications

LEARNING OBJECTIVES

After reading this chapter, you will be able to explain:

1. Explain the use of RS logix software to program and develop some of PLC applications like liquid level control, process control, drink dispenser.
2. Understand the SAP and ladder diagram fundamentals specific to some of PLC applications.

8.1 INTRODUCTION

Since its inception, the PLC has been successfully applied in virtually every segment of industry, including steel mills, paper plants, food-processing plants, chemical plants, and power plants. PLCs made by different manufacturers can be used for wide variety of application wherever we need automation. In this chapter we will use RS logix software to program some PLC applications listed below :

1. Electrical appliances control
2. Liquid level control
3. Process control
4. Main door control
5. Parking control
6. Bottling plant
7. Drink dispenser
8. Traffic light control

8.2 SWITCHING ON-OFF LIGHT

This application is used for an automatic control of light and fan in a room. The static application panel for this application is shown in Fig. 8.1.

8.2.1 SAP Explanation

In this SAP, whenever a person enters into the room, fan and tube light should be ON according to the following conditions:

(*i*) Summer day: Fan ON and tube light OFF.
(*ii*) Summer night: Fan ON and tube light ON.
(*iii*) Winter day: Fan OFF and tube light OFF.
(*iv*) Winter night: Fan OFF and tube light ON.

When a person leaves the room then fan and tube light should go OFF. So fan is related with summer and tube light is related with night. The ladder diagram to program this application is shown in Fig. 8.2.

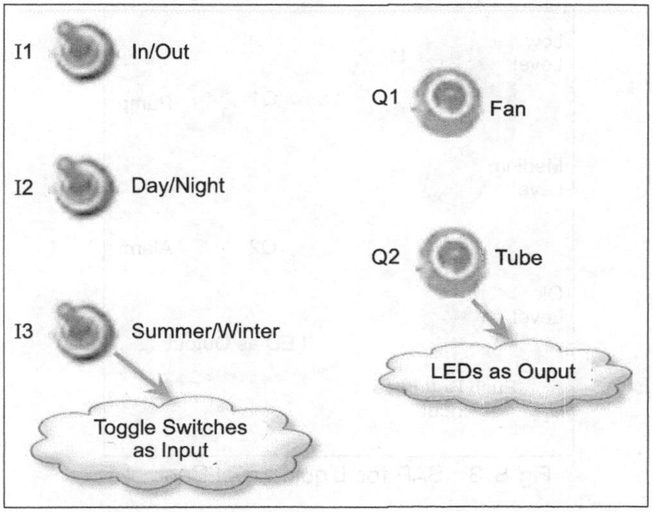

Fig. 8.1 SAP for Electrical Appliances Control

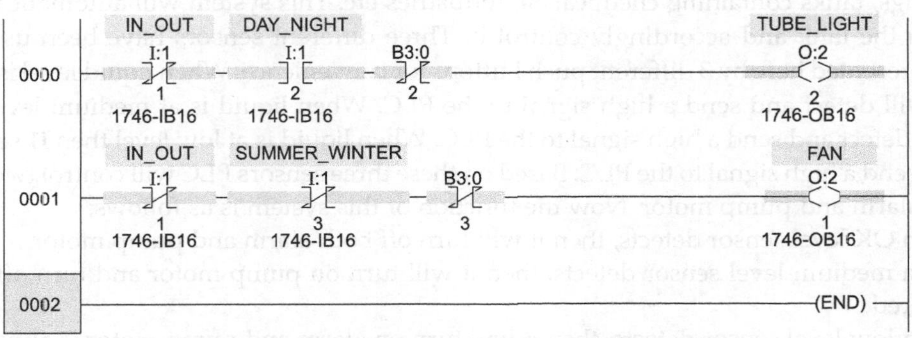

Fig. 8.2 Ladder Diagram Electrical Appliances Control

8.2.2 Ladder Diagram Explanation

Rung 0: In Ladder diagram when person is in means NC input I:1/1 is low and it is night means no input I:1/2 is high then it will latch the output O:2/2 *i.e.,* tube light. When either the person is out means NC input I:1/1 is high or it is day means no input I:1/2 is low then it will unlatch the output O:2/2 *i.e.,* tube light. The bit B3:0/2 is used for SCADA programming.

Rung 1: When the person is in means NC input I:1/1 is low and it is summer means NO input I:1/2 is low then it will latch the output O:2/1 *i.e.,* fan. When either the person is out means NC input I:1/1 is high or it is winter means NO input I:1/2 is high then it will unlatch the output O:2/1 *i.e.,* fan.

8.3 LIQUID LEVEL CONTROL

This application is used for an automatic control of liquid in a tank at remote place. The static application panel for this application is shown in Fig. 8.3.

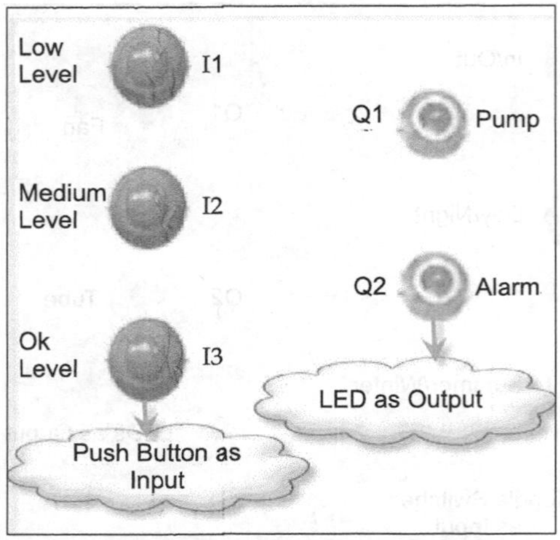

Fig 8.3 SAP for Liquid Level Control

8.3.1 SAP Explanation

This system is used to control the liquid level in a tank. For example overhead water tank in houses and buildings, tanks containing chemicals in industries etc. This system will automatically sense the level in the tank and accordingly control it. Three different sensors have been used in the system, represented here by 3 different push buttons for convenience. When liquid level is OK then I3 sensor will detect and send a high signal to the PLC. When liquid is at medium level then I2 sensor will detect and send a high signal to the PLC. When liquid is at low level then I1 sensor will detect and send a high signal to the PLC. Based on these three sensors PLC will control two outputs which are alarm and pump motor. Now the function of this system is as follows:

1. When OK level sensor detects, then it will turn off both alarm and pump motor.
2. When medium level sensor detects, then it will turn on pump motor and turn off alarm if required.
3. When low level sensor detects, then it will turn on alarm and pump motor is already on.

To make the system work in the above-mentioned way, we first see which of the three push buttons are pressed. There can be three main cases:

1. In case only push button for low level is pressed *i.e.,* I1 high, then the alarm for low level connected to output Q2 and the water pump connected to output Q1 should be high.
2. In case push button for medium level is pressed *i.e.,* I2 high, then it will turn on pump motor connected with Q1 and turn off alarm connected to Q2 if required.
3. In case push button for OK level is pressed *i.e.,* I3 high, then it will turn off both pump motor connected with Q1 and alarm connected to Q2 if required. The ladder diagram to program this application is shown in Fig. 8.4.

8.3.2 Ladder Diagram Explanation

Rung 1: If input I:1/1 *i.e.,* water level extra low sensor or I:1/2 water level medium sensor is high then it will latch the output O:2/1 *i.e.,* water pump.

Rung 2: If input I:1/1 *i.e.,* water level extra low sensor is high then it will latch the output O:2/2 *i.e.,* alarm.

Rung 3: If input I:1/2 water level medium sensor or I:1/3 *i.e.,* water level OK sensor is high then it will unlatch the output O:2/2 *i.e.,* alarm.

Rung 4: If input I:1/3 *i.e.,* water level OK sensor is high then it will unlatch the output O:2/1 *i.e.,* water pump.

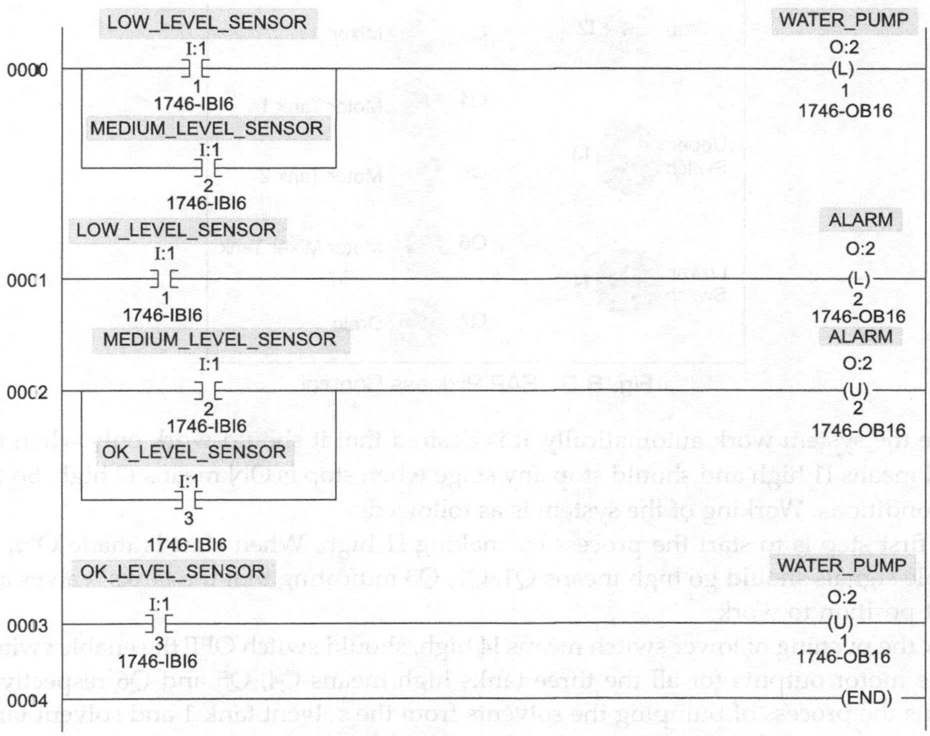

Fig. 8.4 Ladder Diagram for Liquid Level Control

8.4 PROCESS CONTROL

This application is used for an automatic process control in a chemical industry. The static application panel for this application is shown in Fig. 8.5.

8.4.1 SAP Explanation

This SAP is used to control the mixing of two solvents kept in two different tanks and mixed in a mixing tank by the use of motor pumps in each tank. The three motors indicated by three outputs which are Q4, Q5, and Q6 respectively. These outputs will be switched ON when the lower sensor in the mixer tank senses lower level. The drain indicated by output Q7 would be ON when the upper sensor in the mixer tank senses upper level. The two toggle switches are used to start and stop the process and two push buttons are used for lower sensor and upper sensor. The seven outputs available are used to indicate that solvent tank 1 ready, solvent tank 2 ready, mixer tank ready, motor of solvent tank 1 ON, motor of solvent tank 2 ON, motor of mixer tank ON and for the drain ON.

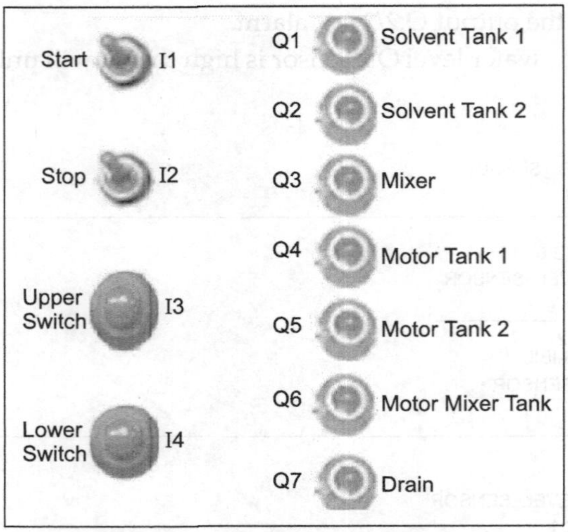

Fig. 8.5 SAP Process Control

To make the system work automatically, it is desired that it should work only when the main start is ON means I1 high and should stop any stage when stop is ON means I2 high. So these are the basic conditions. Working of the system is as followeds:

1. The first step is to start the process by making I1 high. When start is made ON, the tank enable signals should go high means Q1, Q2, Q3 indicating that the tanks' valves are in the right position to work.
2. Now the pushing of lower switch means I4 high, should switch OFF the enable switches and make motor outputs for all the three tanks high means Q4, Q5 and Q6 respectively. This means the process of pumping the solvents from the solvent tank 1 and solvent tank 2 into mixer tank and its mixing in the mixer tank has started.
3. In case the upper switch is pushed means I3 high, then all outputs should go low and drain *i.e.,* Q7 should high.
4. In case stop is made high means I2 high at any stage mentioned above or in the end all outputs should go low.

8.4.2 Ladder Diagram Explanation

Rung 0: Here NO start input I:1/1 is used to ON the tank enable signals indicated by three outputs (Q1, Q2, Q3). These outputs should go OFF if stop input switch is high or if lower switch is high or if upper switch is high. That is why we have taken here NC inputs. In this we are using two bits in order to latch or unlatch the inputs from two sensors.

Rung 1: High input from lower sensor *i.e.,* I:1/4 will latch one bit *i.e.,* B3:0/0 and unlatch the other bit *i.e.,* B3:0/1 as shown in Fig. 8.6.

Rung 2: This latched bit B3:0/0 will switch ON three motors connected to Q4, Q5 and Q6 outputs. The NC input I:1/2 is used here to stop the process in between.

Rung 3: High input from upper sensor *i.e.,* I:1/3 will latch one bit *i.e.,* B3:0/1 and unlatch the other bit *i.e.,* B3:0/0.

Rung 4: This latched bit B3:0/1 will switch ON drain connected to Q7 output. The NC input I:1/2 is used here to stop the process in between.

Rung 5: The NO input I:1/2 is used to unlatch both the bits if latched in order to start the operation again.

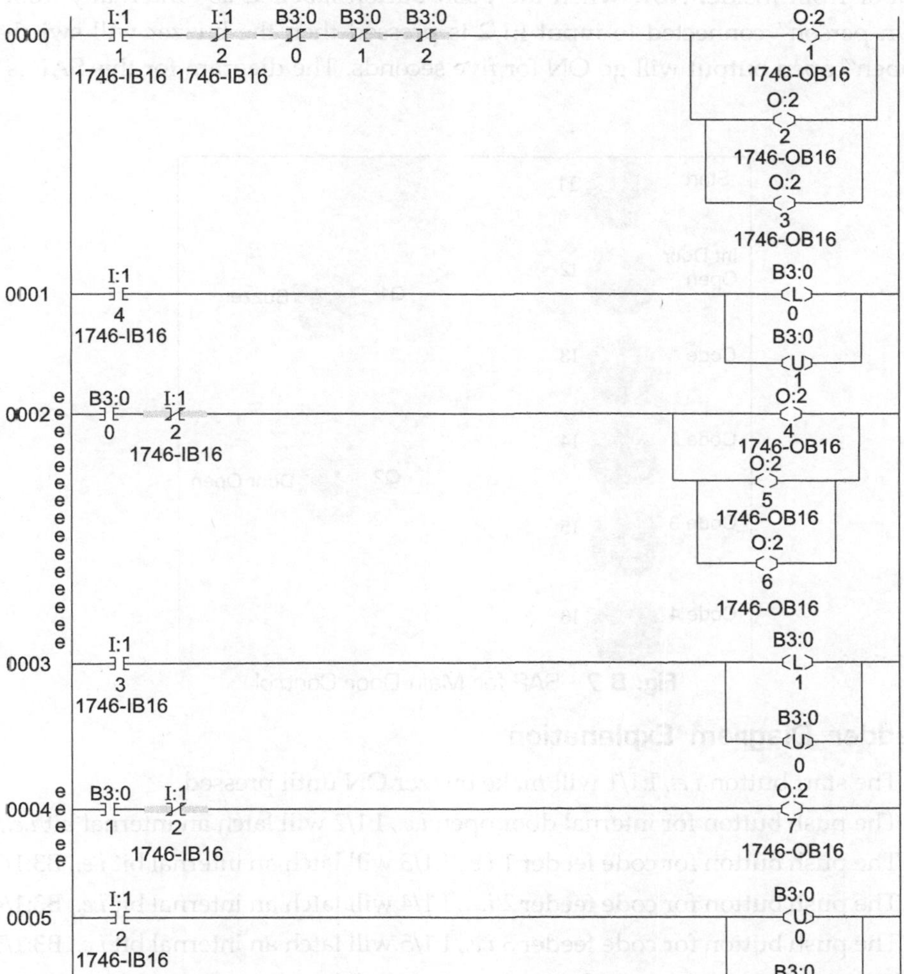

Fig. 8.6 Ladder Diagram for Process Control

8.5 MAIN DOOR CONTROL

This application is used for an automatic control of main door. The static application panel for this application is shown in Fig. 8.7.

8.5.1 SAP Explanation

When a known person wants to open the door then he has to feed the correct code at the door entrance *i.e.,* 1011 through code feeders I3, I4, I5, I6. After feeding the correct code the door will open for five seconds which is connected to output O:2/2.

Now when an unknown person wants to enter then he has to press the start button *i.e.,* I:1/1 which makes Buzzer connected to output O:2/1 ON. He will be allowed to enter when some one opens the door from inside. Now when the push button labelled as "Internally door open for the unknown person" connected to input I:1/2 is pressed then the buzzer will switch OFF and the "Door open" open output will go ON for five seconds. The diagram for this SAP is shown in Fig. 8.8.

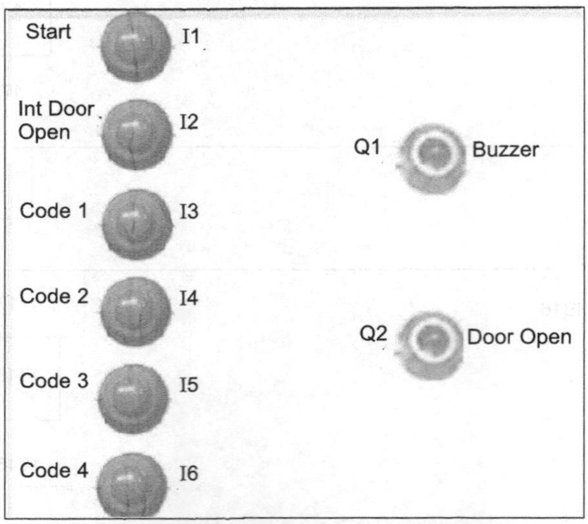

Fig. 8.7 SAP for Main Door Control

8.5.2 Ladder Diagram Explanation

Rung 0: The start button *i.e.,* I:1/1 will make buzzer ON until pressed.

Rung 1: The push button for internal door open *i.e.,* I:1/2 will latch an internal bit *i.e.,* B3:1/2.

Rung 2: The push button for code feeder 1 *i.e.,* I:1/3 will latch an internal bit *i.e.,* B3:1/3.

Rung 3: The push button for code feeder 2 *i.e.,* I:1/4 will latch an internal bit *i.e.,* B3:1/4.

Rung 4: The push button for code feeder 3 *i.e.,* I:1/5 will latch an internal bit *i.e.,* B3:1/5.

Rung 5: The push button for code feeder 4 *i.e.,* I:1/6 will latch an internal bit *i.e.,* B3:1/6.

Rung 6: The timer T4:1 will be activated if bit B3:0/2 *i.e.,* internal door open bit is high or if bit B3:0/3 and bit B3:0/5 and bit B3:0/6 are high to make exact code combination *i.e.,* 1011.

Rung 7: The timer timing bit of timer will open the door connected with Q2 output five seconds.

Rung 8: The done bit of the timer T4:1 will unlatch the bit B3:0/2 in order to repeat the process.

Rung 9: The done bit of the timer T4:1 will unlatch the bits B3:0/3 B3:0/4 B3:0/5 B3:0/6 in order to repeat the process. Due to the limitation of software maximum four bits can be used in branch for unlatching. So 2 different rungs are used to unlatch five bits.

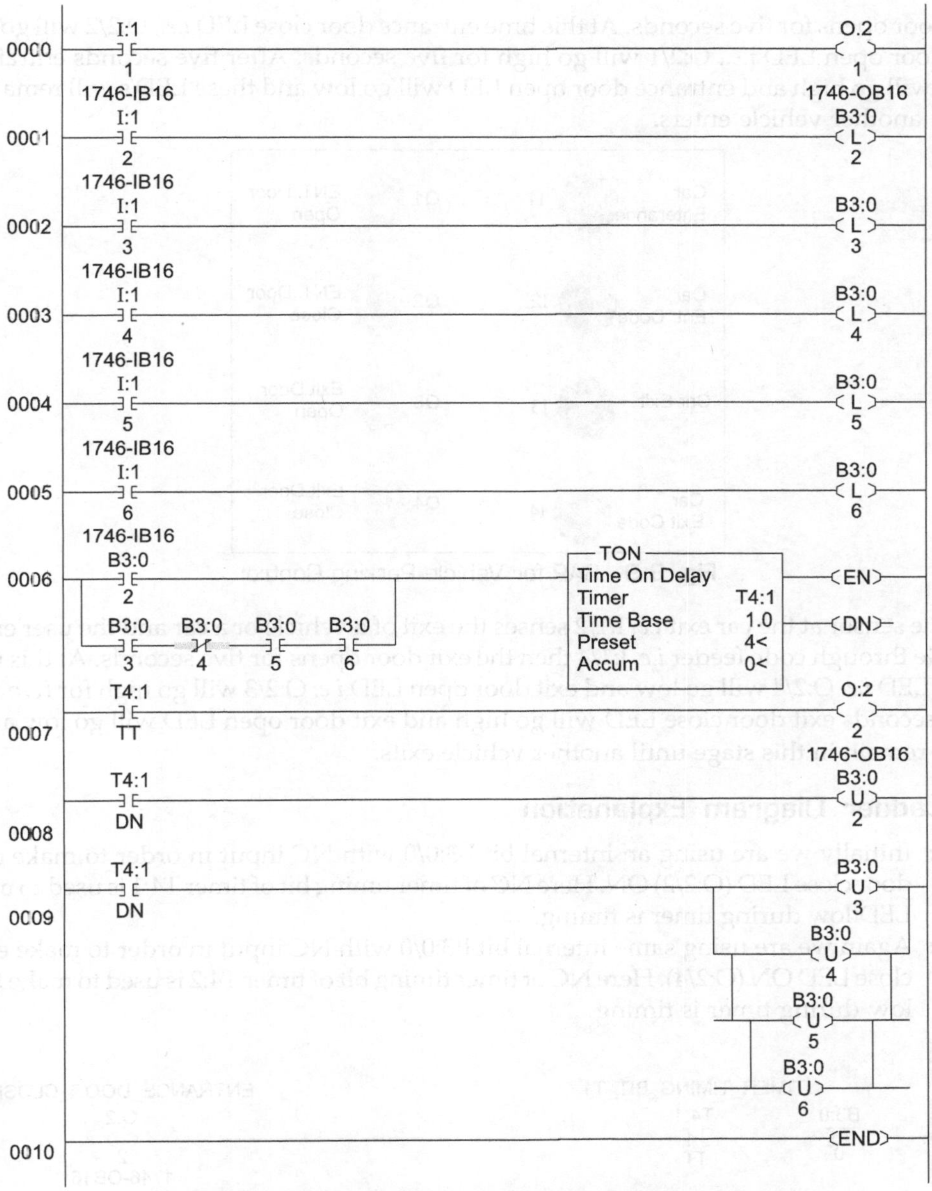

Fig. 8.8 Ladder Diagram for Main Door Control

8.6 VEHICLE PARKING CONTROL

This application is used for an automatic control to park the vehicles. The static application panel for this application is shown in Fig. 8.9.

8.6.1 SAP Explanation

Initially the LED showing the car entrance door closed indicated by 0:2/2 and LED showing the car exit door closed indicated by 0:2/4 are high. When the sensor at the entrance *i.e.*, I:1/1 senses the entry of a vehicle or a car and the user enters the correct code through code feeder *i.e.*, I:1/3 then the entrance door opens for five seconds. At this time entrance door close LED *i.e.,* O:2/2 will go low and entrance door open LED *i.e.,* O:2/1 will go high for five seconds. After five seconds entrance door close LED will go high and entrance door open LED will go low and these LED's will remain in this stage until another vehicle enters.

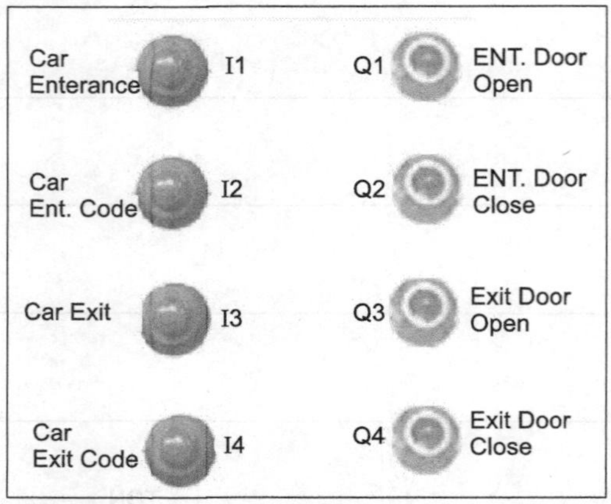

Fig. 8.9 SAP for Vehicle Parking Control

When the sensor at the car exit *i.e.* I:1/2 senses the exit of a vehicle or a car and the user enters the correct code through code feeder *i.e.* I:1/4 then the exit door opens for five seconds. At this time exit door close LED *i.e.* O:2/4 will go low and exit door open LED *i.e.* O:2/3 will go high for five seconds. After five seconds exit door close LED will go high and exit door open LED will go low and these LED's will remain in this stage until another vehicle exits.

8.6.2 Ladder Diagram Explanation

Rung 0: Initially we are using an internal bit B3:0/0 with NC input in order to make entrance door close LED (O:2/2) ON. Here NC of timer timing bit of timer T4:1 is used to make this LED low during timer is timing.

Rung 1: Again we are using same internal bit B3:0/0 with NC input in order to make exit door close LED ON (O:2/4). Here NC of timer timing bit of timer T4:2 is used to make this LED low during timer is timing.

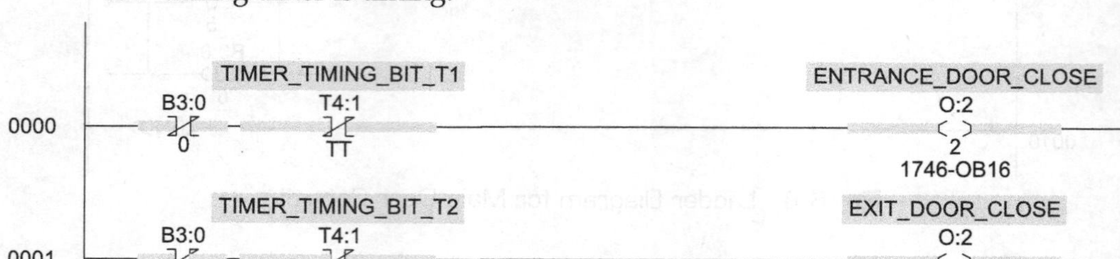

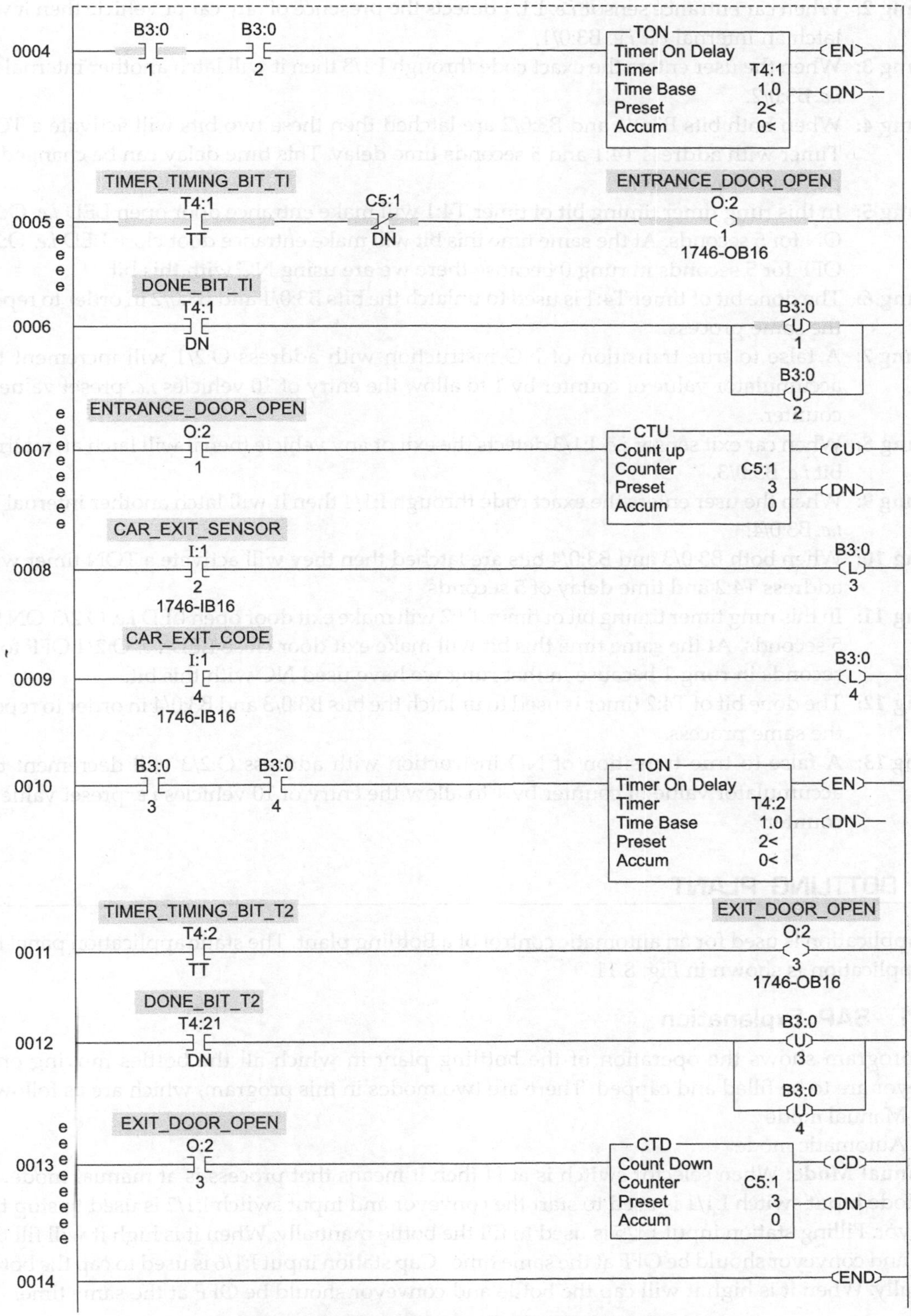

Fig. 8.10 Ladder Diagram for Car Parking Garage

Rung 2: When car entrance sensor *i.e.* I:1/1 detects the presence of any car or vehicle then it will latch an internal bit *i.e.* B3:0/1.

Rung 3: When the user enters the exact code through I:1/3 then it will latch another internal bit *i.e.* B3:0/2.

Rung 4: When both bits B3:0/1 and B3:0/2 are latched then these two bits will activate a TON Timer with address T4:1 and 5 seconds time delay. This time delay can be changed as per user requirement.

Rung 5: In this rung timer timing bit of timer T4:1 will make entrance door open LED *i.e.* O:2/1 ON for 5 seconds. At the same time this bit will make entrance door close LED *i.e.* O:2/2 OFF for 5 seconds in rung 0 because there we are using NC with this bit.

Rung 6: The done bit of timer T4:1 is used to unlatch the bits B3:0/1 and B3:0/2 in order to repeat the same process.

Rung 7: A false to true transition of NO instruction with address O:2/1 will increment the accumulator value of counter by 1 to allow the entry of 10 vehicles *i.e.* preset value of counter.

Rung 8: When car exit sensor *i.e.* I:1/3 detects the exit of any vehicle then it will latch an internal bit *i.e.* B3:0/3.

Rung 9: When the user enters the exact code through I:1/4 then it will latch another internal bit *i.e.* B3:0/4.

Rung 10: When both B3:0/3 and B3:0/4 bits are latched then they will activate a TON timer with address T4:2 and time delay of 5 seconds.

Rung 11: In this rung timer timing bit of timer T4:2 will make exit door open LED *i.e.* O:2/3 ON for 5 seconds. At the same time this bit will make exit door close LED *i.e.* O:2/4 OFF for 5 seconds in rung 1 because in that rung we have used NC with this bit.

Rung 12: The done bit of T4:2 timer is used to unlatch the bits B3:0/3 and B3:0/4 in order to repeat the same process.

Rung 13: A false to true transition of NO instruction with address O:2/3 will decrement the accumulator value of counter by 1 to allow the entry of 10 vehicles *i.e.* preset value of counter.

8.7 BOTTLING PLANT

This application is used for an automatic control of a Bottling plant. The static application panel for this application is shown in Fig. 8.11.

8.7.1 SAP Explanation

This program shows the operation of the bottling plant in which all the bottles moving on a conveyor are to be filled and capped. There are two modes in this program, which are as follows:

1. Manual mode
2. Automatic mode

Manual Mode: When selector switch is at I4 then it means that process is at manual mode. In this mode input switch I:1/1 is used to start the conveyor and input switch I:1/2 is used to stop the conveyor. Filling station input I:1/5 is used to fill the bottle manually. When it is high it will fill the bottle and conveyor should be OFF at the same time. Cap station input I:1/6 is used to cap the bottle manually. When it is high it will cap the bottle and conveyor should be OFF at the same time.

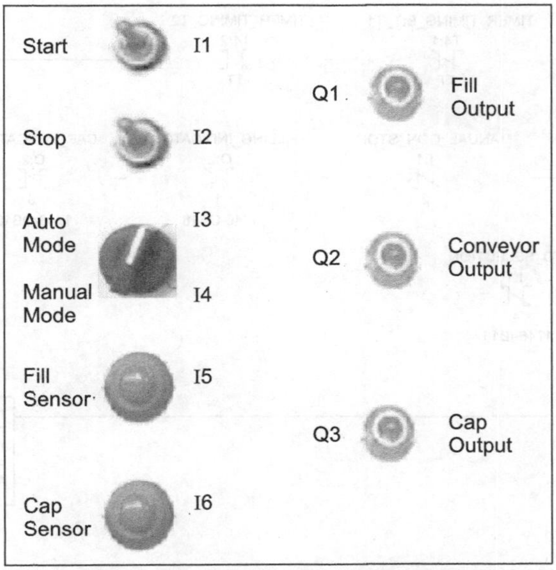

Fig. 8.11 SAP for Bottling Plant

Automatic Mode: For automatic operation means selector switch is not at input I:1/4 and conveyor should start automatically in this mode. When the filling station sensor senses the bottle, it stops the conveyor and keeps Q1 ON for the filling time. After the filling process, conveyor starts again and when cap sensor senses the bottle, it again stops the conveyor and Q3 becomes ON till the capping process is going on. The whole process will go on continuously. The ladder diagram for this SAP is shown in Fig. 8.12.

8.7.2 Ladder Diagram Explanation

Rung 0: Here conveyor will be ON when selector switch is not ON input I:1/4 *i.e.* auto mode or when input I:1/1 is high *i.e.* manual mode. This conveyor will stop when input I:1/2 or filling station output *i.e.* O:2/1 or capping station output *i.e.* O:2/3 is high.

Rung 1: In auto mode when filling sensor *i.e.* I:1/5 detects the presence of a bottle then it will latch a bit B3:0/1. This bit will be latched only if selector switch is at auto *i.e.* input I:1/4 is low.

Rung 2: The latched bit B3:0/1 will activate one timer with preset value of 5 seconds. The done bit of this timer T4:1 is used to reset it in order to repeat the process.

Rung 3: The filling station output O:2/1 will be high if timer timing bit of timer T4:1 is high in auto mode or input I:1/5 is high in manual mode.

Rung 4: The done bit of the timer T4:1 is used to unlatch the bit B3:0/1.

Rung 5: In auto mode when capping sensor *i.e.* I:1/6 detects the bottle then it will latch a bit B3:0/2. This bit will be latched only if selector switch is at auto *i.e.* input I:1/4 is low.

Rung 6: The latched bit B3:0/2 will activate one timer with preset value of 2 seconds. The done bit of this timer T4:2 is used to reset it in order to repeat the process.

Rung 7: The capping station output O:2/3 will be high if timer timing bit of timer T4:2 is high in auto mode or input I:1/6 is high in manual mode.

Rung 8: The done bit of the timer T4:2 is used to unlatch the bit B3:0/1.

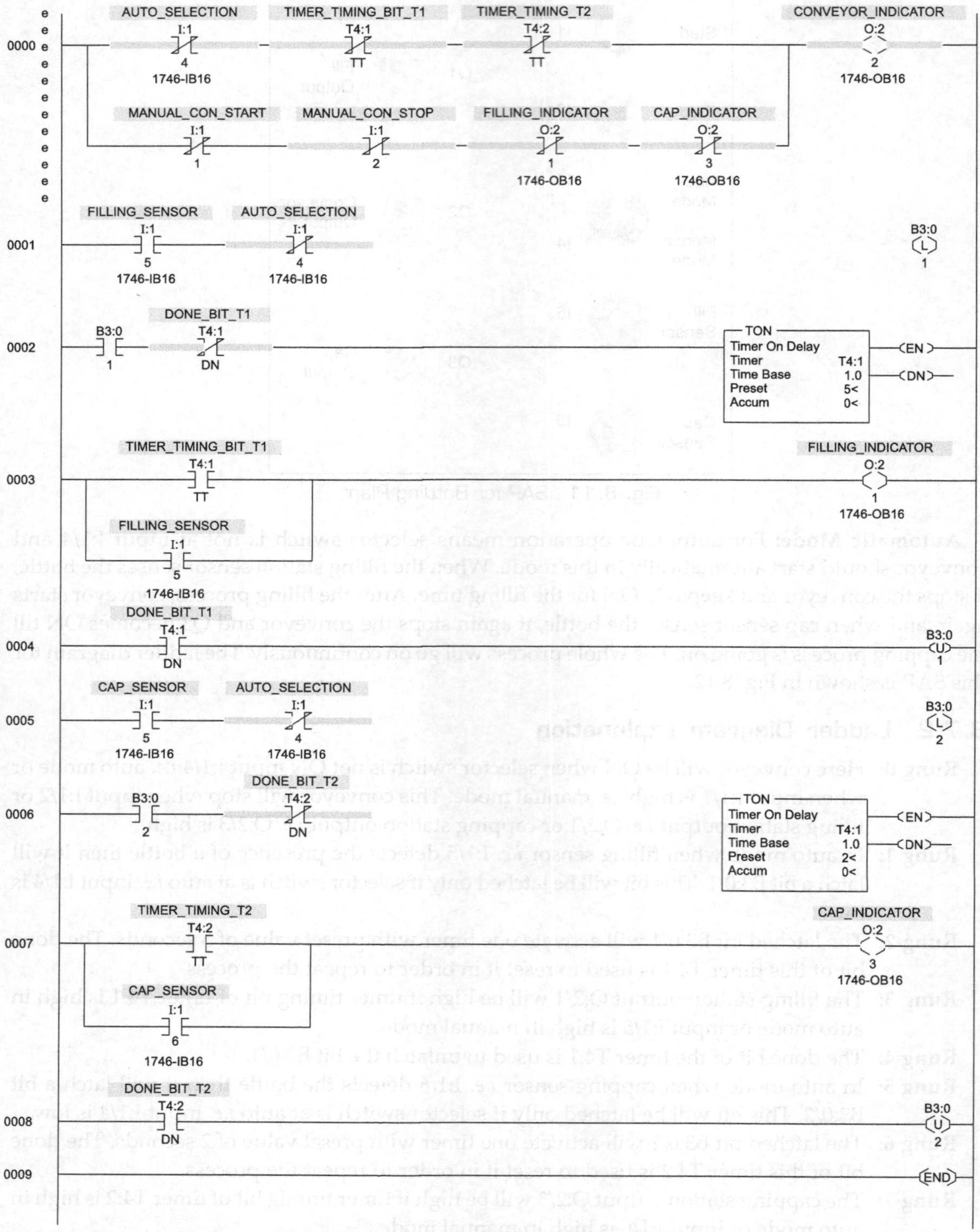

Fig. 8.12 Ladder diagram for Bottling Plant

8.8 DRINK DISPENSER

This application is used for an automatic control of a Drink Dispenser. The static application panel for this application is shown in Fig. 8.13.

8.8.1 SAP Explanation

In this automatic process the user has to put the coin into the drink-dispensing machine can get the required soft drink. The user first presses the push button labelled as "Coin" and then presses any other push button such as "Thumsup", "Coke", "Pepsi", "Fanta", etc. to get the required soft drink. The corresponding LED glows for five seconds indicating the opening of the valve of the required soft drink for the preset time of five seconds. This preset time can be set as per user requirement. The ladder diagram for this SAP is shown in Fig. 8.14.

8.8.2 Ladder Diagram Explanation

Rung 0: When input I:1/1 *i.e.* thums up is pressed then it will latch an internal bit *i.e.* B3:0/1

Rung 1: When input I:1/2 *i.e.* coke is pressed then it will latch an internal bit *i.e.* B3:0/2.

Rung 2: When input I:1/3 *i.e.* pepsi is pressed then it will latch an internal bit *i.e.* B3:0/3.

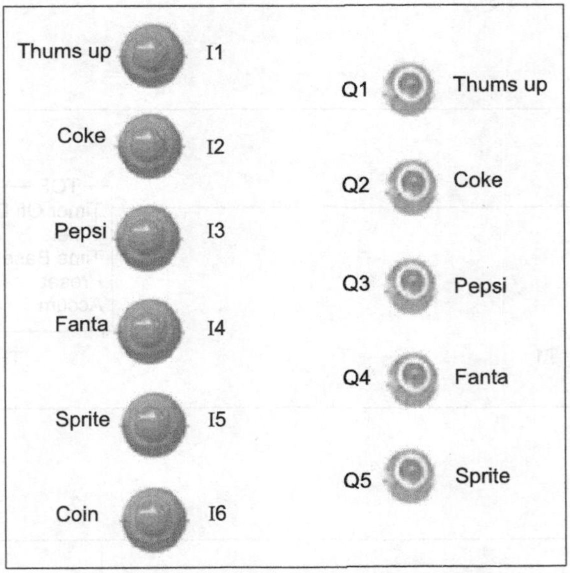

Fig. 8.13 SAP for Drink Dispenser

Rung 3: When input I:1/4 *i.e.* fanta is pressed then it will latch an internal bit *i.e.* B3:0/4.

Rung 4: When input I:1/5 *i.e.* sprite is pressed then it will latch an internal bit *i.e.* B3:0/5.

Rung 5: When input I:1/6 *i.e.* coin is pressed then it will latch an internal bit *i.e.* B3:0/6.

Rung 6: When coin bit *i.e.* B3:0/6 and thums up bit *i.e.* B3:0/1 are high then a timer T4:1 will be activated with preset value of 5 seconds.

Rung 7: The timer timing bit of T4:1 will make thums up output high *i.e.* O:2/1. It means that the thums up valve will open for 5 seconds then it will be stopped.

Rung 8–9: The done bit of T4:1 timer will unlatch all the latched bits in order to repeat the process.

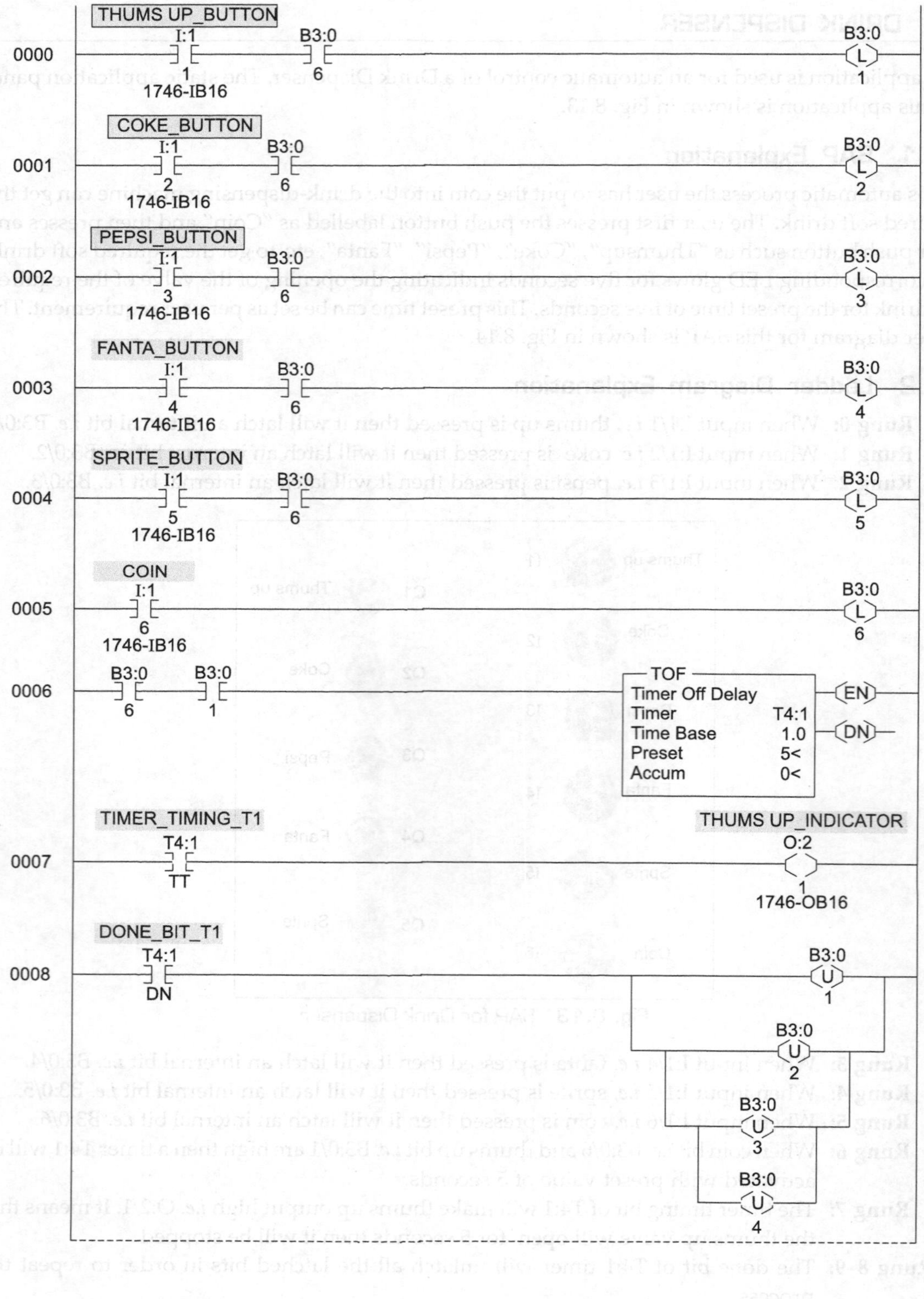
THUMS UP_BUTTON
I:1
B3:0
0000
1
1746-IB16
6
B3:0
L
1
COKE_BUTTON
I:1
B3:0
0001
2
1746-IB16
6
B3:0
L
2
PEPSI_BUTTON
I:1
B3:0
0002
3
1746-IB16
6
B3:0
L
3
FANTA_BUTTON
I:1
B3:0
0003
4
1746-IB16
6
B3:0
L
4
SPRITE_BUTTON
I:1
B3:0
0004
5
1746-IB16
6
B3:0
L
5
COIN
I:1
0005
6
1746-IB16
B3:0
L
6
B3:0
B3:0
0006
6
1
TOF
Timer Off Delay
Timer T4:1
Time Base 1.0
Preset 5<
Accum 0<
EN
DN
TIMER_TIMING_T1
T4:1
0007
TT
THUMS UP_INDICATOR
O:2
1
1746-OB16
DONE_BIT_T1
T4:1
0008
DN
B3:0
U
1
B3:0
U
2
B3:0
U
3
B3:0
U
4

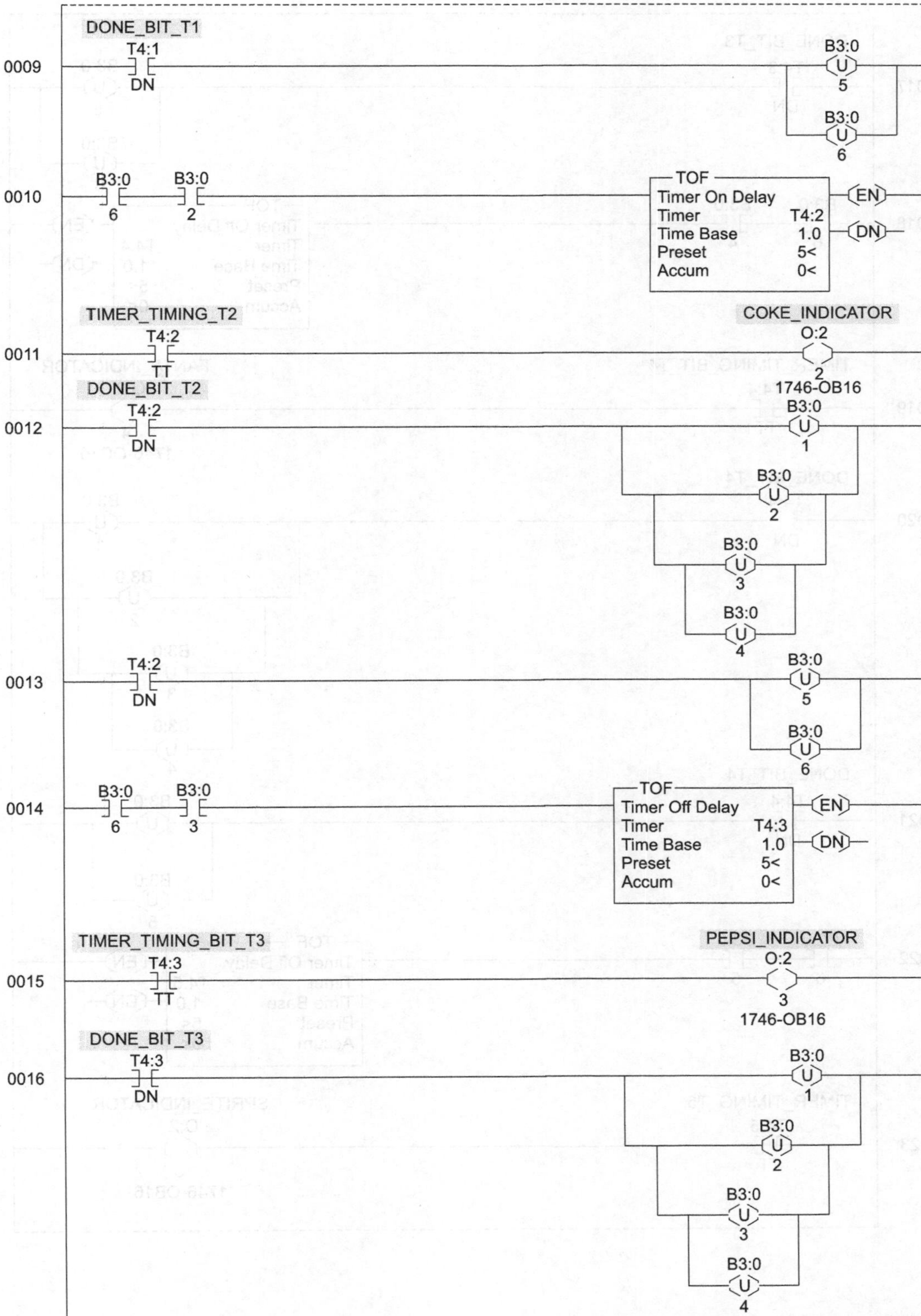

0009
DONE_BIT_T1
T4:1
DN
B3:0
U
5
B3:0
U
6
0010
B3:0
6
B3:0
2
TOF
Timer On Delay
Timer T4:2
Time Base 1.0
Preset 5<
Accum 0<
EN
DN
0011
TIMER_TIMING_T2
T4:2
TT
COKE_INDICATOR
O:2
2
1746-OB16
0012
DONE_BIT_T2
T4:2
DN
B3:0
U
1
B3:0
U
2
B3:0
U
3
B3:0
U
4
0013
T4:2
DN
B3:0
U
5
B3:0
U
6
0014
B3:0
6
B3:0
3
TOF
Timer Off Delay
Timer T4:3
Time Base 1.0
Preset 5<
Accum 0<
EN
DN
0015
TIMER_TIMING_BIT_T3
T4:3
TT
PEPSI_INDICATOR
O:2
3
1746-OB16
0016
DONE_BIT_T3
T4:3
DN
B3:0
U
1
B3:0
U
2
B3:0
U
3
B3:0
U
4

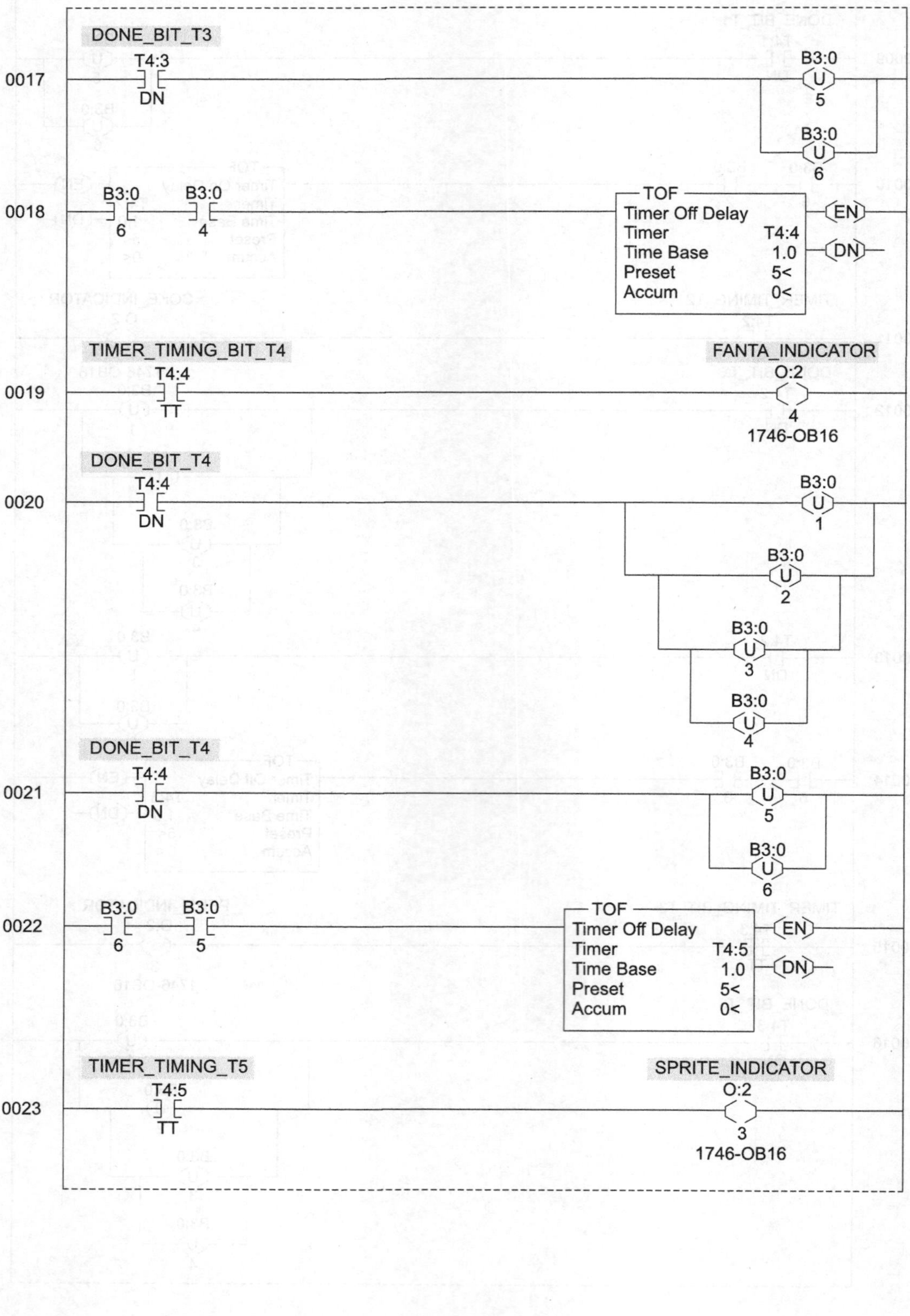
DONE_BIT_T3
T4:3
0017
DN
B3:0
U
5
B3:0
U
6
B3:0
B3:0
0018
6
4
TOF
Timer Off Delay
Timer T4:4
Time Base 1.0
Preset 5<
Accum 0<
EN
DN
TIMER_TIMING_BIT_T4
T4:4
0019
TT
FANTA_INDICATOR
O:2
4
1746-OB16
DONE_BIT_T4
T4:4
0020
DN
B3:0
U
1
B3:0
U
2
B3:0
U
3
B3:0
U
4
DONE_BIT_T4
T4:4
0021
DN
B3:0
U
5
B3:0
U
6
B3:0
B3:0
0022
6
5
TOF
Timer Off Delay
Timer T4:5
Time Base 1.0
Preset 5<
Accum 0<
EN
DN
TIMER_TIMING_T5
T4:5
0023
TT
SPRITE_INDICATOR
O:2
3
1746-OB16

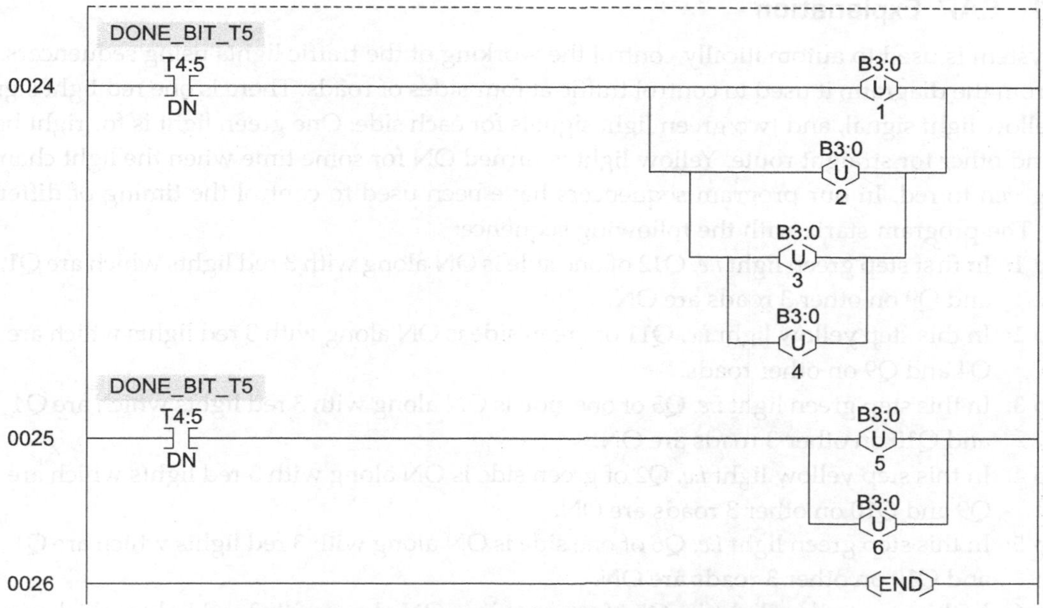

Fig. 8.14 Ladder diagram for Drink Dispenser

Rung 10: When coin bit *i.e.* B3:0/6 and coke bit *i.e.* B3:0/2 is high then a timer T4:2 will be activated with preset value of 5 seconds.

Rung 11: The timer timing bit of T4:2 will make coke output high *i.e.* O:2/2. It means that the coke drink valve will open for 5 seconds then it will be stopped.

Rung 12-13: The done bit of T4:2 will unlatch all the latched bits in order to repeat the process.

Rung 14: When coin bit *i.e.* B3:0/6 and pepsi bit *i.e.* B3:0/3 is high then a timer T4:3 will be activated with preset value of 5 seconds.

Rung 15: The timer timing bit of T4:3 will make pepsi output high *i.e.* O:2/3. It means that the pepsi drink valve will open for 5 seconds then it will be stopped.

Rung 16-17: The done bit of T4:3 will unlatch all the latched bits in order to repeat the process.

Rung 18: When coin bit *i.e.* B3:0/6 and fanta bit *i.e.* B3:0/4 is high then a timer T4:4 will be activated with preset value of 5 seconds.

Rung 19: The timer timing bit of T4:4 will make fanta output high *i.e.* O:2/4. It means that the fanta drink valve will open for 5 seconds then it will be stopped.

Rung 20-21: The done bit of T4:4 will unlatch all the latched bits in order to repeat the process.

Rung 22: When coin bit *i.e.* B3:0/6 and sprite bit *i.e.* B3:0/5 is high then a timer T4:5 will be activated with preset value of 5 seconds.

Rung 23: The timer timing bit of T4:5 will make sprite output high *i.e.* O:2/5. It means that the sprite drink is coming out for 5 seconds then it will be stopped.

Rung 24-25: The done bit of T4:5 will unlatch all the latched bits in order to repeat the process.

8.9 TRAFFIC LIGHT CONTROL

This application is used for an automatic traffic light control. The static application panel for this application is shown in Fig. 8.15.

8.9.1 SAP Explanation

This system is used to automatically control the working of the traffic lights using sequencers. As seen from the diagram, it used to control traffic at four sides of roads. There is one red light signal, one yellow light signal, and two green light signals for each side. One green light is for right hand turn and other for straight route. Yellow light is turned ON for some time when the light changes from green to red. In our program sequencers have been used to control the timing of different lights. The program starts with the following sequence:

Step 1: In first step green light *i.e.* Q12 of one side is ON along with 3 red lights which are Q1, Q4 and Q9 on other 3 roads are ON.

Step 2: In this step yellow light *i.e.* Q11 of green side is ON along with 3 red lights which are Q1, Q4 and Q9 on other roads.

Step 3: In this step green light *i.e.* Q3 of one side is ON along with 3 red lights which are Q4 , Q9 and Q10 on other 3 roads are ON.

Step 4: In this step yellow light *i.e.* Q2 of green side is ON along with 3 red lights which are Q4, Q9 and Q10 on other 3 roads are ON.

Step 5: In this step green light *i.e.* Q6 of one side is ON along with 3 red lights which are Q1 , Q9 and Q10 on other 3 roads are ON.

Step 6: In this step yellow light *i.e.* Q5 of green side is ON along with 3 red lights which are Q1, Q9 and Q10 on other 3 roads are ON.

Step 7: In this step green light *i.e.* Q7 of one side is ON along with 3 red lights which are Q1, Q4 and Q10 on other 3 roads are ON.

Step 8: In this step yellow light *i.e.* Q8 of green side is ON along with 3 red lights which are Q1, Q4 and Q10 on other 3 roads are ON.

Then the whole sequence of steps will be repeated to continuously run the process.

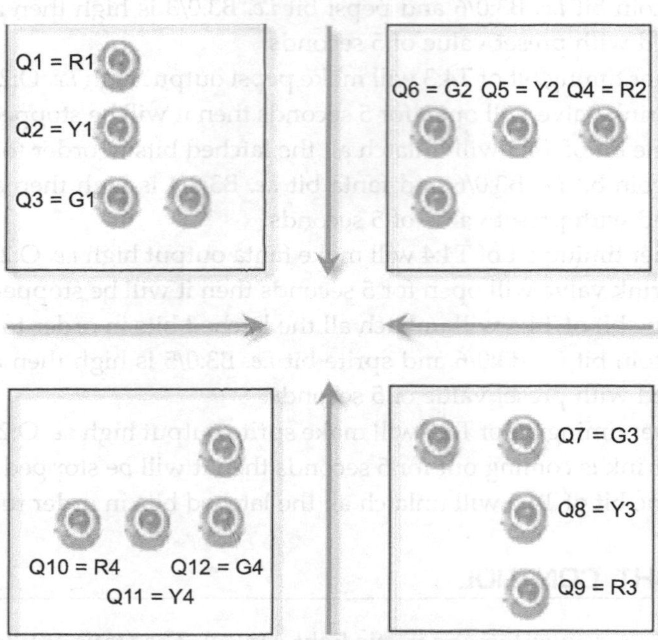

Fig. 8.15 SAP for Traffic Light Control

8.9.2 Ladder Diagram Explanation

Rung 0 In this rung a timer T4:0 is activated with 3 seconds time delay. The done bit of the T4:1 timer is used to reset itself. The resetting of timer will advances the sequencer to next step.

Rung 1 Here timer timing bit of T4:0 is used to drive the steps of sequencer as shown in Fig. 8.16. The steps of sequencer are shown in Fig. 8.17.

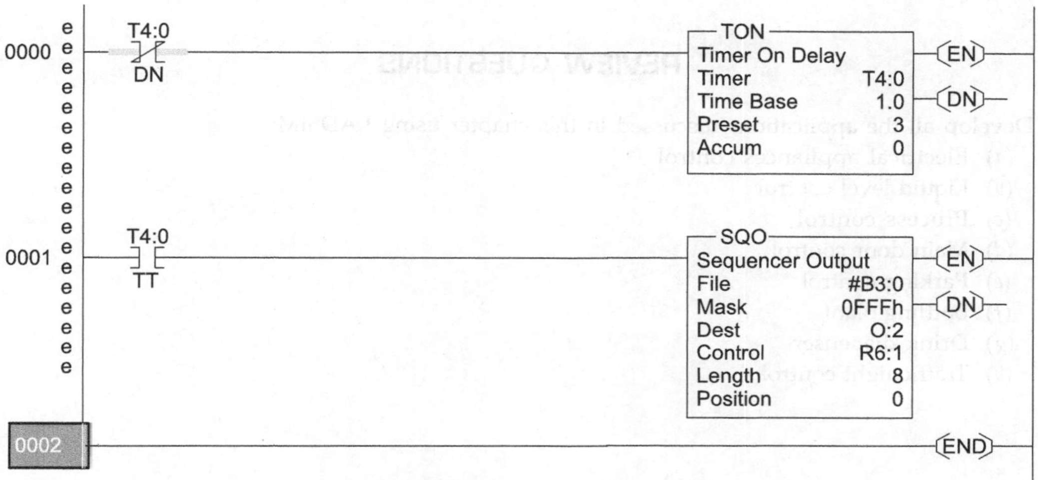

Fig. 8.16 Ladder Diagram for Traffic Light Control

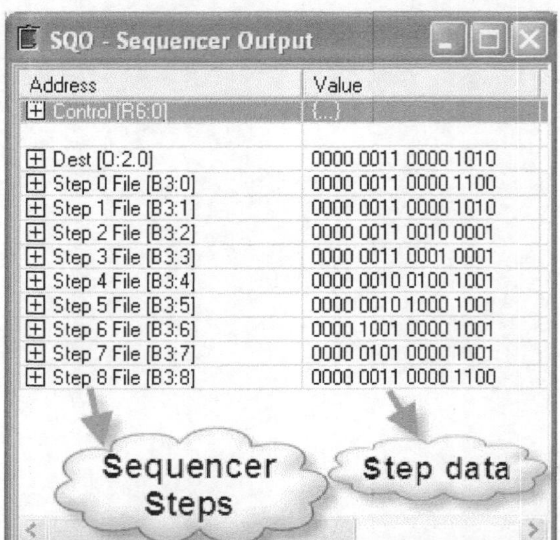

Fig. 8.17 Sequencer Step Data

SUMMARY

PLCs perform a great variety of control tasks, from repetitive ON/OFF control of simple machines to sophisticated manufacturing and process control. In this chapter, we have used the foundation knowledge learned in Chapter 2 and use it to build an understanding of PLC programming. The programming method used in this chapter is graphical method, which uses schematic symbols for relay coils and contacts. The next thing to be considered is how to get the ladder diagram into the programmable controller. In higher order controllers, this can be accomplished through the use of dedicated personal computer software that allows the programmer to enter the ladder diagram as drawn. The software then takes care of translating the ladder diagram into the code required by the controller. In the lower order, more basic controllers, this has to be performed by the programmer and entered by hand into the controller. A variety of PLC applications are developed and programmed in this chapter using RS logix software.

REVIEW QUESTIONS

1. Develop all the applications discussed in this chapter using LADSIM:

(*a*) Electrical appliances control
(*b*) Liquid level control
(*c*) Process control
(*d*) Main door control
(*e*) Parking control
(*f*) Bottling plant
(*g*) Drink dispenser
(*h*) Traffic light control

9

PLC & SCADA INTERFACE

LEARNING OBJECTIVES

After reading this chapter, you will be able to explain:

1. SCADA software Installation
2. SCADA project development
3. Basics of PLC interfacing with SCADA
4. Working with developed projects

9.1 INTRODUCTION

In this chapter we will discuss how a PLC can be interfaced with a SCADA that stands for **"Supervisory Control and Data Acquisition"**. Our discussion will begin with SCADA software installation, followed by project development aspects of SCADA.

9.2 SCADA SOFTWARE INSTALLATION

The RSView32 is used as SCADA software for Allen Bradley PLCs. As part of the installation process, RSView32 creates program folders and modifies registry entries. To make these modifications, the currently logged in user must have administrative rights on the computer on which RSView32 is to be installed. The steps to install the software are as follows:

1. Close all open Windows programs.
2. Place the RSView32 CD-ROM into your CD-ROM drive. The CD-ROM should run automatically. If the CD-ROM does not start automatically. Then run D:\Setup.exe, where D is the drive containing the CD-ROM. Then a new window will appear as shown in Fig. 9.1.

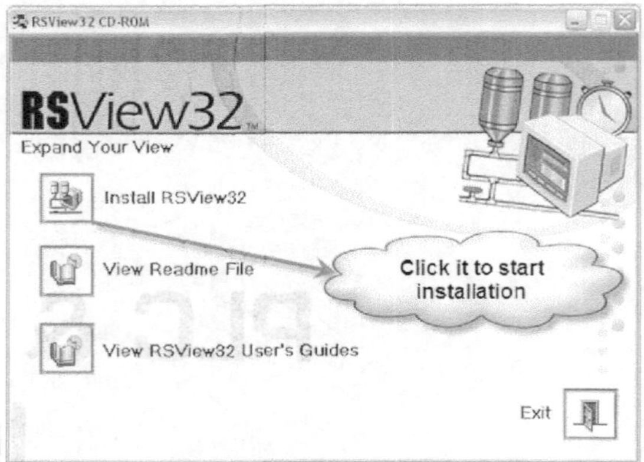

Fig. 9.1 RSView32 Install Window

3. In the RSView32 install window, click the button beside Install RSView32 Runtime/Works.
4. Follow the on-screen instructions which are:
   (*a*) Click Next as shown in Fig. 9.2.

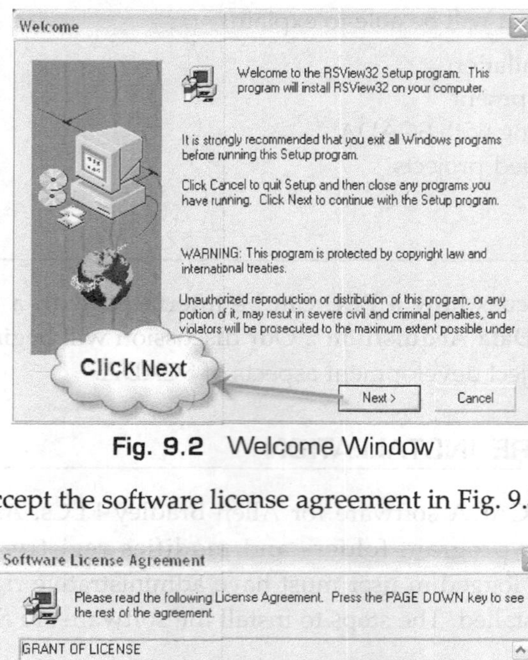

Fig. 9.2 Welcome Window

(*b*) Click Yes to accept the software license agreement in Fig. 9.3.

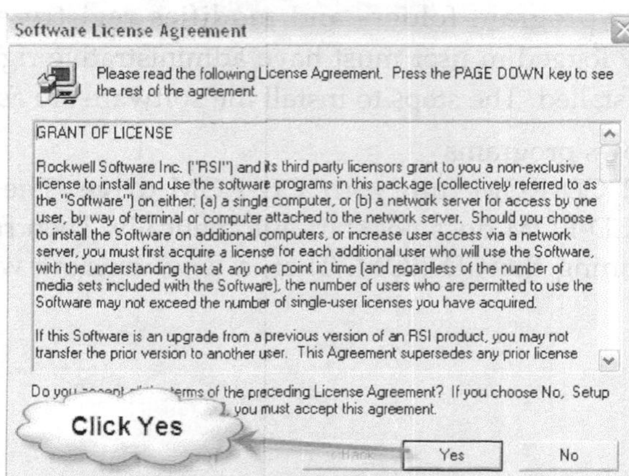

Fig. 9.3 License Agreement Window

(*c*) Enter a name, company name of your choice and 10 digit serial number written on the activation disk. Click **Next** as indicated in Fig. 9.4.

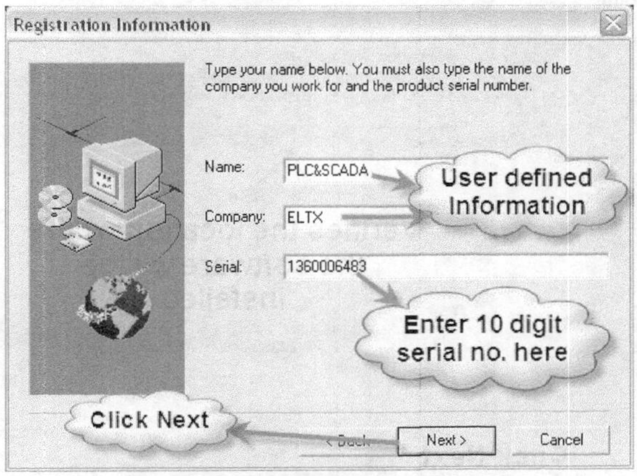

Fig. 9.4 Registration Window

(*d*) Select the installation path by using browse option. Click Next in Fig. 9.5.

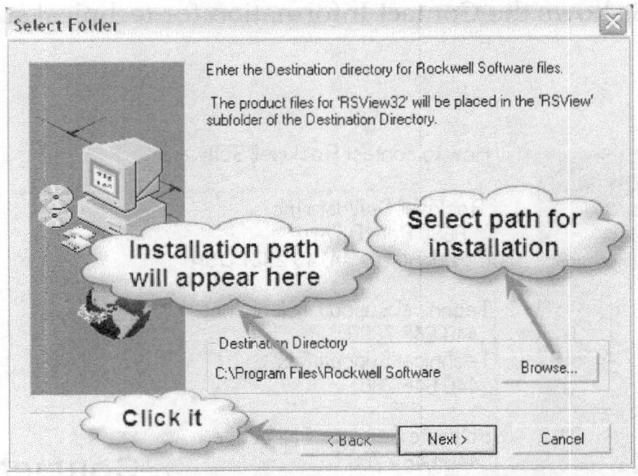

Fig. 9.5 Folder Selection Window

(*e*) Select the required software, you want to install then click Next in Fig. 9.6.

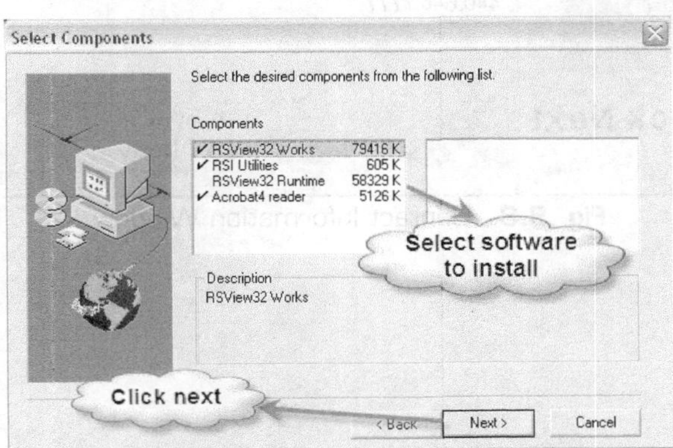

Fig. 9.6 Component Selection Window

(*f*) Figure 9.7 verifies the location where the software will be installed, click Next.

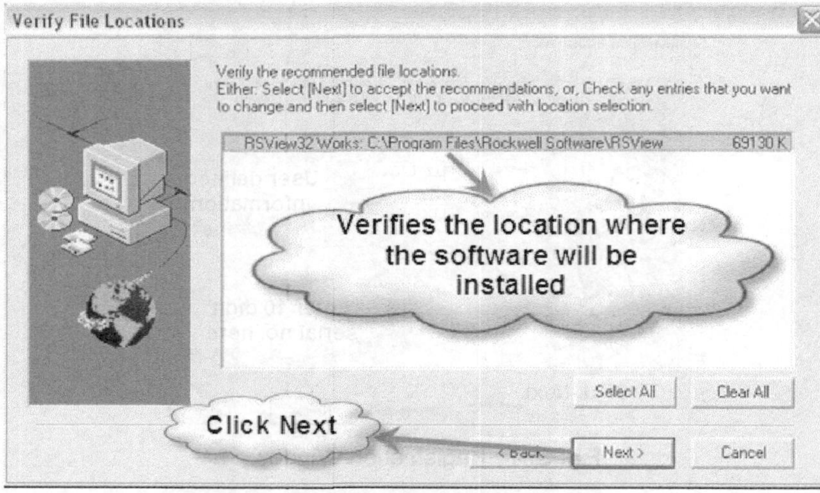

Fig. 9.7 File Location Verification Window

(*i*) Figure 9.8 shows the Contact Information for technical support, click Next.

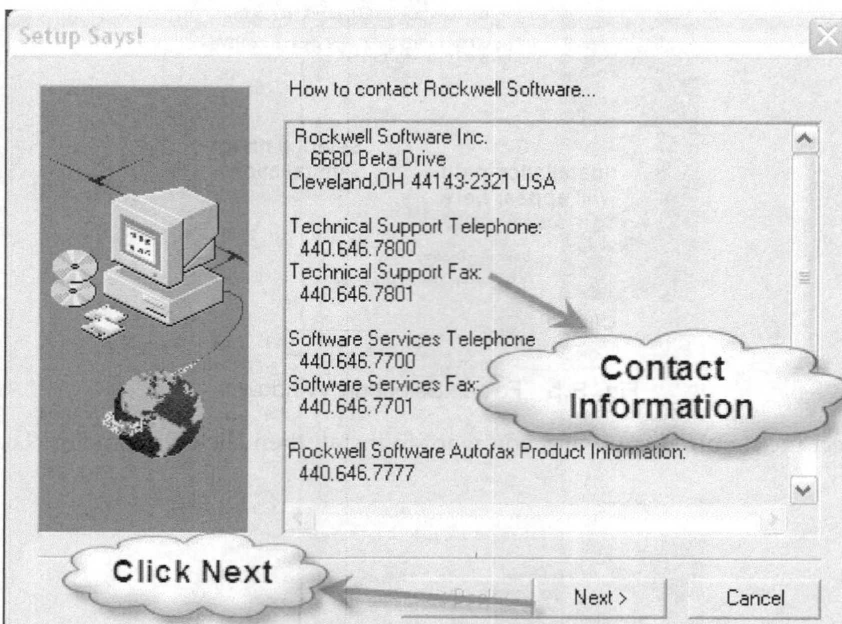

Fig. 9.8 Contact Information Window

(*ii*) Un-check the box "View read me file now" and check "Activate RSview" box as shown in Fig. 9.9 and click finish.

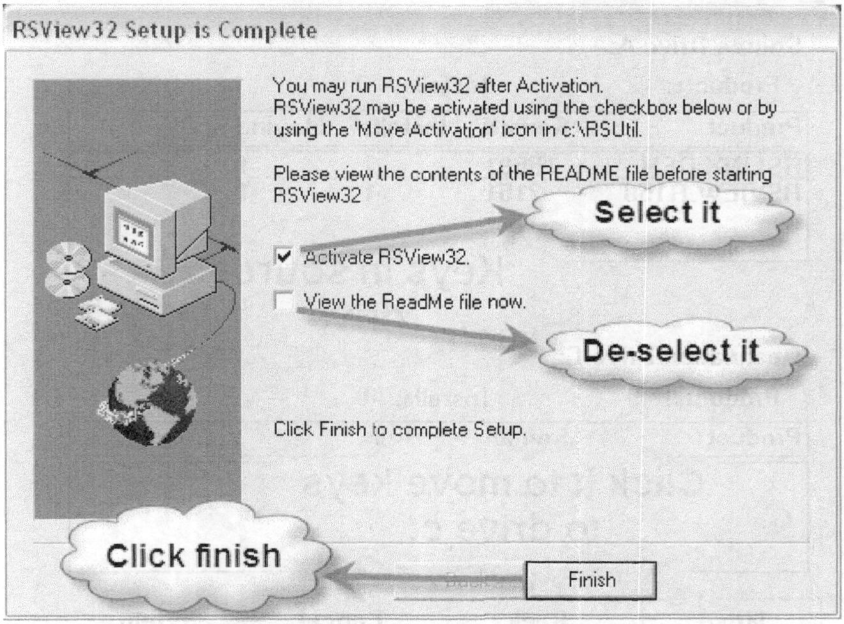

Fig. 9.9 Setup Complete Window

(*iii*) Select source and destination drives to move activation as shown in Fig. 9.10. Click Finish again.

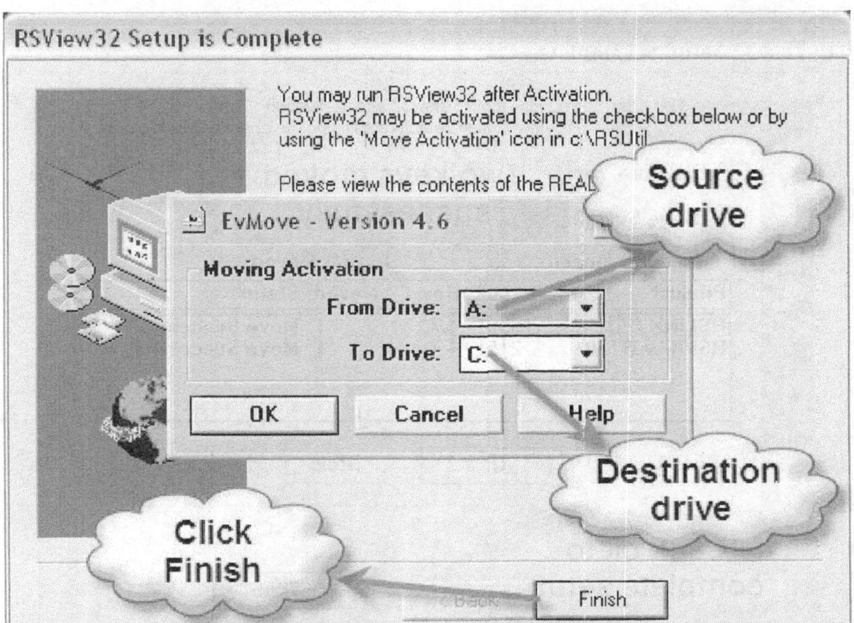

Fig. 9.10 Move Activation Window

(*iv*) Click move in Fig. 9.11 to move the activation keys from drive A: to drive C:

Fig. 9.11 Activation Move Window

(*v*) A message with move successful will appear as shown in Fig. 9.12, click OK.

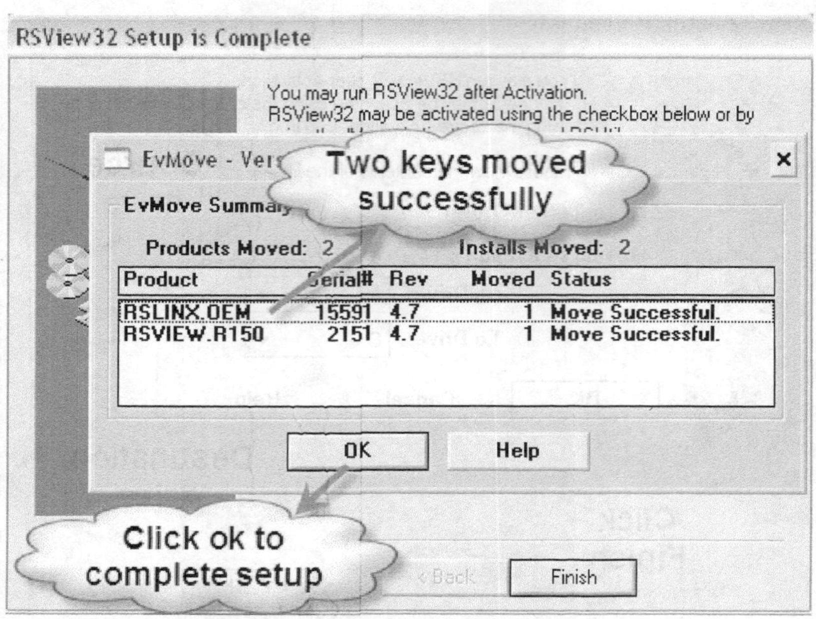

Fig. 9.12 Setup Complete Window

(*vi*) When installation is complete, restart Windows as shown in Fig. 9.13. If you choose not to restart Windows now, be sure to restart it before running RSView32 for the first time.

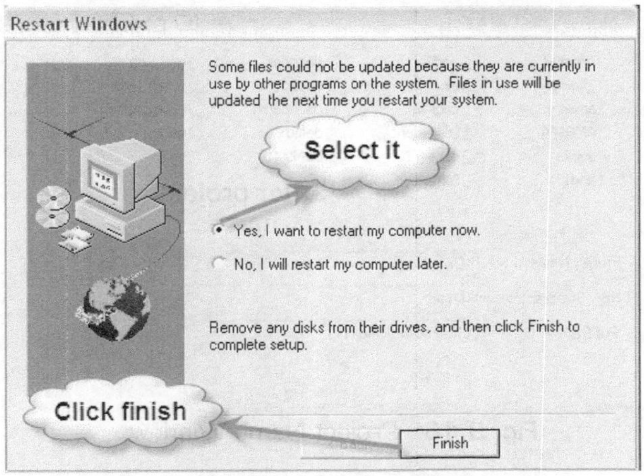

Fig. 9.13 Restart Window

9.3 SCADA PROJECT DEVELOPMENT

To start editing a SCADA project open RSView32 works as shown in Fig. 9.14.

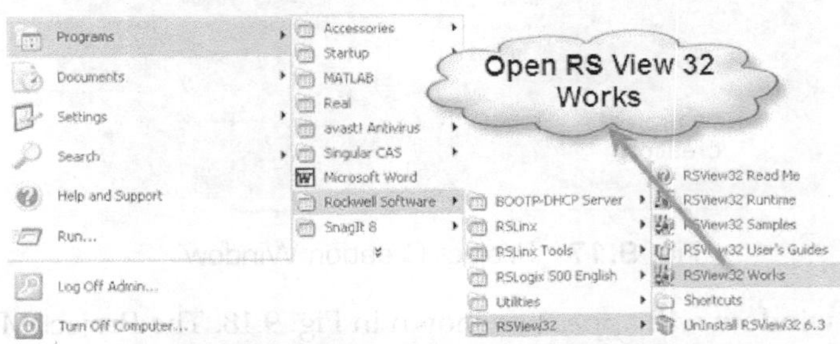

Fig. 9.14 RSView32 Works Open Window

A new window will appear for SCADA project editing as shown in Fig. 9.15. Select create new project in this window.

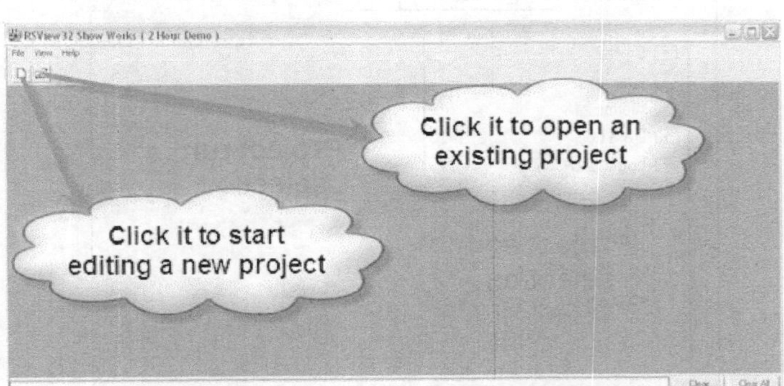

Fig. 9.15 RSView32 Start Window

Enter a project name of your choice *e.g*. "RKDemo" in the new window as shown in Fig. 9.16.

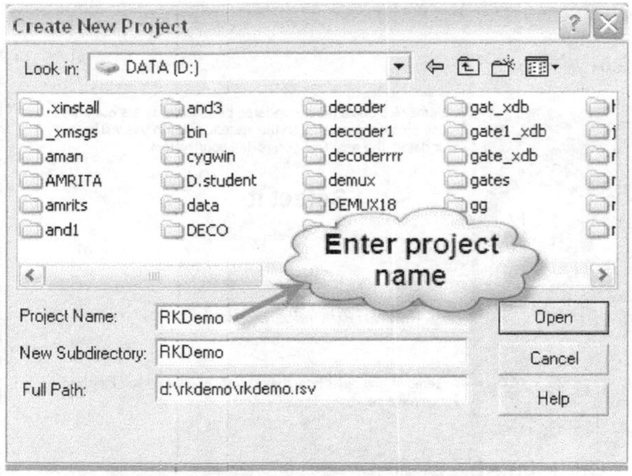

Fig. 9.16 Project Name Window

Then RSView32 will start creating a new project with given name as shown in Fig. 9.17.

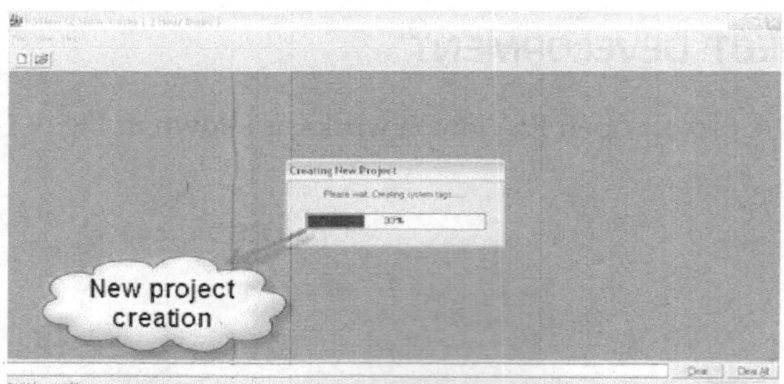

Fig. 9.17 Project Creation Window

A project manager window will appear as shown in Fig. 9.18. The Project Manager is the main tool for working with RSView32.

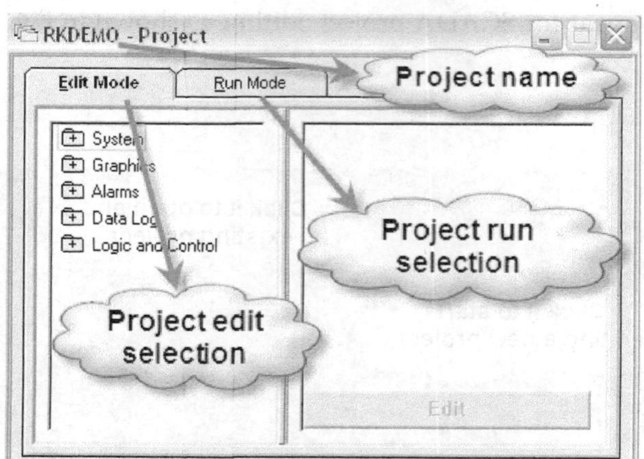

Fig. 9.18 Project Manager Window

The Project Manager window is divided into two panes. When you first create a project, the left pane appears with a series of folders and the right pane is empty as shown in Fig. 9.18. As you create project components, they are displayed in the right pane as shown in Fig. 9.19.

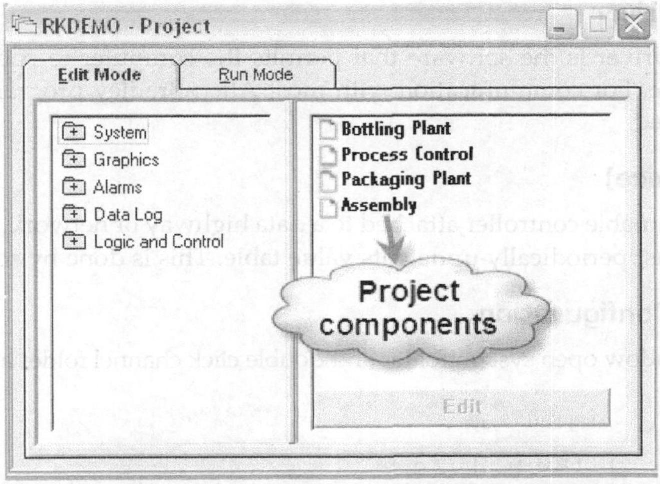

Fig. 9.19 Project Components in Project Manager Window

9.3.1 Direct Driver Communications

RSView32 uses a direct connection to the drivers in RSLinx to communicate with most Allen-Bradley devices. Communication through direct drivers involves:

1. Channels
2. Communication devices
3. Communication driver software
4. Nodes (control devices)

The RSView32 station with its channel and nodes is shown in Fig. 9.20.

Communication Channel

The communication channel is the connection from the RSView32 station to the network the programmable controllers are attached to.

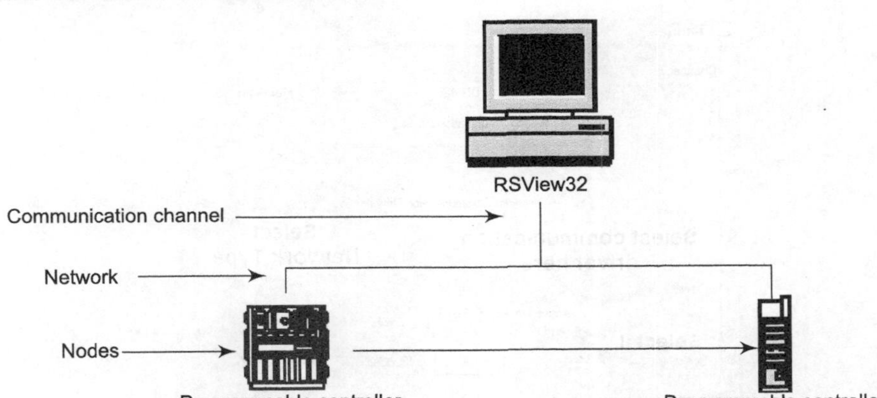

Fig. 9.20 RSView32 Station with Channels and Nodes

Communication Device

The communication device connects the communication channel to the computer. You can use internal devices, such as the 1784-KT/KTX, or you can use external devices connected through the serial port.

Communication Driver

The communication driver is the software that permits the computer to communicate with the communication device. For communication with most Allen-Bradley programmable controllers, RSLinx software is used.

Node (control device)

The node is a programmable controller attached to a data highway or network. Once the RSView32 station is set up, it must periodically update its value table. This is done by scanning its nodes.

9.3.2 Channel Configuration

In project manager window open system folder and double click channel folder as shown in Fig. 9.21.

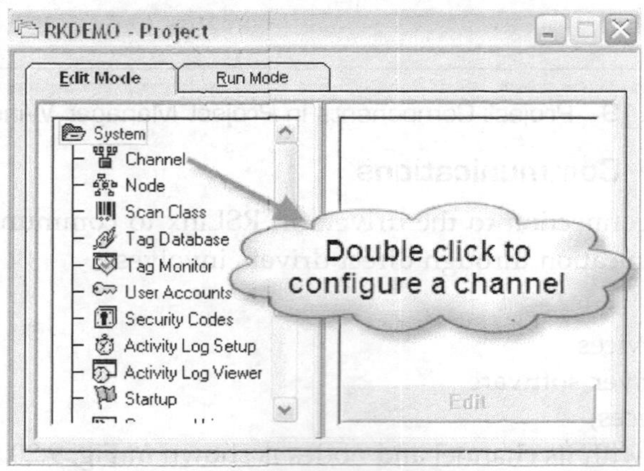

Fig. 9.21 Channel Configuration Window

A new window will appear for configuring the channel as shown in Fig. 9.22. Fill the required information as shown in the figure. After completing channel configuration close the window.

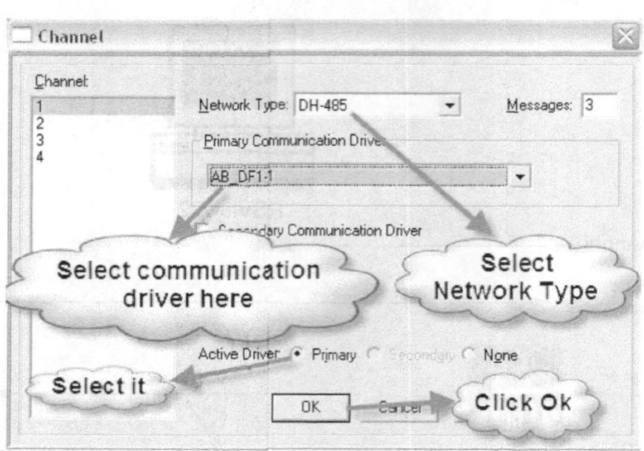

Fig. 9.22

Network Type

A network type can be selected by using the information given in Table 9.1.

Table 9.1 Network Types

To	Select
Communicate with programmable controllers on a ControlNet network	ControlNet
Communicate with programmable controllers on local or remote DH networks	DH
Communicate with programmable controllers on local or remote DH+™ networks *Or* Communicate with SLC ™ controllers on remote DH-485 networks *Or* Connect to Channel O of an Enhanced PLC®	DH+
Communicate with SLC controllers on local DH-485 networks *Or* Connect to an SLC with an RS-232 port	DH-485
Communicate with Ethernet® programmable controllers directly Connected to the TCP/IP network *Or* Communicate with SoftLogix 5 controllers *Or* Communicate with SLC 5/05™ controllers	TCP/IP
Communicate with programmable controllers through the Ethernet interface of a Pyramid Integrator™, or RSLinx Gateway	TCP/IP Bridge
Not use a channel *Or* Clear a channel configuration	None Loaded

Primary and Secondary Communication Drivers

The primary driver is the one your system will use most of the time. By default, this driver is the active driver. The secondary driver is a backup. Primary and secondary drivers must use the same network type. To specify a driver, select the appropriate driver from the list. This list contains the drivers installed and running in RSLinx that are valid for the selected network type. To configure communications without the appropriate hardware or software installed, select None Loaded as the driver.

Active Driver

The active driver can be Primary, Secondary, or None. By default, the Primary driver is active. To disable a channel temporarily without removing its configuration, click None.

9.3.3 Node Creation

In project manager window open system folder and double click node folder as shown in Fig. 9.23.

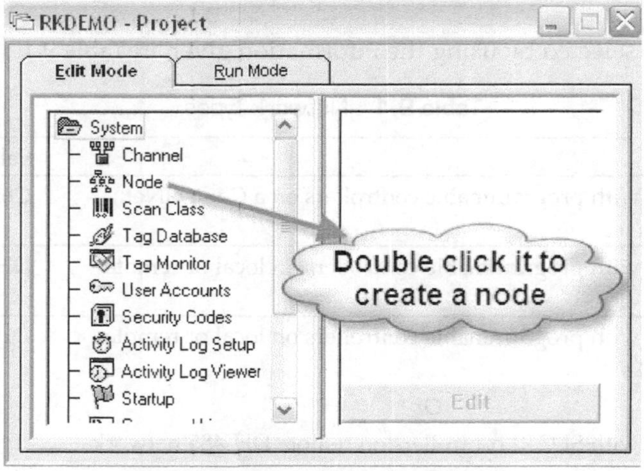

Fig 9.23 Node Creation Window

A new window will appear for creating the node. Fill the required information as shown in Fig. 9.24. After completing node creation close the window.

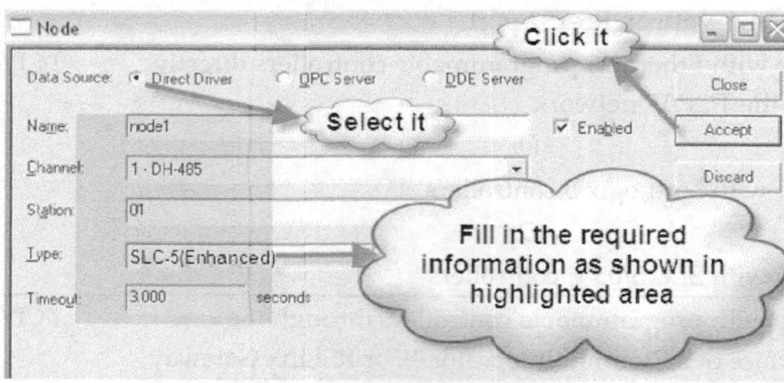

Fig. 9.24 Node Creation Window

Click accept button then created node will appear below as shown in Fig. 9.25.

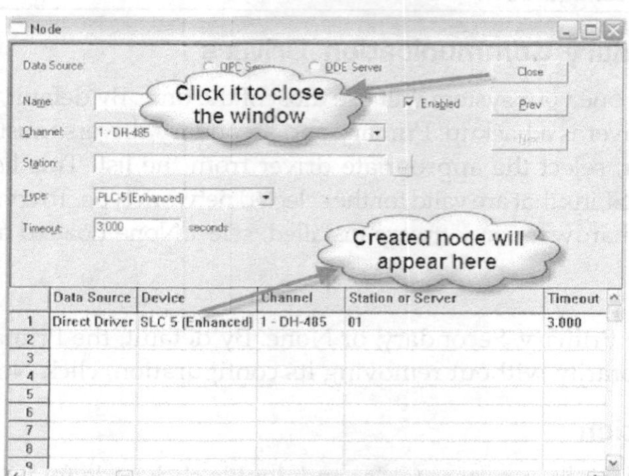

Fig. 9.25 Window with Created Node

Data Source

Select Direct Driver.

Name

Type a name of your choice up to 40 characters long. This name represents the programmable controller. The name can have upper and lower case letters, numbers, dashes, and underscores. Spaces are not permitted.

Enabled

Normally nodes are enabled, allowing collection of values from the programmable controller. However, during setup or troubleshooting you might want to disable a node to prevent communication timeouts or invalid data. When a node is disabled, tag values can still be read and written, but the values are read from and written to the value table instead of the programmable controller. When a tag uses a disabled node at runtime, the tag's state will change from valid to disabled. A tag's state can be seen by viewing the tag in the tag monitor. To disable nodes, clear the enabled check box.

Channel

Select the channel number on which this node resides.

Station

Type the station address of the programmable controller.

Type

Select the type of device from the list as shown in Table 9.2.

Table 9.2 Device Types

For this device type	Select
PLC-2®	PLC-2
PLC-3®	PLC-3
PLC-5/10, PLC-5/12, PLC-5/15, PLC-5/25	PLC-5
PLC-5/11, PLC-5/20, PLC-5/30, PLC-5/40, PLC-5/60, PLC-5/80	PLC-5 [Enhanced]
SLC 5, SLC 5/01™, SLC 5/02™, SLC 5/03™ [OS 300]	SLC 5
SLC 5/03 [OS 301], SLC 5/04™, SLC 5/05	SLC 5 [Enhanced]
SoftLogix 5	SoftLogix 5

Timeout

Type the number of seconds you want RSView32 to wait before reporting a communication error. A timeout period of three seconds is usually enough.

9.3.4 Device Tag Creation

In the tag database, we define the data that we want RSView32 to monitor. Each entry in the database is called a tag. A tag is a logical name for a variable in a device or in local memory (RAM).

For example, a tag can represent a process variable in a programmable controller. The current value of a tag, when required, is updated from the device it is connected to and stored in computer memory which is referred as the value table. This value is immediately accessible to all parts of RSView32. For example, graphic displays use tag values to control animation or update a trend, alarm monitoring compares current tag values to pre-defined limits, and data logging stores tag values to create a historical record.

Tag Types

RSView32 uses the following types of tags shown in Table 9.3:

Table 9.3 Tag Types

Tag	Type of data stored
Analog	Range of values These tags can represent variable states such as temperature or the position of rotary controls.
Digital	0 or 1 These tags can represent devices that can only be on or off, such as switches, contacts, and relays.
String	ASCll string, series of characters, or whole words (maximum of 82 characters). These tags can represent devices that use text, such as a bar code scanner which uses an alphanumeric product code.

Data Sources

When defining an analog, digital, or string tag, you must specify a data source. The data source determines whether the tag receives its values externally or internally.

Device

A tag with Device as its data source receives its data from a source external to RSView32. The data can come from a direct programmable controller driver or from an OPC or DDE server. Tags with Device as the data source count toward the total tag limit which are 150, 300, 1,500, and so on.

Memory

A tag with Memory as its data source receives its data from the RSView32 internal value table. A memory tag can be used to store values internally. Tags with Memory as the data source do not count toward the total tag limit.

In this chapter we will create device tags to interface PLC and SCADA. From next chapter onwards we will create tags in memory to discuss various features of SCADA. In project manager window open system folder and double click tag database as shown in Fig. 9.26. Here we will create device tags for "Electrical appliances control" application discussed in chapter 8. For that application we require to create 5 digital tags, one each for In/Out, Day/Night, Summer/Winter, Fan and Tube inputs and outputs.

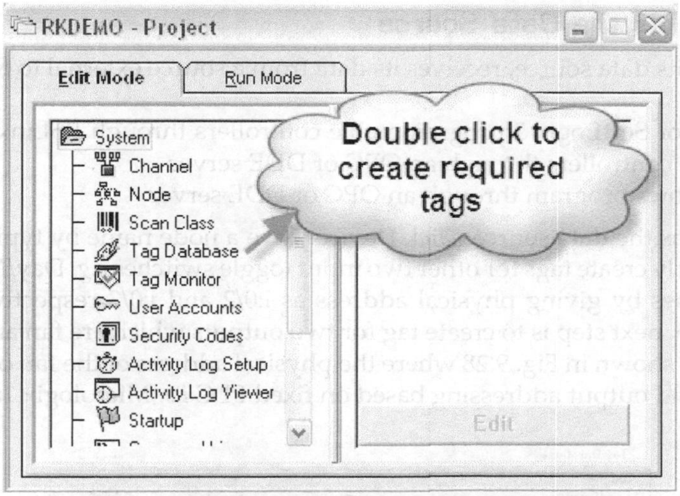

Fig. 9.26 Tag Database in Project Manager Window

A new window will appear to create required tags in the tag data base. Fill in the required information as shown in Fig. 9.27. To create tag for In/Out input toggle switch. Once the tag creation is complete click the accept button.

Security

To restrict access to this tag, select a security code. If access is restricted, operators cannot change a tag value without the proper security code.

Description

Type a description of this tag, up to 128 characters long.

Off Label and On Label

Type text up to 20 characters long that describes the off state when value = 0 and on state when value = 1 of the tag. The off and on labels must be different. One or the other, but not both, can be blank.

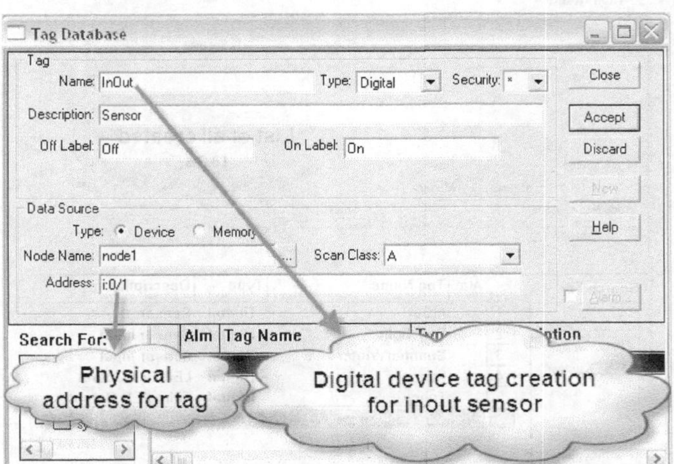

Fig. 9.27 Digital Device Tag Creation Window

Specifying Device as the Data Source

A tag with device as its data source, receives its data from a source external to RSView32. The data can come from:

(i) Allen-Bradley or SoftLogix 5 programmable controllers through RSLinx direct drivers
(ii) Programmable controllers through an OPC or DDE server
(iii) Another Windows program through an OPC or DDE server

To specify device as the data source Click Device. Give a node name by typing the name of the existing node. Similarly create tags for other two input toggle switches *e.g*. Day/Night and Summer/Winter toggle switches by giving physical address as i:0/2 and i:0/3 respectively in the address field of Fig. 9.27. Now next step is to create tag for two outputs which are fan and tube. The device tag creation for fan is shown in Fig. 9.28 where the physical address of the fan output is o:0/1. Here we are using input and output addressing based on fixed PLC *i.e.* Micrologix 1000.

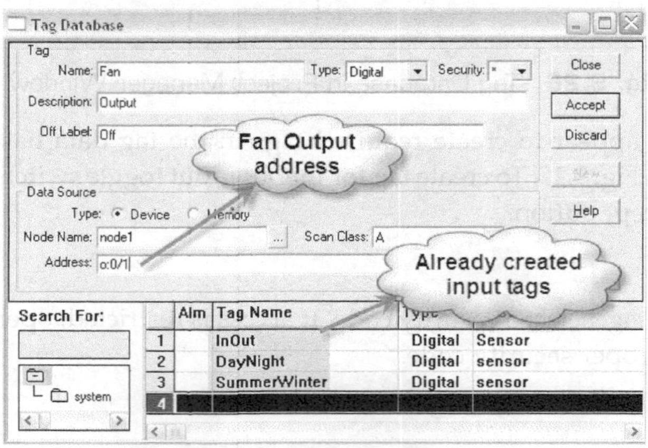

Fig. 9.28 Output Tag Creation Window

Similarly create tag for tube output by changing the address to o:0/2 in the address field of Fig. 9.28. Accept it to reflect the settings. Once all the tags are created they can be viewed in created tag list as shown in Fig. 9.29. Now close the tag database.

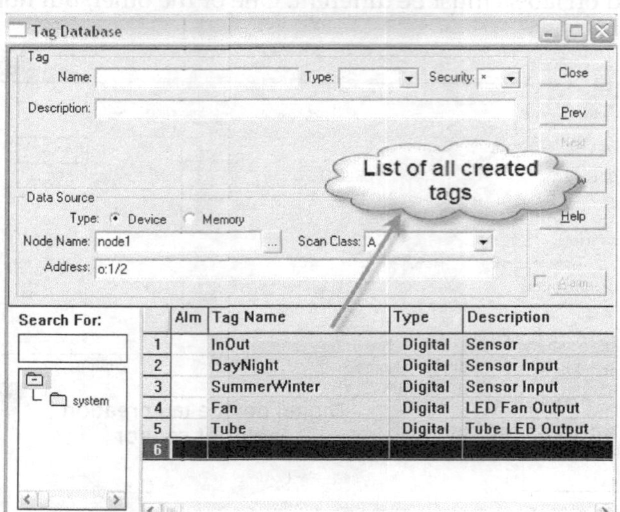

Fig. 9.29 All Created I/O Tags

9.3.5 Graphic Display Creation and Animation

The next step is to create graphic for the required application and animate it. Here we will create graphic for "Electrical appliances control" application whose SAP is shown in Fig. 6.1. To start editing the graphic for the said application Open the graphic folder in project manager window and click display. An untitled display window will appear. Place mouse on this display window and right click to get the list of options as shown in Fig. 9.30.

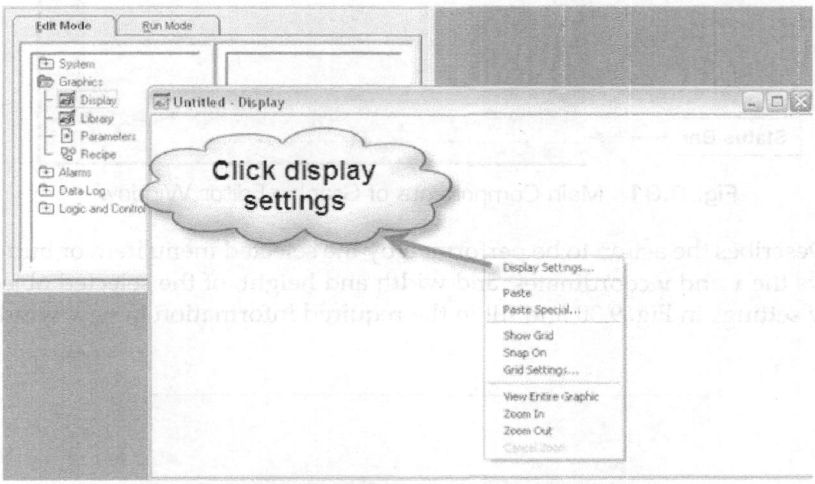

Fig. 9.30 Graphic Display Window

A graphic display represents the operator's view of plant activity. The display can show system or process data and provide operators with a way to write values to an external device such as programmable controller. Operators can also print the display at runtime to create a visual record of tag values. The components that make up a graphic display are called graphic objects. Objects can be:

1. Created in the Graphic Display editor
2. Dragged and dropped from a graphic library
3. Copied and pasted from another Windows application
4. Created by another Windows application and inserted in the graphic display using OLE *i.e.* Object Linking and Embedding.
5. ActiveX Z objects embedded in the graphic display

The Fig. 9.31 shows the main components of the Graphic Display editor. The Graphic Display editor's main components are:

Toolbar: Contains buttons for commonly used menu items. The figure shows only one toolbar but there are several, including toolbars for drawing tools, line and fill colors, and fill patterns. You can hide or show toolbars using the View menu, and you can move the toolbars anywhere on the screen.

The toolbars are a convenient way to quickly perform an action. You can hide or show them using the items on the View menu. If there is a check mark beside the toolbar name, the toolbar is visible. If there is no check mark, the toolbar is hidden.

Drawing area: Is the area for creating graphic displays. Change the background color of this area in the Display Settings dialog box.

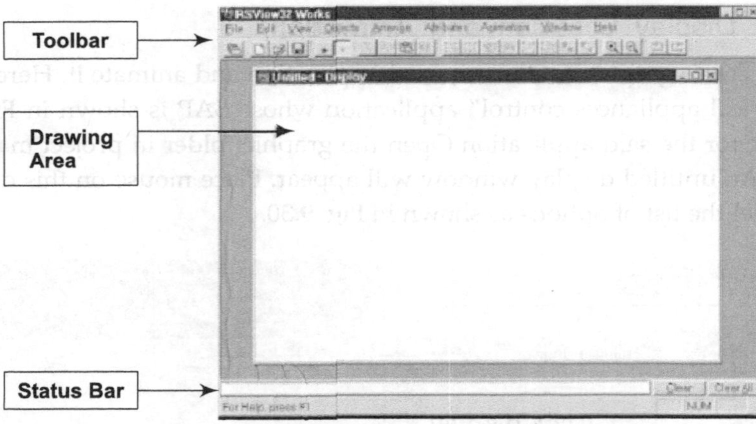

Fig. 9.31 Main Components of Graphic Editor Window

Status bar: Describes the action to be performed by the selected menu item or button. The status bar also displays the *x* and *y* coordinates, and width and height, of the selected object.

Click display settings in Fig. 9.30 and fill in the required information in new window as shown in Fig. 9.32.

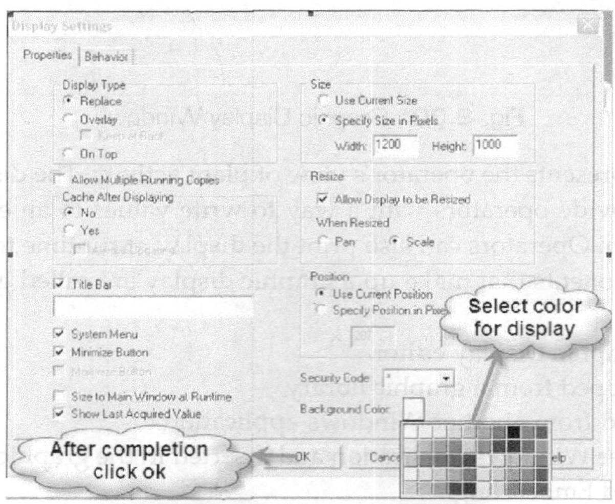

Fig. 9.32 Display Setting Window

Saving the Display Settings

You can save the settings for the current graphic display, or you can save the settings as the default for all new graphic displays. To save the settings for the current display, click OK. To create your own default settings, click Set as Default. When you open a new graphic display, it will use the display settings you have configured. Note that if you click Set as Default the settings for the current display are not saved. You must also click OK to save the settings for the current display.

Display Settings Dialog Box

The Display Settings dialog box has two tabs, Properties and Behavior. Use the Properties tab to specify the display options which are:

1. Display type
2. Title bar and other display attributes
3. Size, resizing, and position
4. Security
5. Color background

Use the Behavior tab to specify these display options:

On-screen keyboard

1. **Display type:** The different available options for display type are shown in Fig. 9.33 which are:

   (*i*) Replace (*ii*) Overlay (*iii*) Keep at Back (*iv*) On Top

(i) Replace

Replace is the default display type. Use this option if you want the graphic display to replace other open graphic displays when it opens. RSView32 will close any graphic display that the newly opened display overlaps. This way you don't need to issue separate commands to close the other displays.

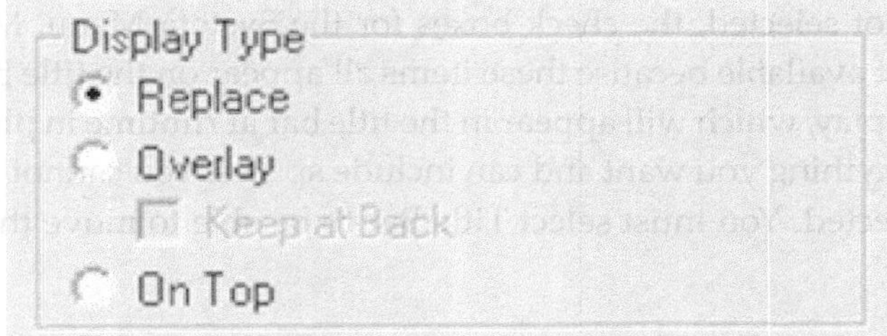

Fig. 9.33 Display Types

(ii) Overlay

Use this option if the graphic display doesn't need to replace others or appear on top. The display will layer with other displays, overlapping some and being overlapped by others as the focus changes between multiple displays. Overlay displays will always appear behind On Top displays, and will be replaced by Replace displays. Use Overlay with care, as keeping multiple displays open can affect system performance adversely.

(iii) Keep at Back

Check this box if you always want this graphic display at the back. It is recommended that you use the On Top display type to control the layering of displays. This option will be activated when you choose the Overlay.

(iv) On Top

Use this option to keep the graphic display on top at all times. It will remain on top even if another display has focus. However, if more than one graphic display of the On Top type is open at once, the display that has focus, or had the most recent focus, appears on top.

2. **Title bar and other attributes:** The figure to configure the settings of the title bar and other display attributes is shown in Fig. 9.34.

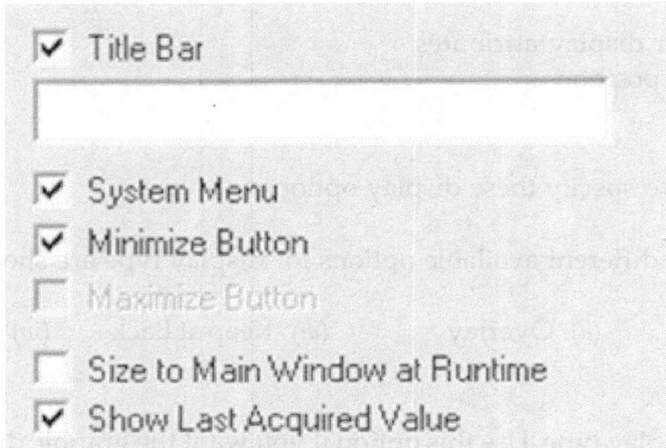

Fig. 9.34 Title Bar and Other Attributes

Title Bar

To have a title bar appear on the graphic display at runtime, select the Title Bar check box. If the Title Bar check box is not selected, the check boxes for the System Menu, Minimize button, and Maximize button are not available because these items all appear on the title bar. You can also type a title for the graphic display, which will appear in the title bar at runtime in place of the component name. The title can be anything you want and can include spaces. You cannot type a title if the Title Bar check box is not selected. You must select Title Bar to be able to move the screen's position at runtime.

Size to Main Window at Runtime

If you select this check box, the graphic display will be the size of the RSView32 main window when the display starts running. The graphic display is panned or scaled, depending on which option is selected under Resize. RSView32 has resolution-independent graphics. This means that no matter what resolution you use to create your graphic displays, they will automatically resize to fit the monitor on which they are displayed.

Show Last Acquired Value

Select this check box to have a graphic displayed with the last known value for each tag in the display until current values arrive from the programmable controller. In many projects, selecting this option will result in graphics displaying more quickly. If you do not select this option, objects with values that have not yet been updated will appear in outline form. The outline indicates that data is not current or is in error.

Selecting Show Last Acquired Value might not affect a graphic display the first time the display starts because the tags used in the display might not be initialized and so might not have any values. To indicate that there are no tag values, the objects appear in outline form. Once the tags have been initialized and tag values arrive, the objects appear in their normal form.

3. **Display size:** The display size setting option window is shown in Fig. 9.35.

Fig. 9.35 Display Size Option Window

To set the display size, do one of the following: select Use Current Size. When you save the graphic display, the size of the window at that time becomes the default size. Select Specify Size in Pixels, and type values in the Width and Height fields. The window size and position you specify here can be overridden by the RSView32 Display command. The Display command accepts size and position parameters, which override the Display Settings options.

4. **Resize display:** The display resize setting option window is shown in Fig. 9.36.

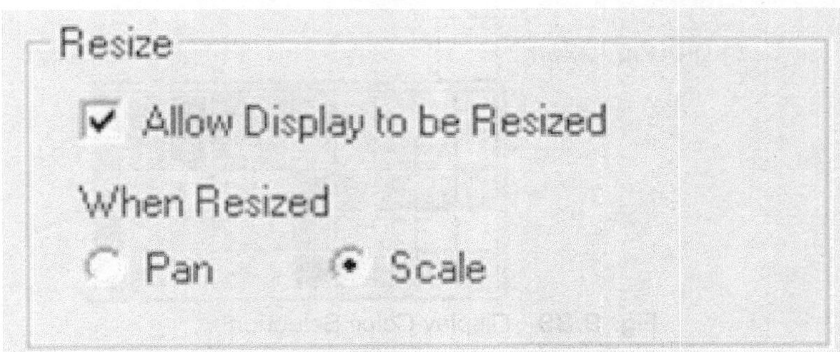

Fig. 9.36 Display Resize Options

Pan

Pan resizes the display so objects retain their original size when the display size changes. For example, if the display was resized to one quarter its original size, only one quarter of the display would be visible. In this case scroll bars appear for viewing the rest of the display.

Scale

Scale resizes the display so objects in the display are magnified or reduced to retain their size in relation to the display size. For example, if the display was resized to one quarter its original size, the entire display would be visible but it would all be one quarter its original size.

5. **Display position:** Figure 9.37 shows the available display position options. Select Use Current Position. When you save the graphic display, the position of the window at that time becomes the default position. Select Specify Position in Pixels and type values in the X and Y fields. These fields specify the position of the window in relation to the top-left corner of the RSView32 main window. The X value positions the window from the left edge of the main window and the Y value positions the window from the top of the main window.

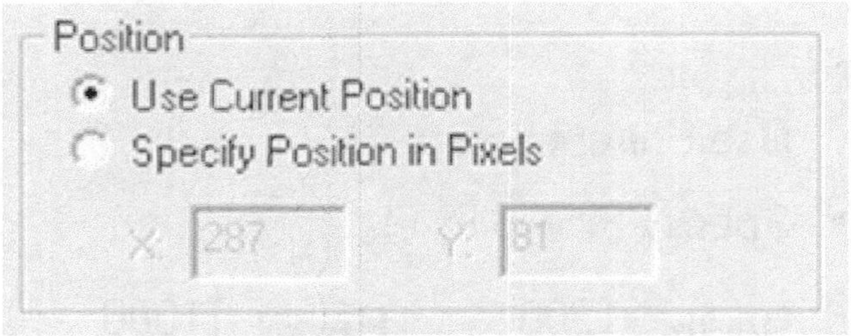

Fig. 9.37 Display Position Window

6. **Security code:** To restrict access to the graphic display, select a security code in Fig. 9.38.

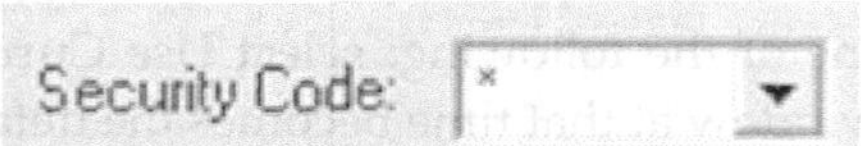

Fig. 9.38 Security Code Selection

7. **Background color:** To select a background color, click the Background Color box, and then click a color on the palette as shown in Fig. 9.39.

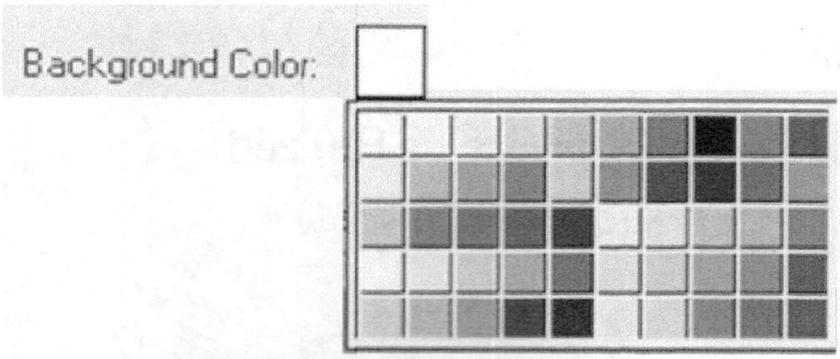

Fig. 9.39 Display Color Selection

Behavior Display Settings

On-screen keyboard: Select behavior button in display setting windows and click the Display On-screen keyboard as shown in Fig. 9.40.

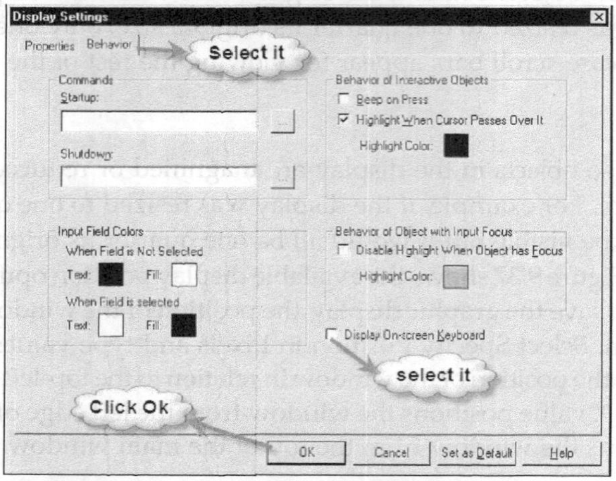

Fig. 9.40 Behavior Display Setting Window

Select this option to display a keyboard on the screen at runtime for numeric input and string input in the graphic display. The keyboard appears only when an operator selects one of these fields for input. The on-screen keyboard is typically used with systems that do not have a hardware keyboard, such as systems that use touch screens only. When the operator clicks or touches the selected field, or presses Enter on a hardware keyboard, the on-screen keyboard appears. The keyboard for string input allows alphanumeric entry as shown in Fig. 9.41.

Fig. 9.41 Alphanumeric Key Board

The keypad for numeric input fields allows numeric entry only as shown in Fig. 9.42.

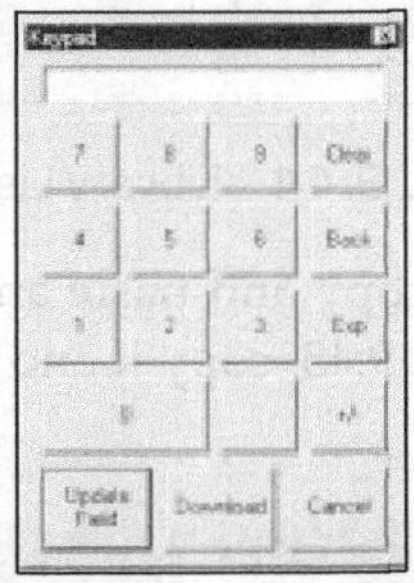

Fig. 9.42 Numeric Key Board

Once the display setting window is completely configured then click OK. The display window will look like the one shown in Fig. 9.43.

Fig. 9.43 Configured Display Window

The next step is to select the required objects from the available libraries of RSView32 and animate them. To select the required components click library in project manager window as shown in Fig. 9.44. Then a list of libraries will appear on right pane of window.

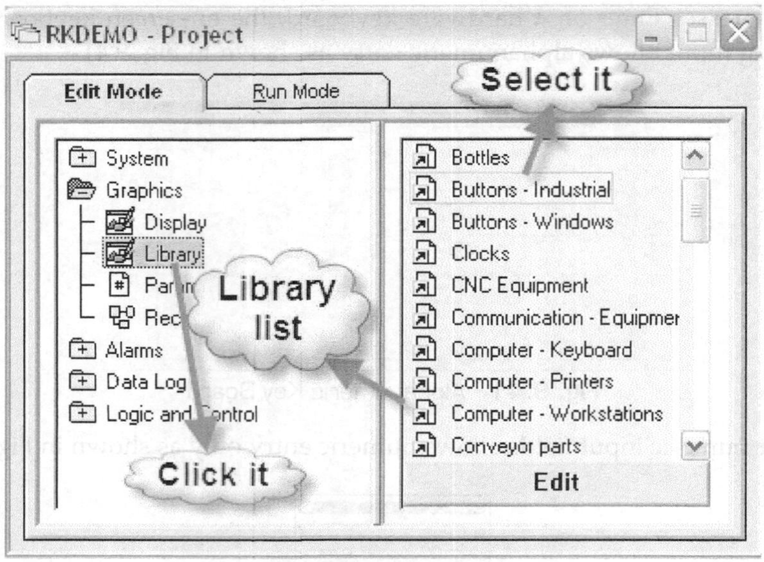

Fig. 9.44 Library List

Select "Buttons-Industrial" library to copy and paste any component from this library. First of all select a toggle switch as shown in Fig. 9.45, copy it and paste it on the display window.

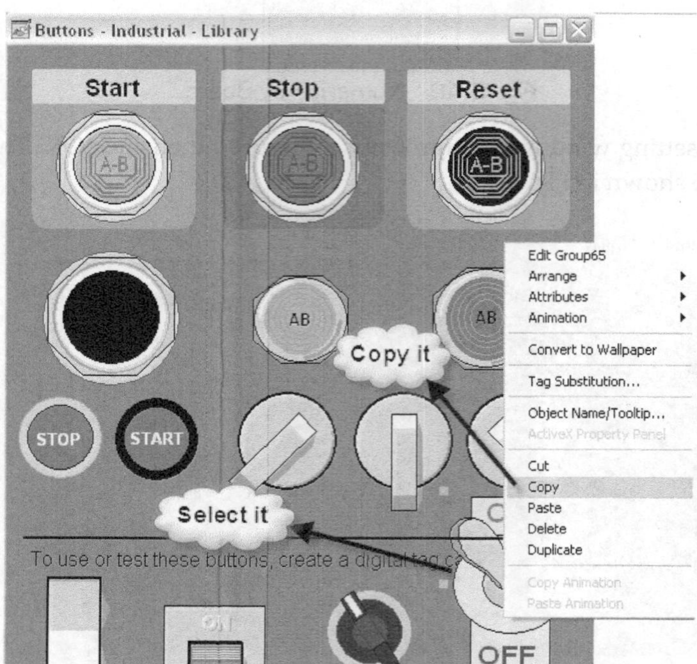

Fig. 9.45 Object Selection Window

Now animate the object by selecting the portion that you want to animation as shown in Fig. 9.46. Here we want to give rotation animation to handle of toggle switch. Select the handle and click right button of mouse to select rotation animation.

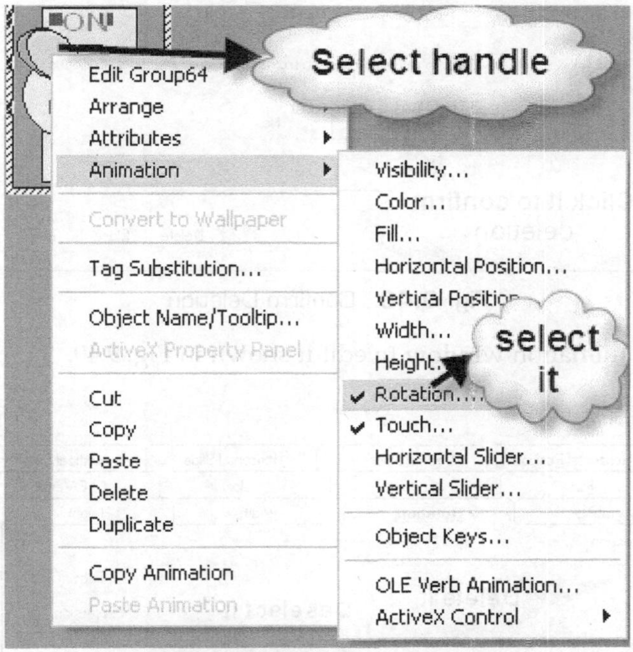

Fig. 9.46 Rotation Animation

Here in above figure we can monitor that touch animation is also selected. First delete this touch animation in as shown in Fig. 9.47.

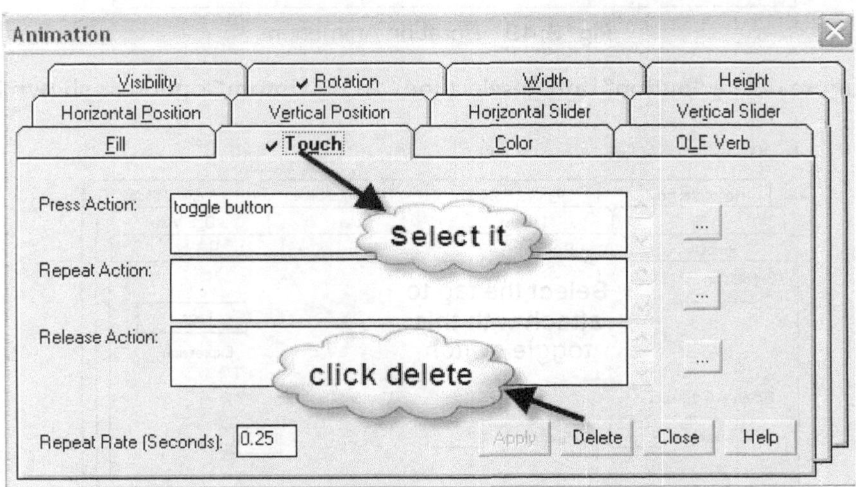

Fig. 9.47 Touch Deletion

A new window will appear with a message to confirm delete process. Click Yes in this window as shown in Fig. 9.48 which will delete the touch animation.

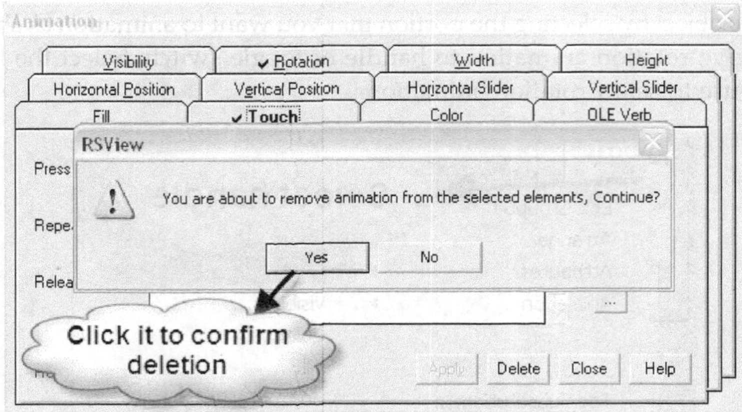

Fig. 9.48 Confirm Deletion

Now select rotation animation window to edit it shown in Fig. 9.49.

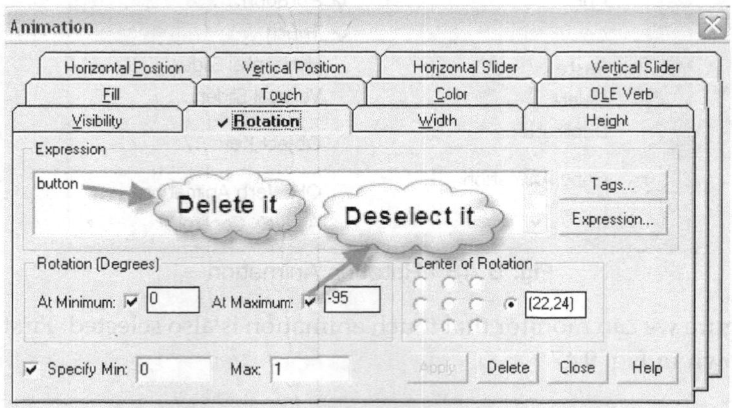

Fig. 9.49 Rotation Animation

In the last figure delete "button" and deselect the "At maximum" option as shown in Fig. 9.50.

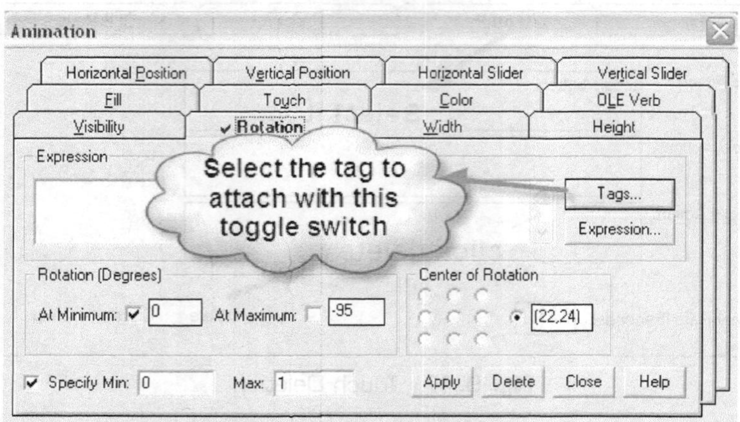

Fig. 9.50 Tag Selection

Click the tags button to get the list of all tags that we created in our tag creation section. You will get a list of tags as shown in Fig. 9.51. Select the tag that you want to attach with this toggle switch *e.g.* InOut tag.

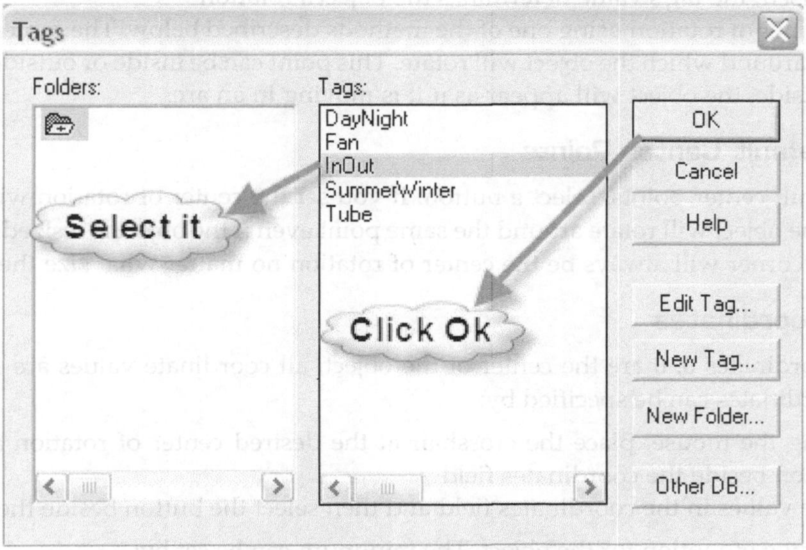

Fig. 9.51 List of Created Tags

Place the marker at the bottom of the handle and rotate the handle up to the desired position as shown in Fig. 9.52. The value in "At maximum" will vary automatically. Check the box with the desired value in it and click apply button to complete rotation animation editing.

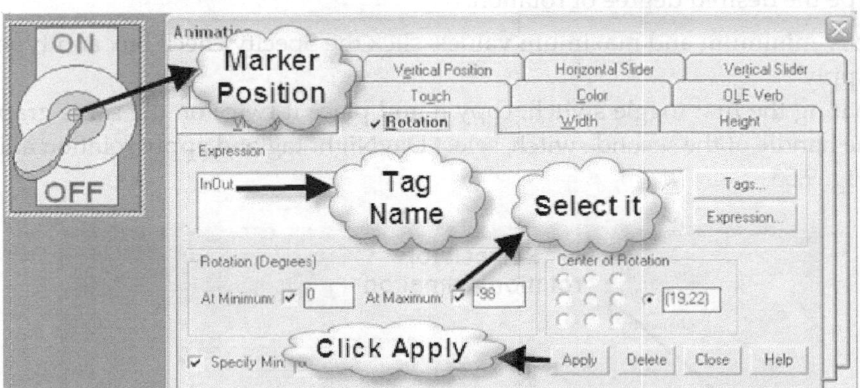

Fig. 9.52 Rotation Animation with First Switch

Rotation Animation

With rotation animation, an object rotates around an anchor point based on the result of an expression in relation to its minimum and maximum values. The degree of rotation is proportional to the value of the expression. For example, if the value of the expression is halfway between the minimum and maximum values, the object will rotate half the specified amount. To configure rotation animation perform the following steps:

1. Select the object.
2. Click the Rotation tab in the Animation dialog box.
3. Create an expression. The expression can be a tag name or a more complex expression. If a tag name is used, the tag's value determines the object's rotation.
4. Set the center of rotation using one of the methods described below. The center of rotation is the point around which the object will rotate. This point can be inside or outside of the object. If it is outside, the object will appear as if it is moving in an arc.

Using the Default Center Points

To use the default center points, select a button. If you set the center of rotation with the default center points, the object will rotate around the same point even if the object is resized. For example, the top-middle corner will always be the center of rotation no matter what size the object is.

Specifying Coordinates

The default coordinates 0, 0 are the center of the object; all coordinate values are relative to the center. The coordinates can be specified by:

(*a*) using the mouse, place the crosshair at the desired center of rotation and select the button beside the coordinates field

(*b*) type values in the coordinates field and then select the button beside the field

5. Set the range of motion for the object. The minimum can be set by:

(*a*) click the Rotate tool and visually set the degree of rotation

(*b*) type the desired degree of rotation.

The maximum can be set by:

(*i*) click the Rotate tool and visually set the degree of rotation

(*ii*) type the desired degree of rotation.

6. To specify minimum and maximum values, click the Specify check box and type the values.
7. Click Apply

After animating the first toggle switch, copy it and place it twice on the same graphic display. Now select the handle of the second switch, select DayNight tag and apply rotation animation to it as shown in Fig. 9.53.

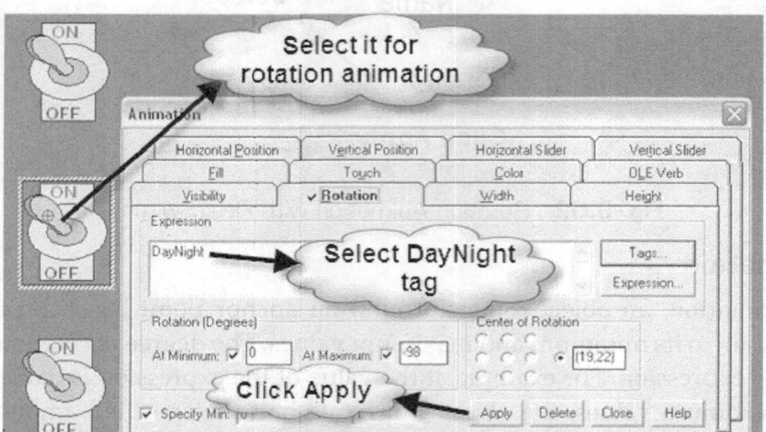

Fig. 9.53 Rotation Animation for Second Switch

Now select the handle of the third switch, assign SummerWinter tag and apply rotation animation to it as shown in Fig. 9.54.

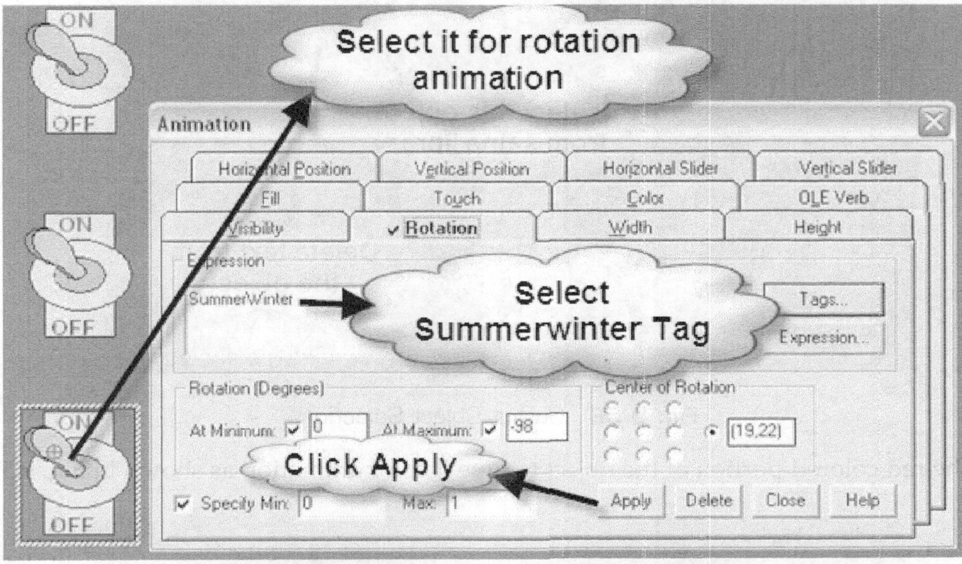

Fig. 9.54 Rotation Animation for Third Switch

Now select text icon, edit text and place on desired places in graphic display as shown in Fig. 9.55.

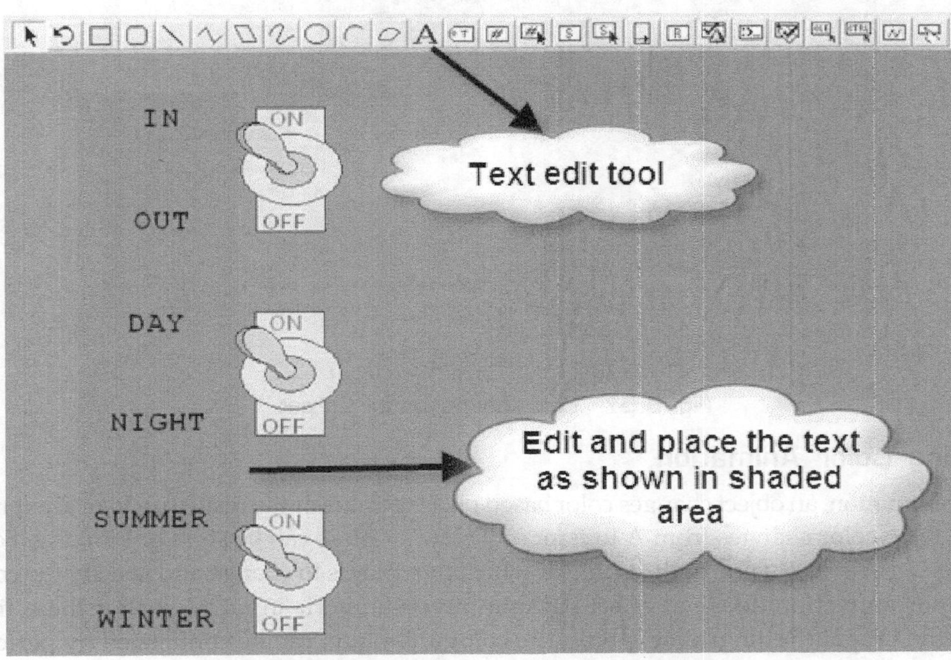

Fig. 9.55 Text Placement Using Text Editor

Select the object shown in Fig. 9.56 from the same library and place it on the display window. Select the text of the object and delete it as shown in figure.

Fig. 9.56 Output Object Selection

Select the red colored portion of the object to give it color animation as shown in Fig. 9.57.

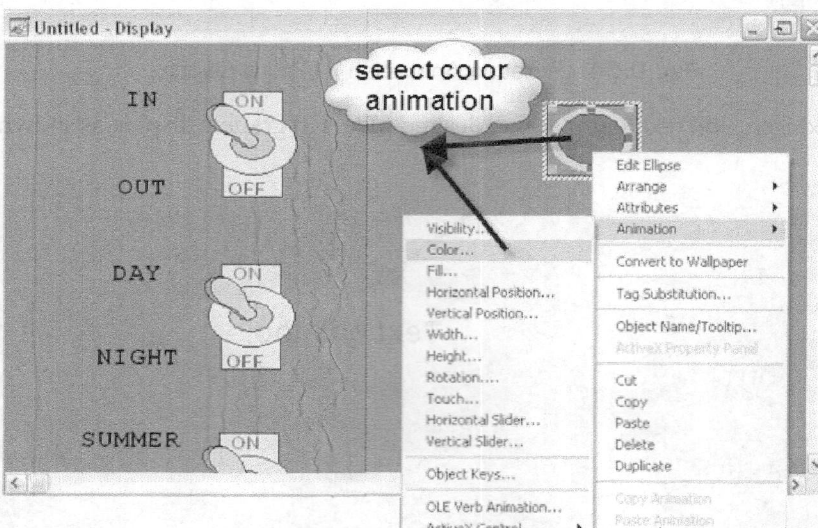

Fig. 9.57 Color Animation for Output

Configuring Color Animation

With color animation, an object changes color based on a tag value or the result of a logical expression. Specify up to 16 color changes from A to P for any object. Colors can be solid or blinking. For each color change, specify the value or threshold at which the color is to change and specify the colors to change. Color animation does not affect string input or numeric input. Color for these fields is defined in the Display Settings dialog box. The color animation can be configured by performing the following steps:

1. Select the object.
2. Click the Color tab in the Animation dialog box.

3. Create an expression. The expression can be a tag name or a more complex expression. If a tag name is used, the tag's value determines the color.
4. In the list box, select a threshold from A through P to configure a value and colors. You can change the default values and colors for thresholds A and B. The other thresholds have no defaults.
5. In the Value field, type the threshold value. When the expression value reaches the threshold then color of the object changes. Enter threshold values in ascending order. That is, A must be lower than B, and B must be lower than C and so on.
6. Click Solid or Blink for the line and fill colors and then click the line and fill colors. Line colors apply to the outline of the object and to text. Fill colors apply to the inside of the object. To select a color, place the cursor over the box, click to open the color palette, and then click the desired color. Solid colors have only one line and one fill color. Blinking colors have two line and two fill colors, which are the line and fill colors, the object will alternate between when blinking. If you select blink, the two colors must be different.
7. In the Blink Rate field, specify how many seconds each blink cycle will be. For example, if you specify two seconds, the object will be one color for one second and the other color for one second.
8. Click Apply

The color animation window is shown in Fig. 9.58. Fill the information as shown in it like select the required tag *i.e.*, fan etc. Once the configuration is complete click Apply.

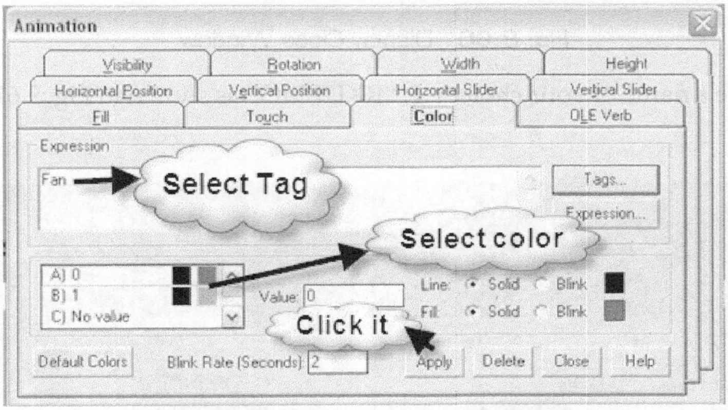

Fig. 9.58 Color Animation Window

Now copy and paste this fan object and change its text as tube. Select again the red portion of pasted object and configure it for color animation by assigning tube tag to it as shown in Fig. 9.59.

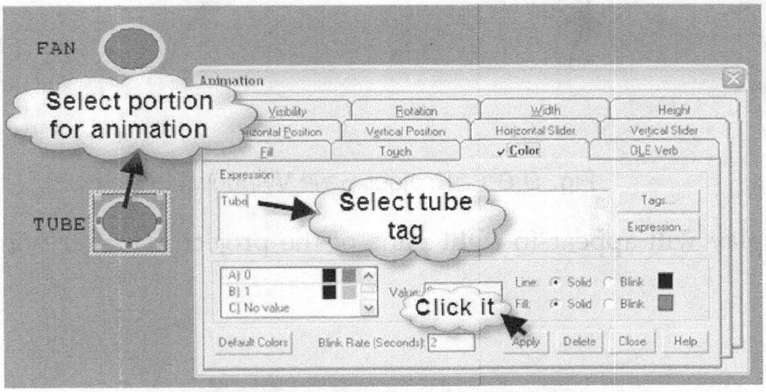

Fig. 9.59 Color Animation for Tube

Once all the required animations in the display are complete then close it then you will notice a window asking to save the display as shown in Fig. 9.60.

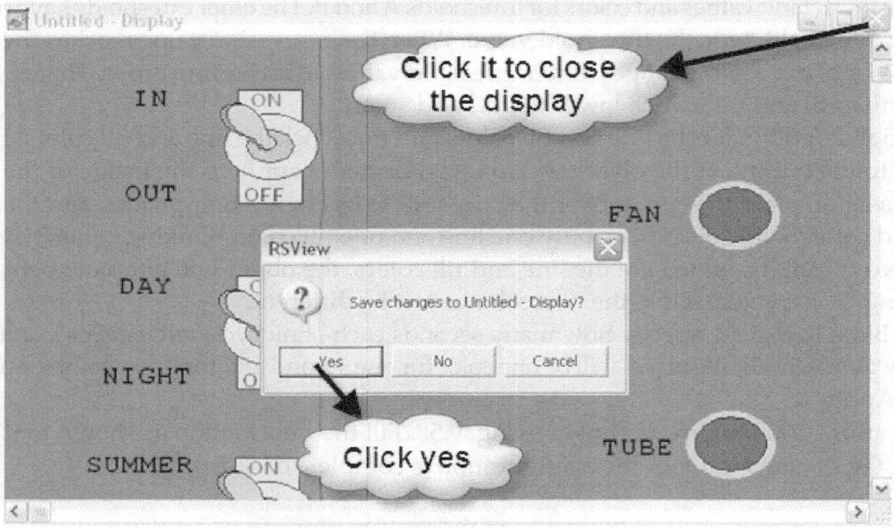

Fig. 9.60 Display Close Window

Click Yes and give a name of your choice *e.g.* "RKDEMO" as shown in Fig. 9.61.

Fig. 9.61 Project Save Window

Now created display will appear in right pane of the project manager window as shown in Fig. 9.62.

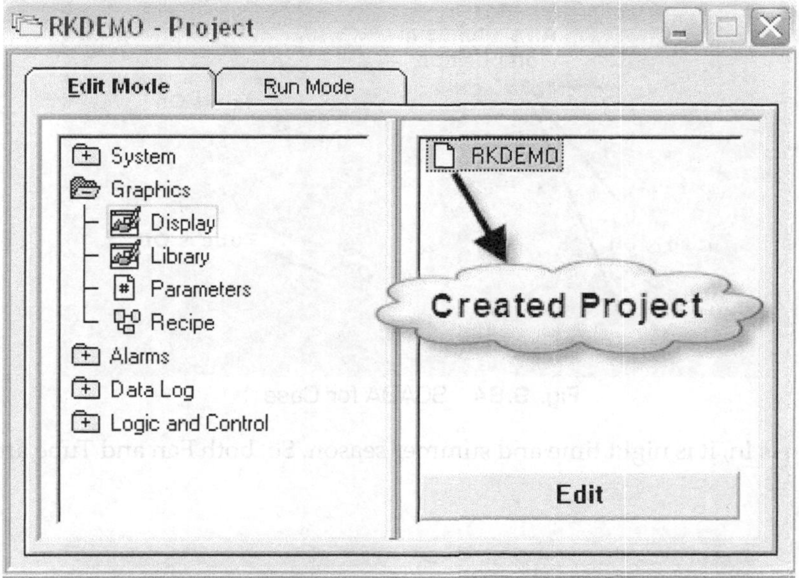

Fig. 9.62 Project Manager with Created Object

The next step is to run both PLC and SCADA projects together. Open the ladder diagram of "Electrical appliances control" SAP developed in chapter 6 and run it as shown in Fig. 9.63.

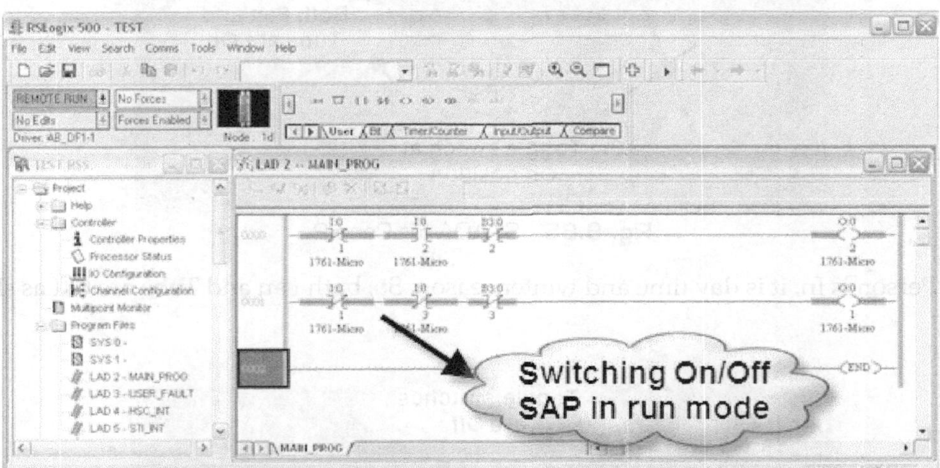

Fig. 9.63 Electrical Appliances Control SAP in Run Mode

In Project Manager Window of RSView32 click Run, select the created project and click run project and then start it as shown in Fig. 9.64. Then the graphic display will show exact position of inputs and outputs as that of physical location in SAP. For example exact handle position of each toggle switch and On/Off state of both the outputs connected to fan and tube.

Case 1: Person is In, it is day time and summer season so only Fan is On and Tube is Off.

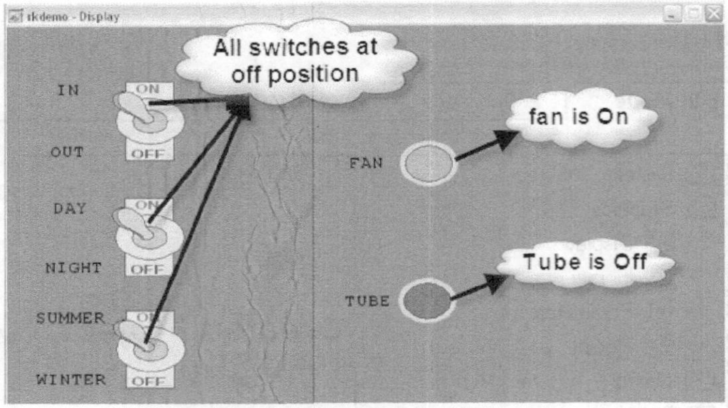

Fig. 9.64 SCADA for Case 1

Case 2: Person is In, it is night time and summer season. So, both Fan and Tube are On as shown in Fig. 9.65.

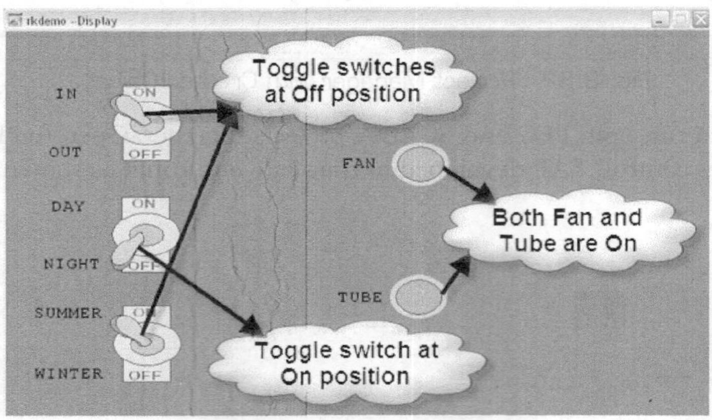

Fig. 9.65 SCADA for Case 2

Case 3: Person is In, it is day time and winter season. So, both Fan and Tube are Off as shown in Fig. 9.66.

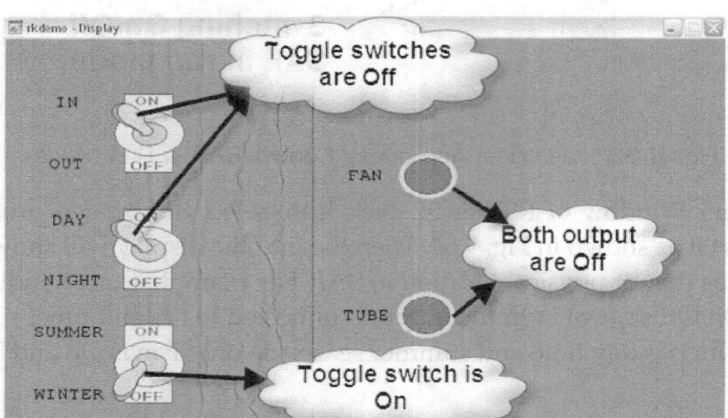

Fig. 9.66 SCADA for Case 3

Case 4: Person is In, it is day time and winter season. So, both Fan and Tube are Off as shown in Fig. 9.67.

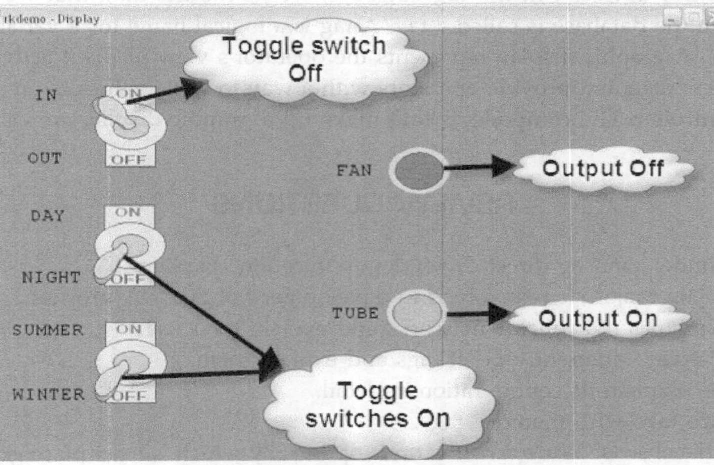

Fig. 9.67 SCADA for Case 4

Case 5: Person is Out then both output must be Off irrespective of and conditions on switch 2 and 3 as shown in Fig. 9.68.

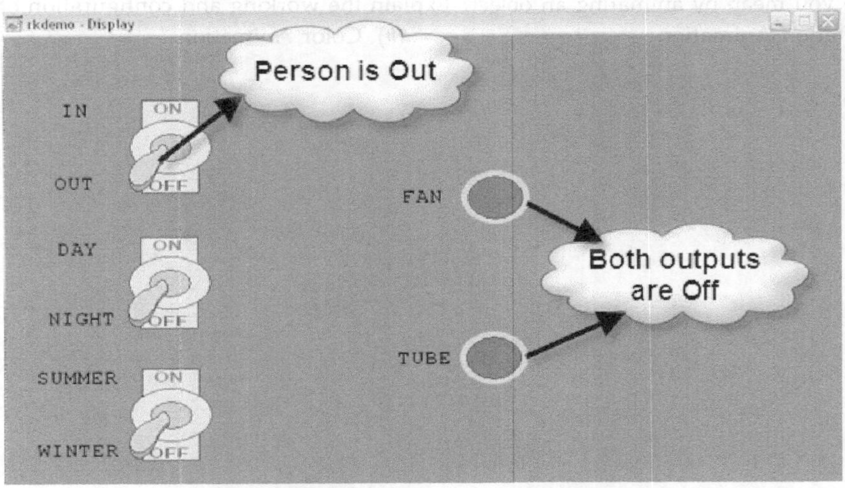

Fig. 9.68 SCADA for Case 5

SUMMARY

SCADA that stands for "Supervisory Control and Data Acquisition". The RSView32 is used as SCADA software for Allen Bradley PLCs. As part of the installation process, RSView32 creates program folders and modifies registry entries. To make these modifications, the currently logged in user must have administrative rights on the computer on which RSView32 is to be installed. The Project Manager is the main tool for working with RSView32. The Project Manager window is divided into two panes. When you first create a project, the left pane appears with a series of folders and the right pane is empty. RSView32 uses a direct connection to the drivers in RSLinx to communicate with most Allen-Bradley devices. Communication through direct drivers involves: Channels , Communication devices , Communication driver software, Nodes (control devices . In the tag database, we define the data that we want RSView32 to monitor. Each entry in the database is called a tag. A tag is a logical name for a variable in a device or in local memory (RAM). A graphic display represents the operator's view of plant activity. The display can show system or process data and provide operators with a way to write values to an external device such as programmable controller. The components that make up a graphic display are called graphic objects.

REVIEW QUESTIONS

1. What do you understand by direct driver communication? Explain it.
2. Name and explain major steps involved in development of SCADA projects.
3. List out various types of available networks and explain their use.
4. What is a tag? Name different type of tags and explain them.
5. What is a node? Explain its configuration in detail.
6. Compare device tags with memory tags.
7. What are graphic objects? Write down various options by which we can place graphic objects on the display.
8. Explain main components of the graphic editor window.
9. Name various display types and explain them.
10. Explain the need of on-screen keyboard in display.
11. What do you mean by animating an object? Explain the working and configuration of:
    (*a*) Rotation Animation (*b*) Color Animation

10

SCADA Animations

LEARNING OBJECTIVES

After reading this chapter, you will be able to explain:

1. SCADA animations
2. Project creation using memory tags
3. Visibility animation, text animation, Numeric display, Numeric input etc.

10.1 INTRODUCTION

In this section we will discuss various types of animation we can attach to graphic objects. After creating graphic objects, we can attach animation that links an object to a tag so that the object's appearance changes to reflect changes in the tag's value.

10.2 ANIMATION DIALOG BOX

To attach animation, use the Animation dialog box. To open the Animation dialog box right-click an object to open the context menu, select Animation, and then click a menu item as shown in Fig. 10.1.

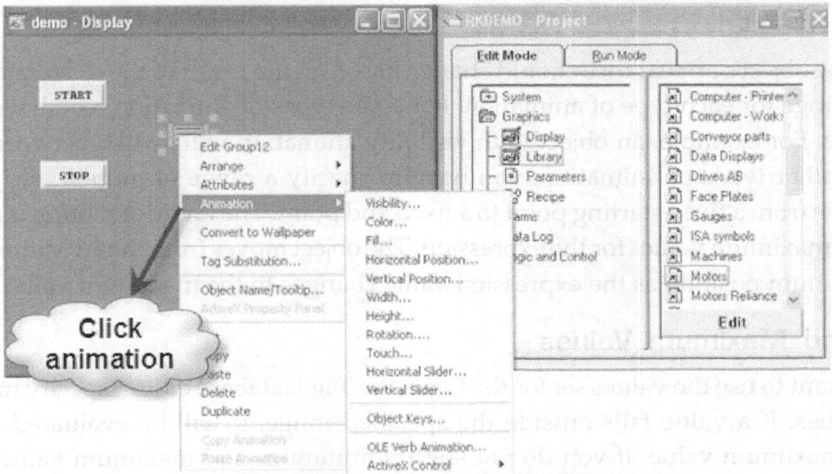

Fig. 10.1 Opening Animation Dialog Box

About the Animation Dialog Box

The Animation dialog box is a floating dialog box, which means you can have it open all the time and can move it around the screen as shown in Fig. 10.2.

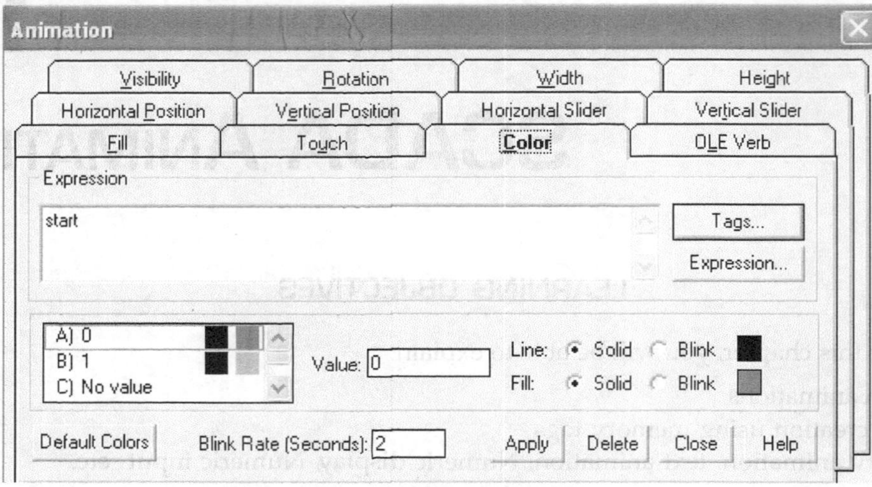

Fig. 10.2 Animation Dialogue Box

Animation Tabs

Click the tab for the type of animation you want to configure.

Expression Area

Create one or more expressions either by typing them, choosing the Expression button, or both. Multiple expressions are evaluated in the order they are listed. To supply a tag name for an expression, click the Tags button and then select a name, or type the name. Enclose tag names that contain dashes or start with a number in braces { } when you use them in an expression. This distinguishes the characters in the tag name from the characters in the expression.

Expression Result

Specify how the object's appearance should change based on the result of the expression. The fields in this area change for each type of animation. For some types of animation, you have to specify a change of state. For example, an object with visibility animation will switch between visible and invisible. For other types of animation, you have to specify a range of motion. For example, an object will move from a fixed starting point to a fixed end point. The range of motion is related to the minimum and maximum values for the expression. The object moves from the At Minimum position to the At Maximum position as the expression value changes to the maximum value.

Minimum and Maximum Values

If you do not want to use the values set for the tag in the Tag Database editor, specify minimum and maximum values. If a value falls outside the specified range, it will be evaluated as either the minimum or maximum value. If you do not specify minimum and maximum values, RSView32 uses the minimum and maximum values for the first tag in the expression.

Apply, Delete, and Close Buttons

These buttons do the following:

Apply: Applies and validates the animation configured for the selected object or group of objects. Choosing another tab is the same as choosing Apply—the animation you have configured is applied to the object and validated.

Delete: Deletes the animation for the selected object.

Close: Prompts to apply changes and closes the dialog box.

Using Tags

When configuring animation for objects, you are linking objects to tags, so you have to specify a tag name or tag placeholder. You can use tag names that you have already added to the tag database or you can use a new tag name. The tag database is used to define the data that we want RSView32 to monitor. Each entry in the database is called a tag. A tag is a logical name for a variable in a device or in local memory *i.e.,* RAM. A tag acquires its value either from device or memory. A tag configured as device tag takes its value from PLC with which it is attached as discussed in previous chapter. Now from this chapter onwards we will create project by creating tags in memory. To discuss various features of RSView32, we will create a new project with a name of own choice *e.g.* rkdemo as shown in Fig. 10.3.

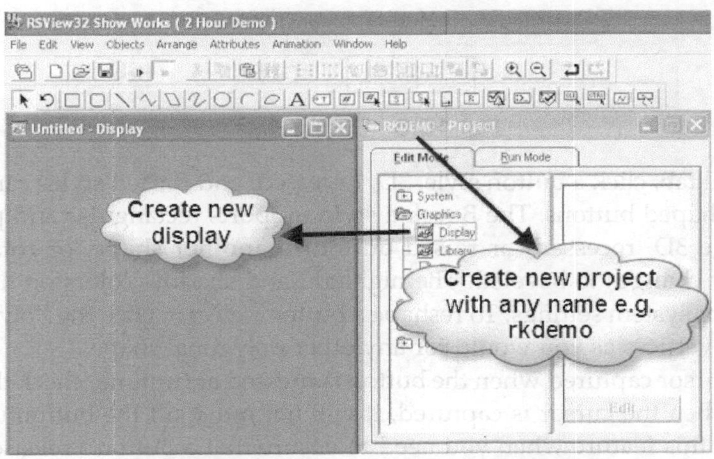

Fig. 10.3 New Project with Name DEMO

10.3 PROJECT CREATION USING MEMORY TAGS

Button Configuration

Use the Button tool to create buttons that work like standard Windows buttons. You can attach RSView32 commands to the buttons so they trigger actions when pressed and released. Select button icon from the toolbox window shown in Fig. 10.4 to create one button on the display window.

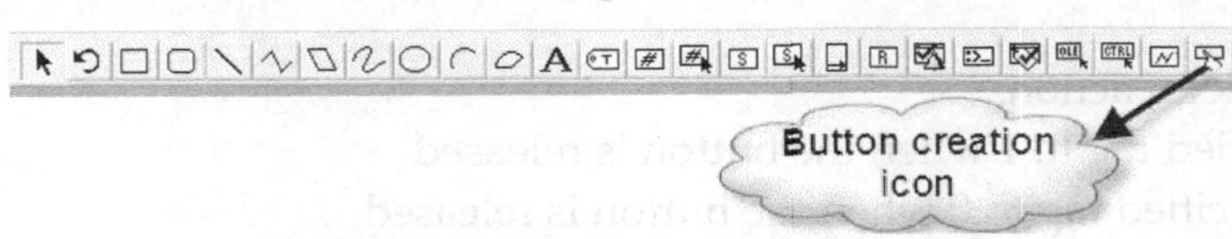

Fig. 10.4 Toolbox Window

After selecting button icon press left mouse button and drag it to create one button of required size and configure button style in general tab window as shown in Fig. 10.5.

General Tab

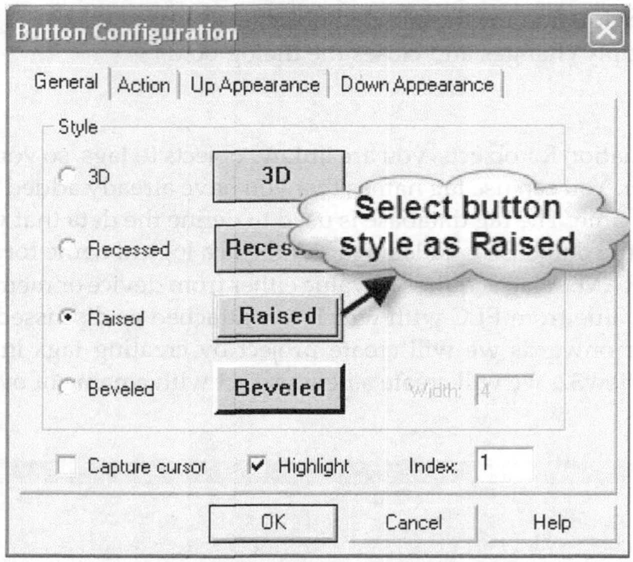

Fig. 10.5 Button Style

1. In the General tab, click a button style. 3D, Recessed, and Raised styles can be used only for rectangular-shaped buttons. The Beveled style supports rectangular and polygonal shapes. If you change 3D, recessed, or raised button to another shape, or rotate it, the button automatically changes to beveled. The highlight and shadow colors on the beveled button use the current system settings. To reshape a button, select it, click the Polygon tool and then move or add vertices as you would for any other polygonal shape.
2. To have the cursor captured when the button is pressed at runtime, check the Capture cursor check box. When the cursor is captured, it will not move off the button until the button is released. Use this feature when you need to ensure that a button's release action is carried out. If the Capture cursor check box is not selected, the button will be have like standard Windows buttons, allowing the user to 'cancel' the release action by holding the mouse and moving off the button. If you want the button to be highlighted when the button has focus, check the Highlight box. The option, Disable Highlight When Object has Focus, on the Behavior tab of the Display Settings dialog box, overrides this Highlight setting.
3. In the Index field, either leaves the default value or type a new value. Index numbers determine the tab sequence for interactive objects such as buttons and input fields and are also used in recipe files.

Action Tab

In the action tab, click an action:

Set: Sets the specified tag to 1 when the button is released.

Reset: Sets the specified tag to 0 when the button is released.

Toggle: Sets the specified tag to 0 or 1 depending on its current state. If the tag's value is 0, the tag is set to 1 when the button is pressed. If the tag's value is any non-zero value, the tag is set to 0 when the button is pressed.

Momentary On: Sets the specified tag to 1 when the mouse button is pressed and held down and sets the tag to 0 when the button is released.

Momentary Off: Sets the specified tag to 0 when the button is pressed and held down and sets the tag to 1 when the button is released.

Command: Runs the specified RSView32 command or macro. In the Press action, Repeat action, and Release action fields, type an RSView32 command or macro, or press the ... button to open the Command Wizard.

You must specify either a press or release action; both are not required. If you specify a repeat action, the action is repeated at the rate specified in the Repeat rate field for as long as the mouse button is held down. Pressing the Enter key does not cause a repeat action to be executed because Windows interprets this as a series of press actions, so the repeat action is not carried out. The Enter key works for the press and release actions.

Now select action tab and configure button action as set and give any tag name *e.g.* start to this button as shown in Fig. 10.6.

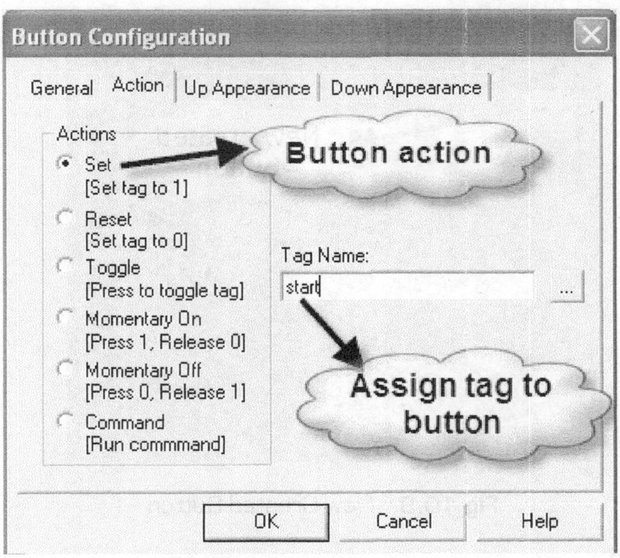

Fig. 10.6 Button Action

A message will appear asking for confirm tag creation as shown in Fig. 10.7. Click Yes in it.

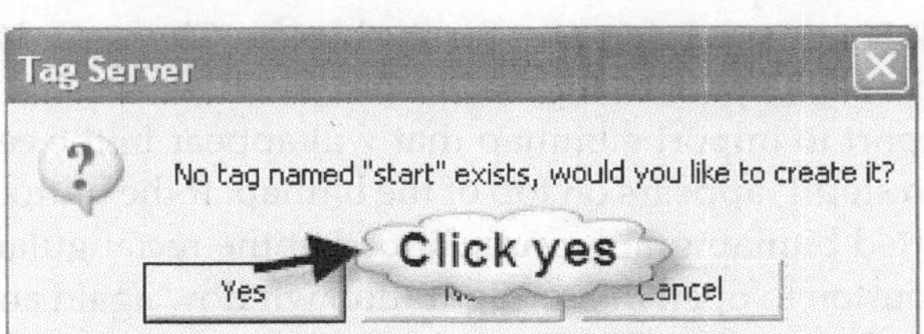

Fig. 10.7 Confirm Tag Creation

Then a tag creation window will exist as shown in Fig. 10.8. Create one digital tag in RSView memory by selecting data source as Memory.

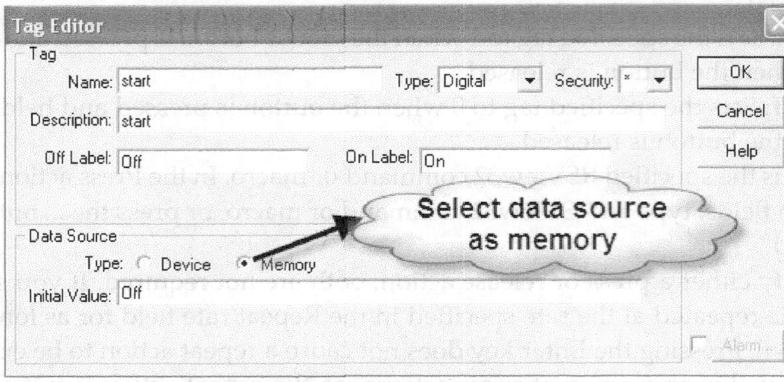

Fig. 10.8 Digital Tag Creation

Once the tag is configured click OK in tag editor window then a new button will be created as shown in Fig. 10.9.

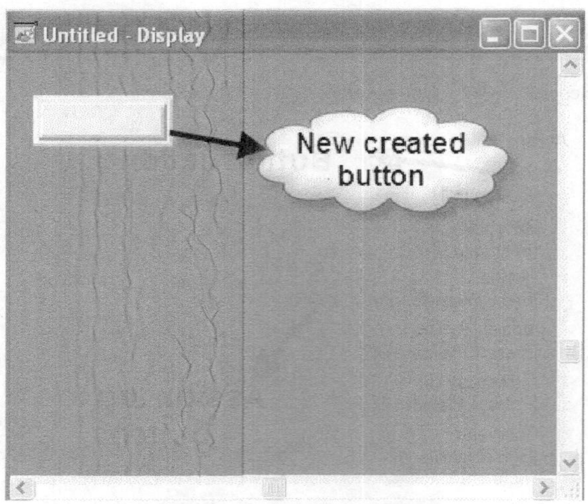

Fig 10.9 New Created Button

Up Appearance Tab

In the Up Appearance tab, specify what the button will look like when it is not pressed.

1. In the Button Label field, type a label for the button. Click the color box to specify the label's color and then click Font to specify a font for the button. Choose only TrueType fonts.
2. In the Fill field, choose a fill pattern for the button.
3. If desired, click Import to import a bitmap that will appear in the center of the button. The label, as specified in step 1, appears on top of the bitmap. If the button is a polygon, the label text and any imported bitmap graphic are clipped at the rectangular extents of the button.

Now double click the button to open the configuration window again and select up appearance tab to type the name of the button as "START" as shown in Fig. 10.10.

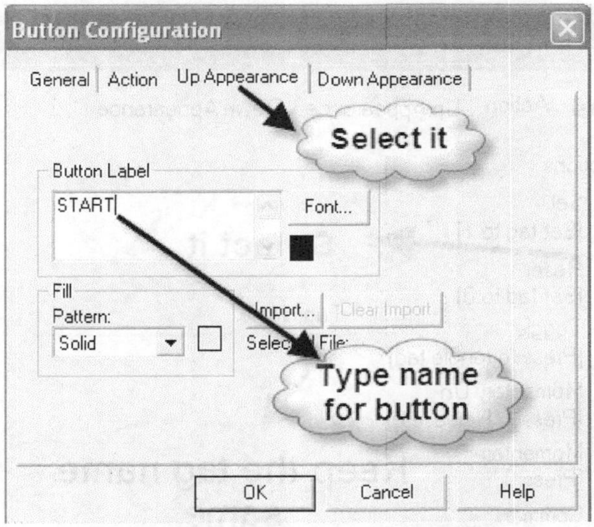

Fig. 10.10 Up Appearance of Button

Down Appearance Tab

In the Down Appearance tab, specify what the button will look like when it is pressed. If you want the button to appear the same when it is not pressed and pressed, click Same as "Up Appearance". If you want the button to appear differently, choose different options. Select down appearance tab of configuration window and check the box against "Same as up appearance" as shown in Fig. 10.11 and click OK to complete the button configuration.

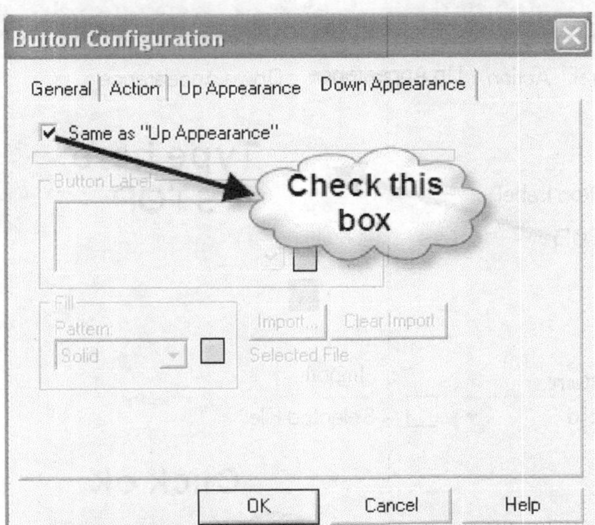

Fig. 10.11 Down Appearance of Button

Then you will notice "START" as up appearance of the created button. Now next step is to copy the created button and paste it on the same display. Again a configuration window will appear for this second button. In action tab select action as "Reset" as shown in Fig. 10.12.

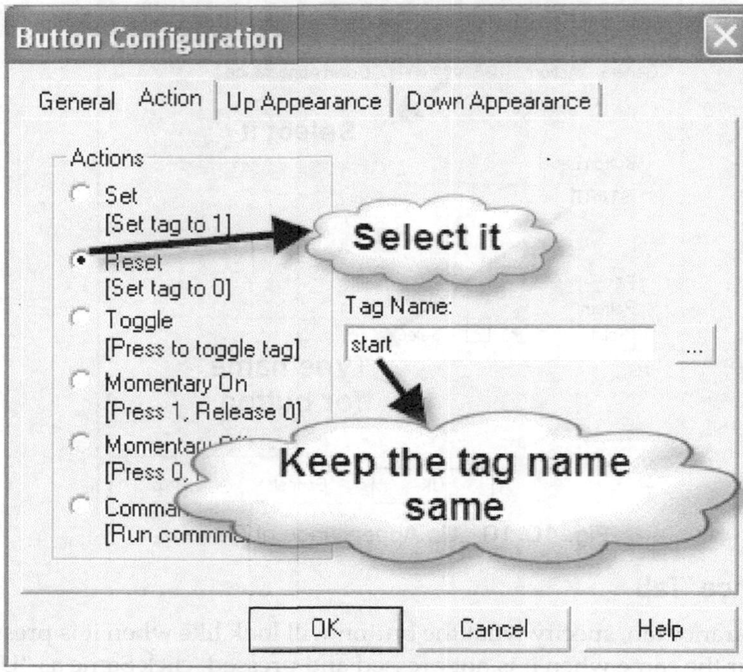

Fig. 10.12 Second Button Configuration

In up appearance tag change the up appearance form "START" to " STOP" shown in Fig. 10.13.

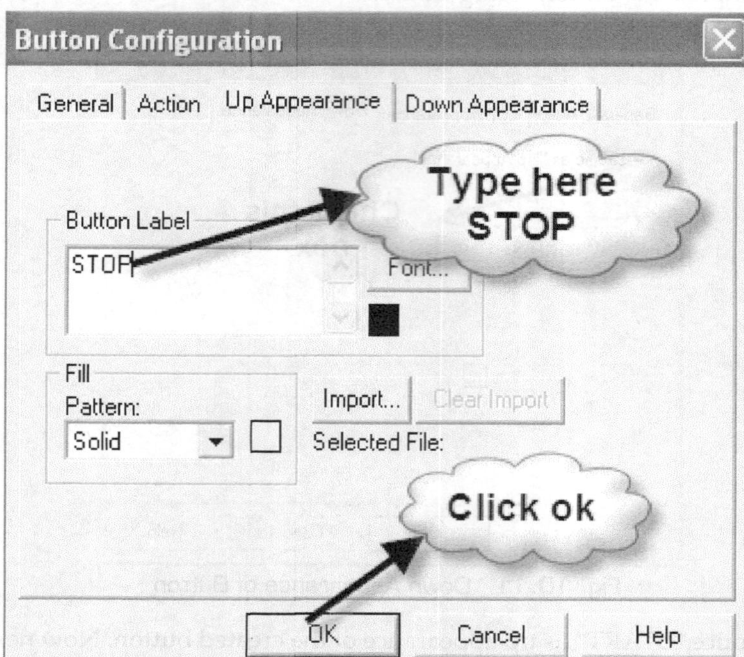

Fig. 10.13 Configuring Up Appearance

Then a new button will be created with up appearance as "STOP" as shown in Fig. 10.14.

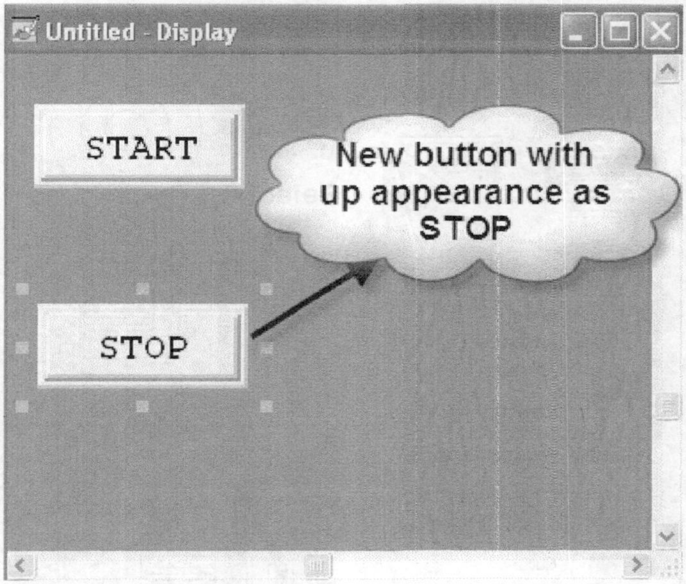

Fig. 10.14 Two Buttons in Display Window

In the project manager window select libraries and open motor library from the list of available libraries as shown in Fig. 10.15.

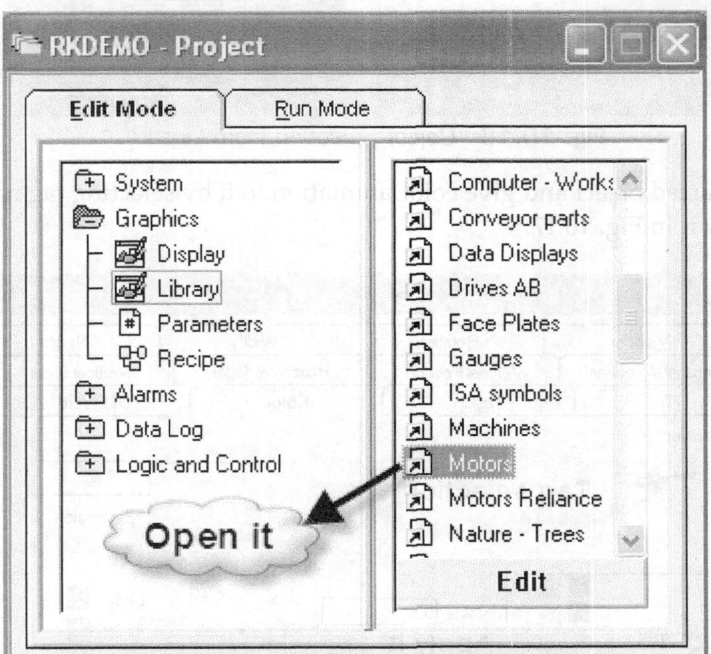

Fig. 10.15 Motor Library Selection

From the motor library select the required object, copy it and paste it on the display window as shown in Fig. 10.16.

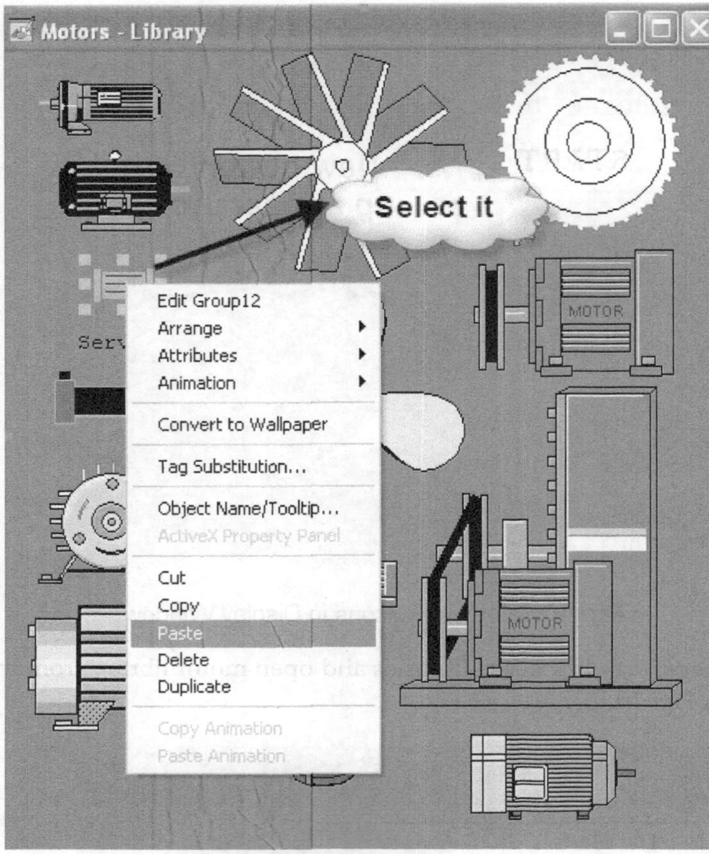

Fig. 10.16 Object Selection from Library

Now select the pasted object and give color animation to it by selecting tag name as start in the tag window as shown in Fig. 10.17.

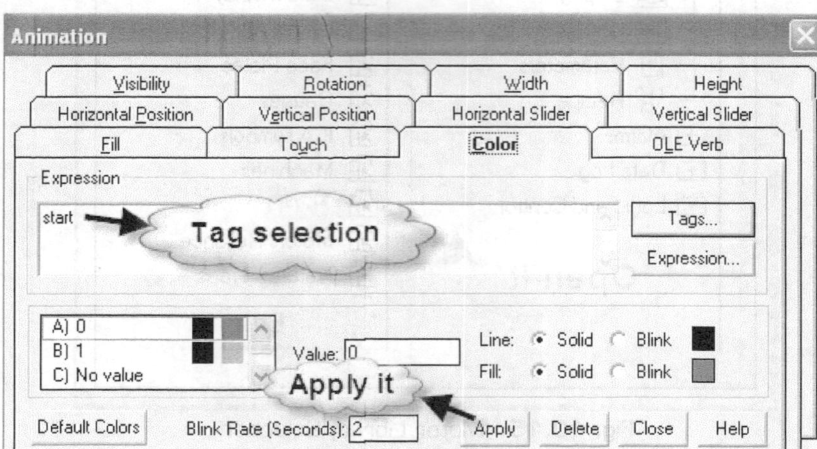

Fig. 10.17 Color Animation for Motor

Click cross sign ☒ to close the display then a message will appear asking to save the display. Click Yes in message window and type a display name of your choice *e.g.* demo as shown in Fig. 10.18.

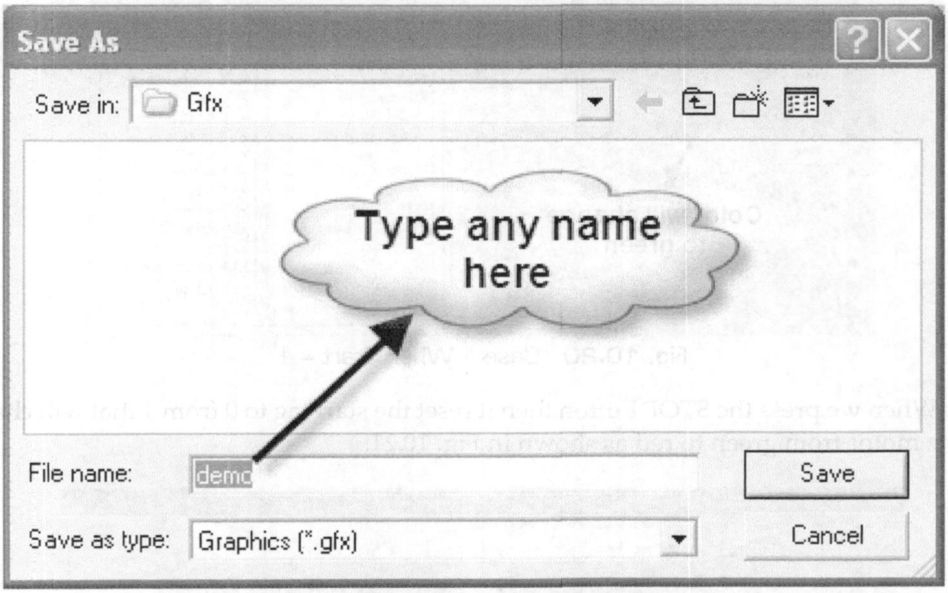

Fig. 10.18 Display Save Window

Now run the project by using run icon ▸ shown in Fig. 10.19. In figure we run the project then initially the color of the motor is red because initial value of start tag is 0.

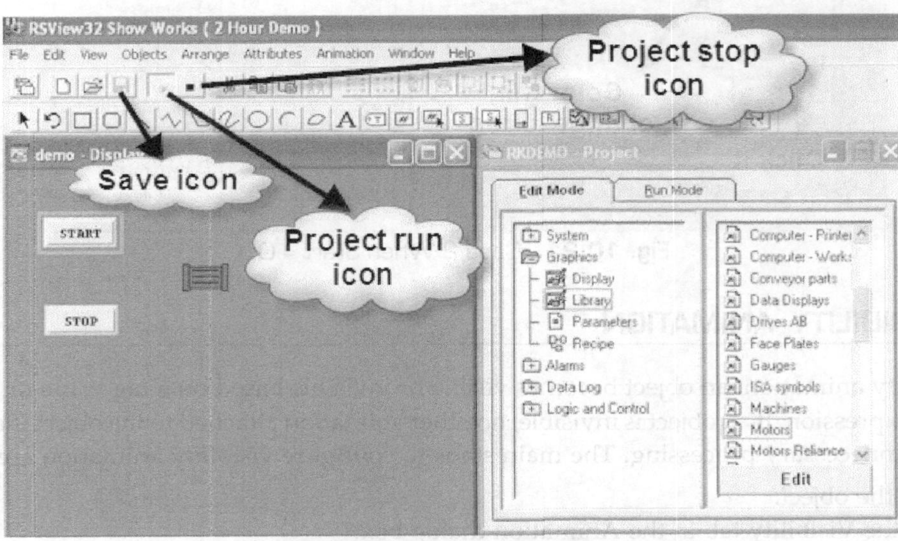

Fig. 10.19 Initial Run State

Case 1: When we press the START button then it set the start tag to 1 from 0 that will change the color of the motor from red to green as shown in Fig. 10.20.

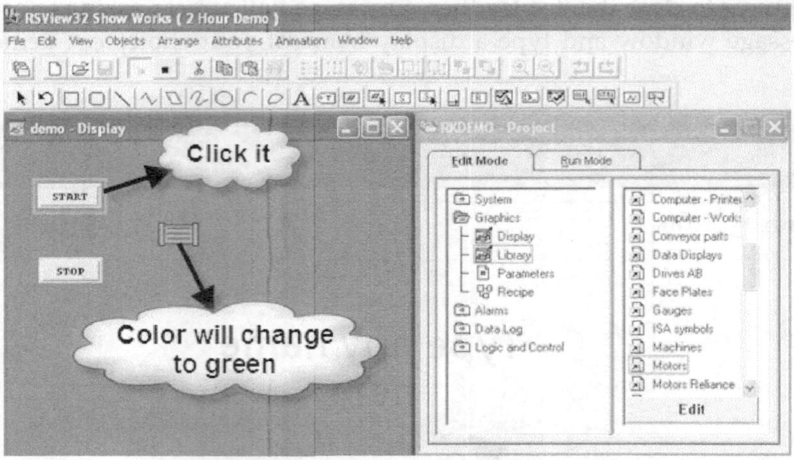

Fig. 10.20 Case 1 When Start = 1

Case 2: When we press the STOP button then it reset the start tag to 0 from 1 that will change the color of the motor from green to red as shown in Fig. 10.21.

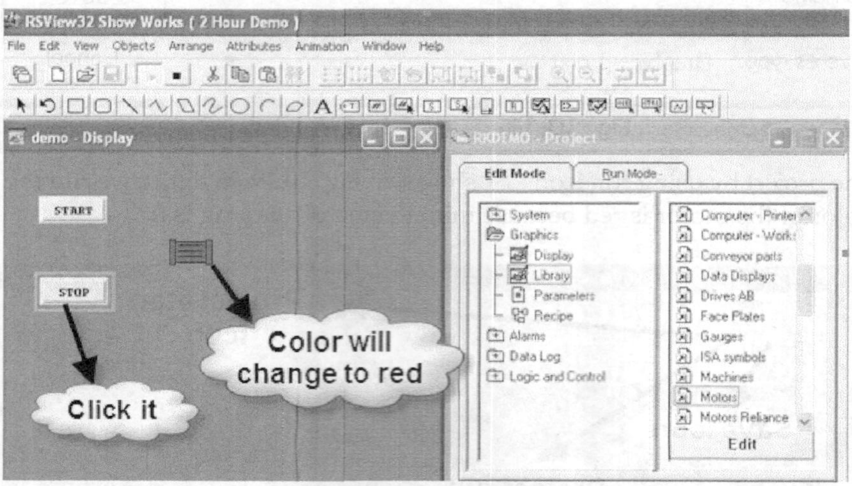

Fig. 10.21 Case 2 When Start = 0

10.4 VISIBILITY ANIMATION

With visibility animation, an object becomes visible or invisible based on a tag value or the result of a logical expression. If an object is invisible, no other animation attached to the object is evaluated to prevent unnecessary processing. The main steps to configure visibility animation are:

1. Select the object.
2. Click the Visibility tab in the Animation dialog box.
3. Create an expression. The expression can be a tag name or a more complex expression. If a tag name is used, the tag's value determines visibility.
4. To make an object invisible when the tag or expression value is true then click Invisible. If you want the object to be visible when the tag or expression value is true then click Visible.

5. Click Apply.

To give visibility animation to start button of demo display, select it and open visibility animation as shown in Fig. 10.22.

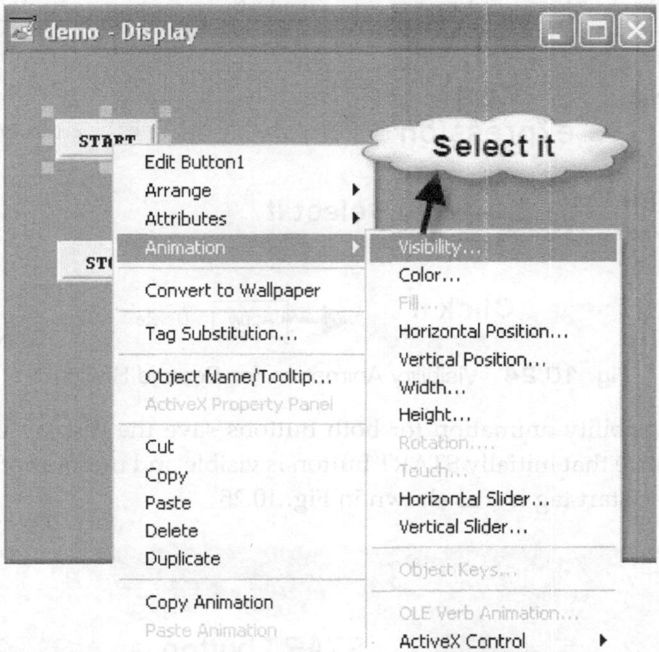

Fig. 10.22 Select Visibility Animation for Button

In the visibility animation editor window write tag expression as "start = 0", select visible and apply it as shown in Fig. 10.23.

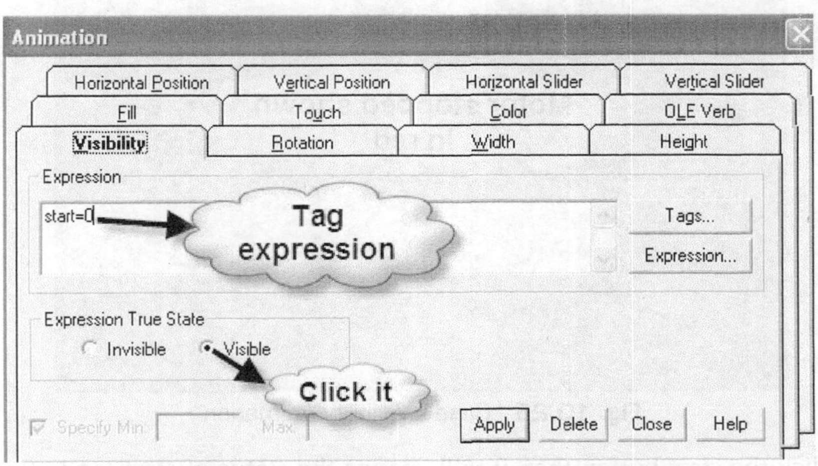

Fig. 10.23 Visibility Animation for First Button

Similarly configure the second button for visibility animation by writing tag expression as "start = 1" and select visible which is shown in Fig. 10.24.

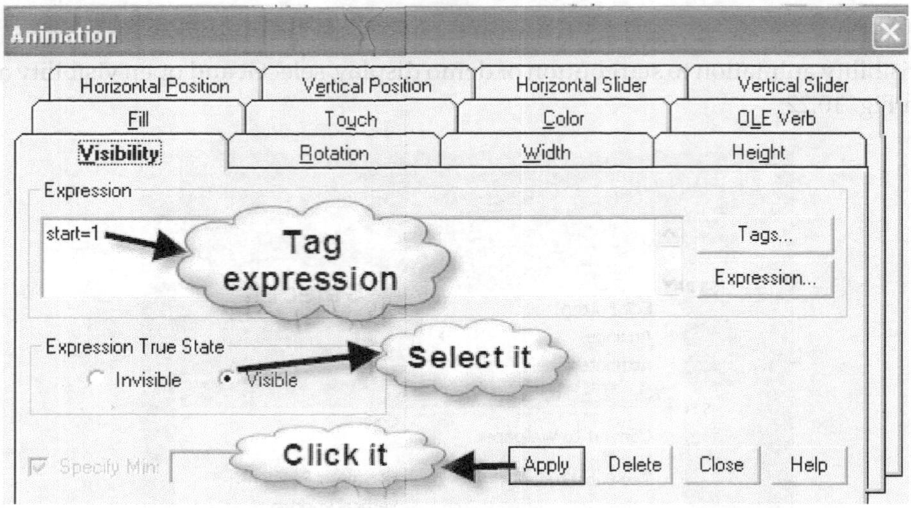

Fig. 10.24 Visibility Animation for Second Switch

After configuring visibility animation for both buttons save the display using save icon and run it. You will notice that initially START button is visible and motor is of red/dark grey color means stopped because start tag is 0 as shown in Fig. 10.25.

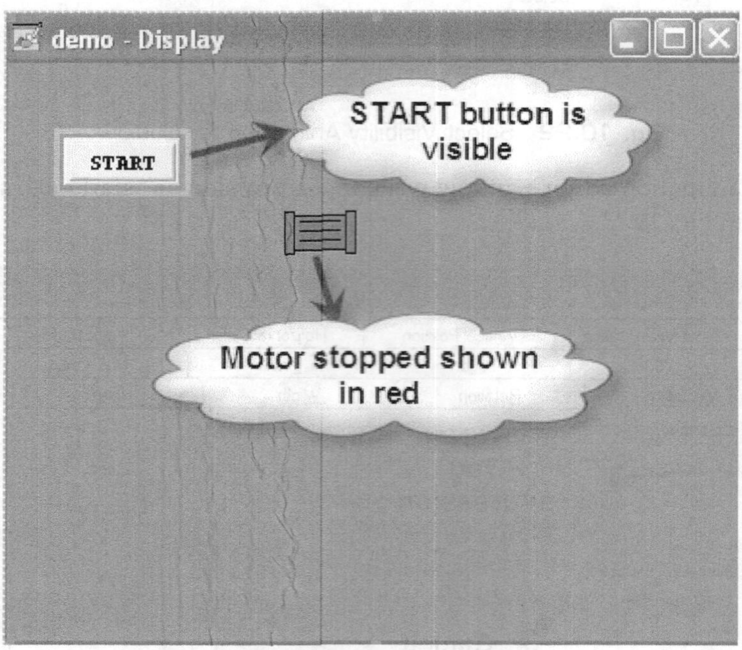

Fig. 10.25 Case1 Visibility Animation

When you click the start button then it will change the status of start tag to 0 which in turn changes the color of the motor to green/light grey color. The change in status of start tag will also make start button invisible and stop button visible as shown in Fig. 10.26.

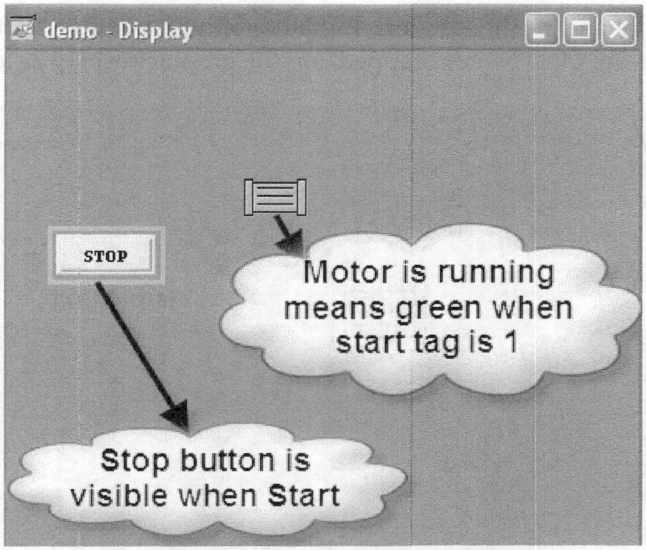

Fig. 10.26 Case 2 Visibility Animation

10.5 TEXT ANIMATION

Select text tool *i.e.* A from tool box window and type "Motor Running" as text. Configure this text for visibility animation in which text will be visible when start tag is 1 as shown in Fig. 10.27.

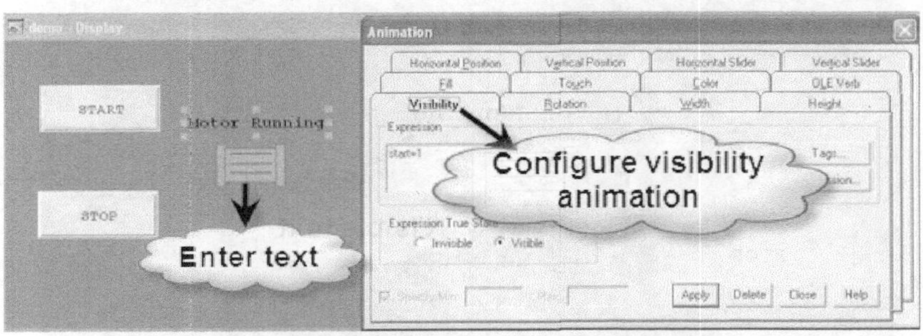

Fig. 10.27 Configuring Text Animation 1

Again select text tool and type "Motor Stopped" as text. Now configure this text for visibility animation in which text will be visible when start tag is 0 as shown in Fig. 10.28.

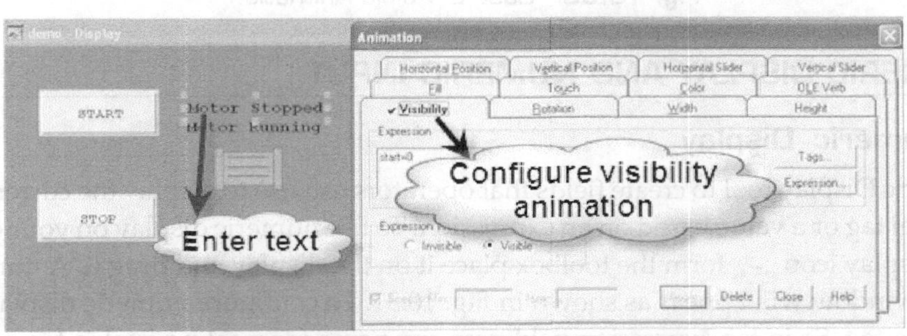

Fig. 10.28 Configuring Text Animation 2

Now save the display and run the project. Initially you will monitor start button, motor with red color and text showing that "Motor Stopped" on the graphic display as shown in Fig. 10.29.

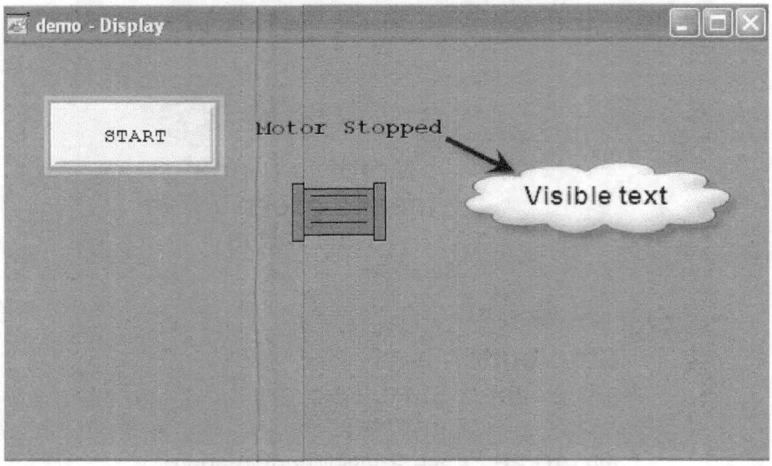

Fig. 10.29 Case1 Visible Animation

When you press start button then you will monitor stop button, motor with green color and text showing that "Motor Running" on the graphic display as shown in Fig. 10.30.

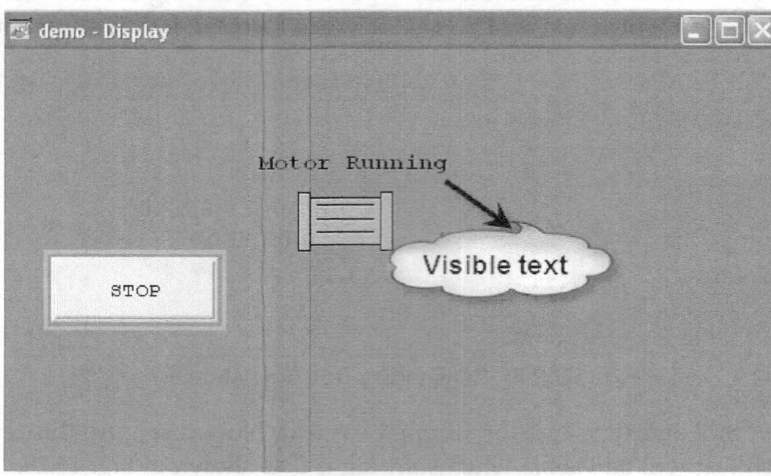

Fig. 10.30 Case 2 Visible Animation

10.6 NUMERIC DISPLAY AND NUMERIC INPUT

10.6.1 Numeric Display

Use the Numeric Display tool to create fields that operators can use to display the current value of an analog or digital tag or a value based on an expression. To use numeric display on your display select the numeric display icon form the toolbox place it on the display and drag it. A numeric display configuration window will appear as shown in Fig. 10.31. To configure numeric display we need to create an analog tag. So in tag expression field type any name for creating analog tag *e.g.* "analog".

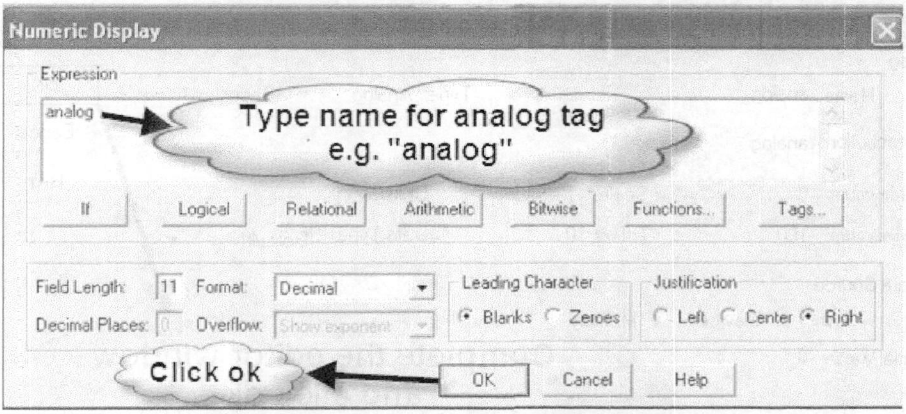

Fig. 10.31 Numeric Display Configuration

Fill in the fields as follows:

Expression: Create an expression. At runtime, the numeric result of this expression is displayed in the Numeric Display field.

Field Length: Type a value to specify the maximum number of characters that will be displayed in this field. Be sure the field is long enough to accommodate the result of the expression.

Format: Select a number base for the value.

Decimal Places: Type the number of places you want after the decimal point. This number must be at least one less than the field width.

Overflow: This field is available only if Floating Point is selected in the format.

Format Field: Use this field to specify how you want the floating point value displayed if it is too long for the field. The value can be shown as an exponent, rounded up to fit in the field, or completely replaced by asterisks.

Leading Character: Select Blanks or Zeroes.

Justification: Select Left, Center, or Right to specify how the values will be aligned within the field.

Once numeric display configuration is complete, click OK then a new message window will appear asking to create new tag with name as analog. Click Yes in window as shown in Fig. 10.32.

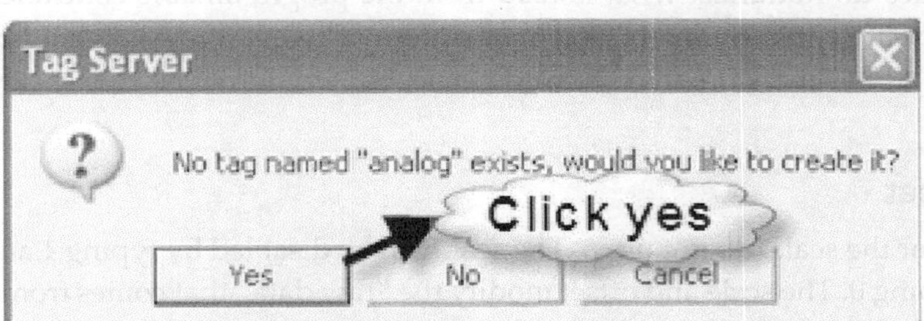

Fig. 10.32 Tag Creation Message Window

Analog Tag Configuration

An analog tag editor window will appear. Now configure analog tag by entering the information shown in Fig. 10.33 and click OK.

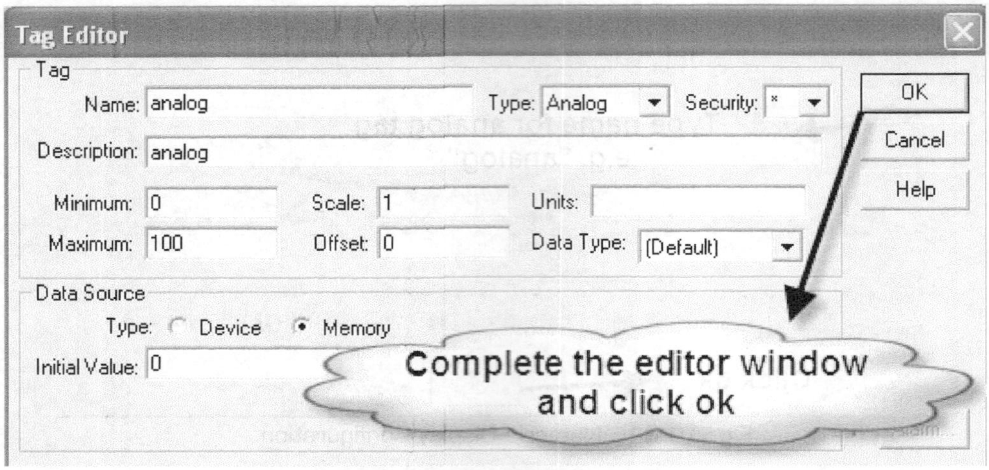

Fig. 10.33 Analog Tag Editor Window

1. If the tag is part of a folder, select that folder in the folder hierarchy. The folder name appears in the Name field and is the first part of the tag name.
2. Type a tag name. If the tag is part of a folder, type the name after the backslash (\).
3. For Type, select Analog.
4. Fill in the fields as outlined below:

Security

To restrict access to this tag, select a security code. If access is restricted, operators cannot change a tag value without the proper security code.

Description

Type a description of this tag, up to 1210 characters long.

Minimum and Maximum

Type the minimum and maximum values that can be written to the programmable controller or server. These values do not affect what is read from the programmable controller or server. For example, if you specify a minimum of 0 and a maximum of 100, RSView32 would be able to read a value of 200 from a PLC and store it in the value table, but would not be able to write this value to the PLC.

Scale and Offset

Type a number. For the scale, do not use 0. The scale can be disabled by typing 1 and the offset can be disabled by typing 0. The scale and offset modify the "raw data" that comes from and goes to the programmable controller before it is saved in the value table. The scale and offset also modify the value specified in RSView32 before it is written to the programmable controller. The scale is a multiplication factor means the value from the PLC is multiplied by the scale. The offset is a fixed value means after the value from the PLC is multiplied by the scale, the offset amount is added. This formula shows the relationship between the PLC value and the amount stored in the RSView32 value table: RSView32 value = (PLC value . scale) + offset.

Units

Type a text label, for display only, up to 20 characters long. This specifies how the tag value is measured for example, gallon, PSI, min, sec.

Data Type

The data types are shown in Table 10.1.

Table 10.1 Data Types

Data type	Description	Range
Default	Depends on the data source and node type: If Device is the data source, for direct driver nodes RSView32 matches the data format specified by the address, for example, N7:0 = Integer, F8:0 = Floating Point. Do not use Default for PC or DDE nodes, as this might produce unexpected results. If Memory is the data source, Default is Floating Point.	See below.
Unsigned Integer	Unsigned 16-bit integer	0 to 65, 535
Integer	Signed 16-bit integer	–32, 768 to 32, 767
Long Integer	Signed 32-bit integer	–2, 147, 483, 648 to 2,147, 483, 647
Floating Point	Single-precision (32-bit) Floating point	$-3.402823E^{+38}$ to $-1.175494E^{-38}$, 0, $1.175494E^{-38}$ to $3.402823E^{+38}$
Byte	Unsigned 8-bit integer	0 to 255
3-Digit BCD	3-digit binary-coded decimal	0 to 999
4-Digit BCD	4-digit binary-coded decimal	0 to 9.999

For tags with Device as the data source, select the data type that matches the format of the data stored in the programmable controller or Windows application. If you are using direct driver nodes, use the Default data type to automatically match the data format specified by the address. Click OK to complete the analog tag creation.

10.6.2 Numeric Input

Use the Numeric Input tool to create fields that operators can use to enter data for analog and digital tags. When the display runs, operators can use these fields to write values to the value table. The values can then be accessed by the programmable controller, DDE device, or RSView32 memory.

To use numeric input on your display select the numeric input icon form the toolbox place it on the display and drag it. A numeric input configuration window will appear as shown in Fig. 10.34. In tag name field select type or already created analog tag.

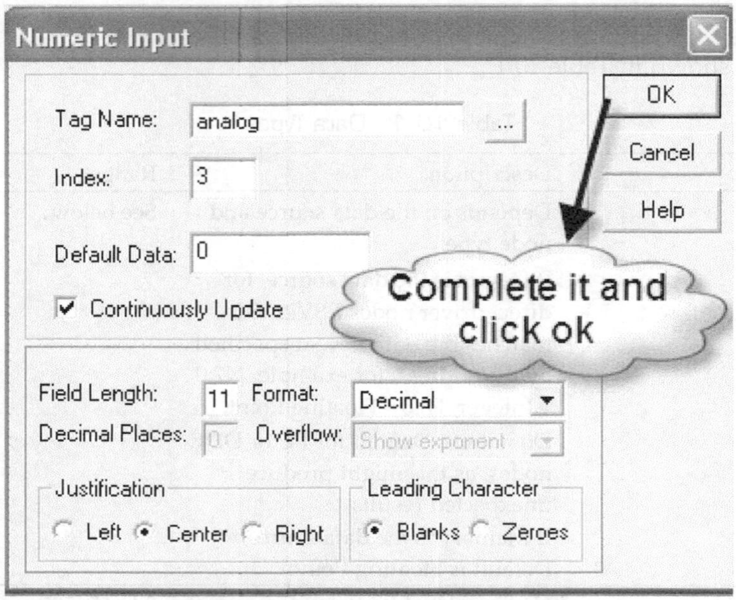

Fig. 10.34 Numeric Input Window

Fill in the fields as follows:

Tag Name: Type a tag name or a placeholder for an analog or digital tag.

Index: Either leave the default as it is or type a number other than the default. Index numbers determine the Tab sequence for interactive objects such as input objects and buttons, and are also used in recipe files.

Default Data: Type a value that will appear in the numeric input field when the graphic is initially displayed.

Continuously Update: Check this box to have tag value changes appear in the field when the operator is not using the field to input data.

Field Length: Type a value to specify the maximum number of characters that will be displayed in this field. Be sure the field is long enough to accommodate all the possible values for the tag.

Format: Select a number base for the value.

Decimal Places: Type the number of places you want after the decimal point. This number must be at least one less than the field width.

Overflow: You can select this field only if Floating Point is selected.

Format Field: Specify how you want the floating-point value displayed if it is too long for the field. The value can be shown as an exponent, rounded up to fit in the field, or completely replaced by asterisks.

Justification: Select Left, Center, or Right to specify how the values will be aligned within the field.

Leading Character: Select Blanks or Zeroes.

Once the numeric input configuration is complete then click OK.

Using the On-screen Keyboard

Displays can be set up so that an operator can use an on-screen keyboard for input entry in numeric and string input fields. The characters the operator types using the on-screen keyboard appear in the selected input field when the operator presses the Download or Update Field button on the on-screen keyboard, or when the operator presses Enter on a hardware keyboard. For on screen keyboard go to display settings select behavior and check the box "Display on screen keyboard" as shown in Fig. 10.35. Click OK.

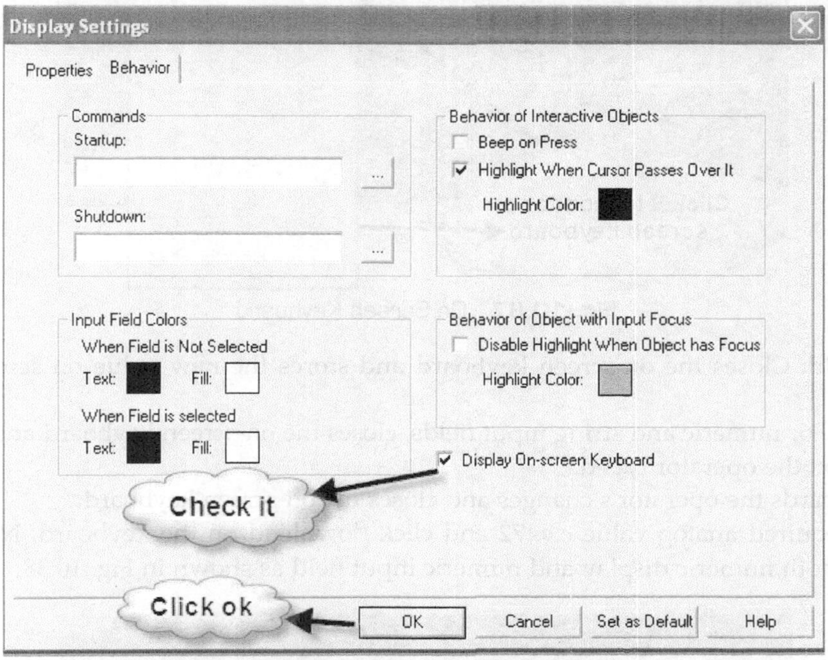

Fig. 10.35 Display Setting Window

Then you will monitor both numeric display and numeric input on your display as shown in Fig. 10.36.

Fig. 10.36 Display with Numeric Display and Numeric Input

Now save the display and run the project. You will observe 0 in both numeric display and numeric input fields. Click the numeric input then a numeric keyboard will appear as shown in Fig. 10.37.

Fig. 10.37 On Screen Keyboard

Update Field: Closes the on-screen keyboard and stores the new value on screen, for later download.

Download: For numeric and string input fields, closes the on-screen keyboard and downloads the value or text the operator typed.

Cancel: Discards the operator's changes and closes the on-screen keyboard.

Enter the required analog value *e.g.* 72 and click download on the keyboard. Now you will monitor 72 in both numeric display and numeric input field as shown in Fig. 10.38.

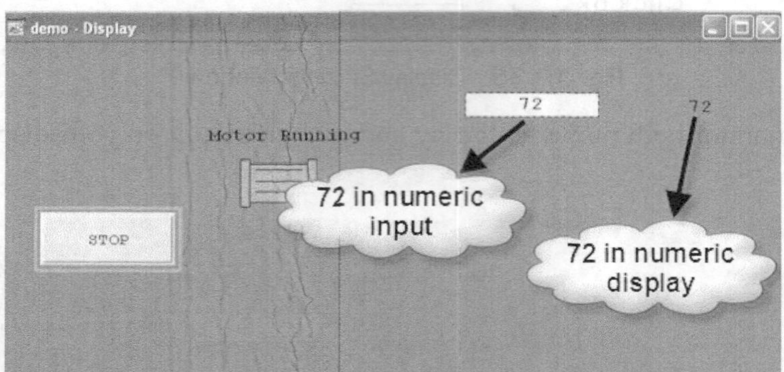

Fig. 10.38 Numeric Display and Numeric Input with 72 as Analog Value

10.7 STRING DISPLAY AND STRING INPUT

10.7.1 String Input

Use the String Input tool to create fields that operators can use to enter data for string tags. When the display runs, operators can use these fields to write values to the value table. The values can then be accessed by the programmable controller, DDE device, or RSView32 memory.

To use string input on your display select the string input icon form the toolbox place it on the display and drag it. A string input configuration window will appear as shown in Fig. 10.39. To configure string input you need to create a string tag. In tag name field type any name for creating string tag *e.g.* "string".

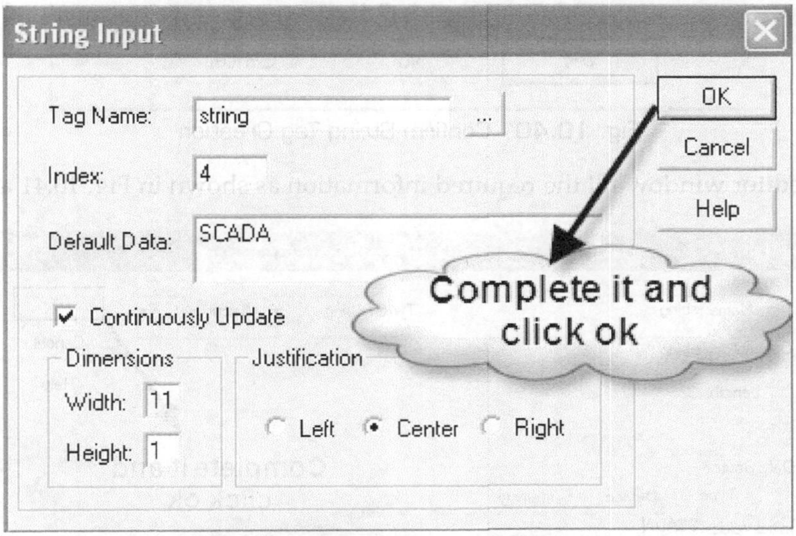

Fig. 10.39 String Input Window

Fill in the fields as follows:

Tag Name: Type a tag name or a placeholder for an analog or digital tag.

Index: Either leave the default as it is or type a number other than the default. Index numbers determine the Tab sequence for interactive objects such as input objects and buttons, and are also used in recipe files.

Default Data: Type a value that will appear in the numeric input field when the graphic is initially displayed.

Continuously Update: Check this box to have tag value changes appear in the field when the operator is not using the field to input data.

Field Length: Type a value to specify the maximum number of characters that will be displayed in this field. Be sure the field is long enough to accommodate all the possible values for the tag.

Format: Select a number base for the value.

Decimal Places: Type the number of places you want after the decimal point. This number must be at least one less than the field width.

Overflow: You can select this field only if Floating Point is selected in the format.

Format Field: Specify how you want the floating-point value displayed if it is too long for the field. The value can be shown as an exponent, rounded up to fit in the field, or completely replaced by asterisks.

Justification: Select Left, Center, or Right to specify how the values will be aligned within the field.

Leading Character: Select Blanks or Zeroes.

Complete the string input window configuration and click OK. You notice a message asking to confirm string tag creation. Click Yes in this window as shown in Fig. 10.40.

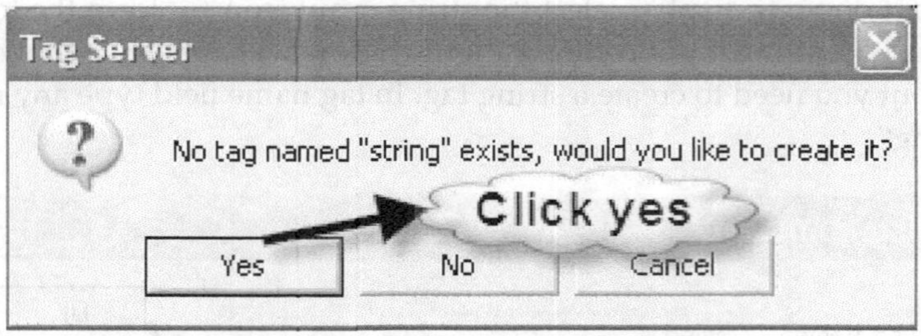

Fig. 10.40 Confirm String Tag Creation

In string tag editor window fill the required information as shown in Fig. 10.41 and click OK.

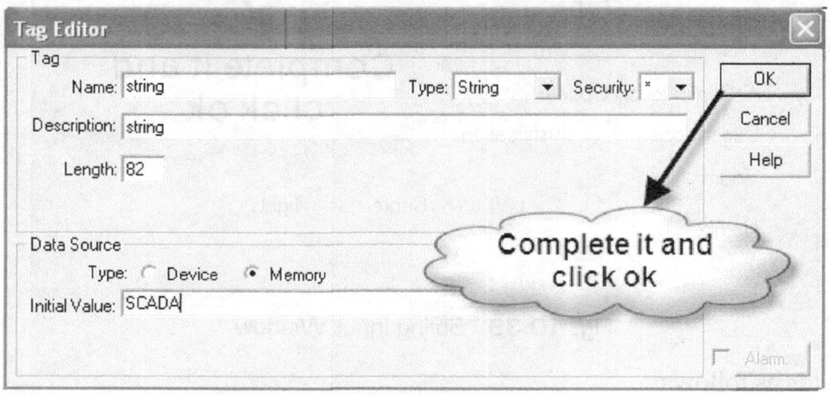

Fig. 10.41 String Tag Configuration

String Tag Configuration

1. If the tag is part of a folder, select the folder in the folder hierarchy. The folder name appears in the Name field and is the first part of the tag name.
2. Type a tag name.
3. For Type, select String.
4. Fill in the fields as outlined below:

Security

Same as described above in analog tag configuration

Description

Same as described above in analog tag configuration

Length

Type a number between 1 and 102 to specify the length of the string tag in characters. RSView32 will accept only a length that is a multiple of the size of the Allen-Bradley programmable controller data element you are addressing.

Specifying a Data Source

The following topics describe how to complete the Data Source fields

Specifying Device as the Data Source

A tag with device as its data source receives its data from a source external to RSView32. The data can come from Allen-Bradley through RSLinx direct driver. To specify device as the data source as shown in Fig. 10.42:

1. Click Device.
2. Supply a node name.

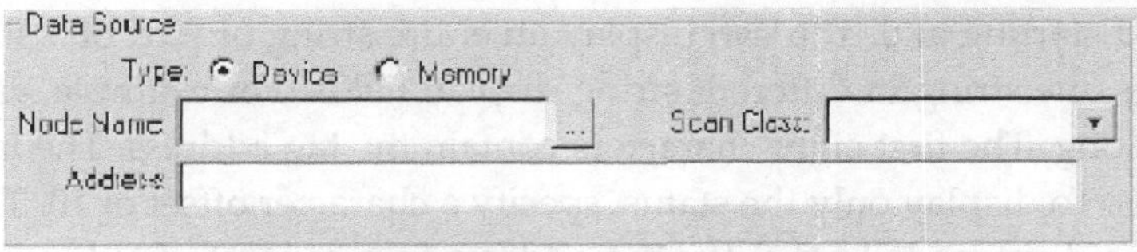

Fig. 10.42 Device Tag Configuration

3. Select a scan class.
4. In the Address field, enter input or output address of the tag.

Specifying Memory as Data Source

Click memory to specify it as the data source as shown in Fig. 10.43.

1. Click Memory.
2. In the Initial Value field, type the tag's starting value. When you first run a project, a memory tag has the value defined in the Initial Value field. All subsequent runs of the project use the tag's last value before the project stopped.

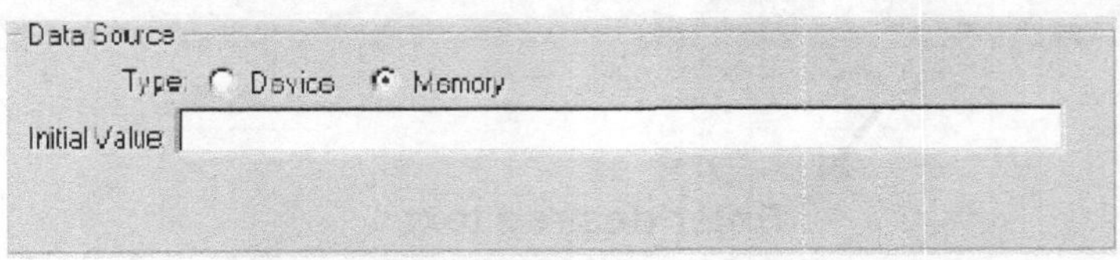

Fig. 10.43 Memory Tag Configuration

10.7.2 String Display

Use the String Display tool to create fields that operators can use to display the current value of a string tag. To use string display on your display select the string display icon from the toolbox then place it on the display and drag it. A string display configuration window will appear as shown in Fig. 10.44. In tag name field select tag *e.g.* "string".

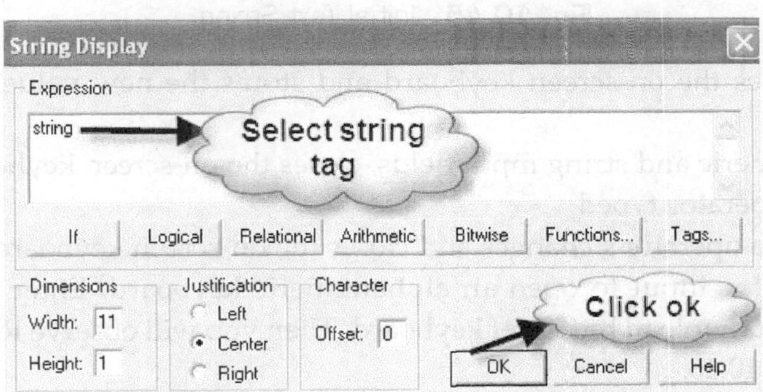

Fig. 10.44 String Display Configuration

Fill in the fields as follows:

Expression: Create an expression. At runtime, the string result of this expression is displayed in the String Display field.

Dimensions: Type a value to specify the maximum number of characters that will be displayed in this field. Be sure the field is long enough to accommodate the result of the expression.

Justification: Select Left, Center, or Right to specify how the values will be aligned within the field.

Character Offset: Select a number of characters. This number determines the first characters to be displayed in the field starting at 0. You can display an entire string or part of a string. You can also display parts of the same string in different string display fields. For example, assume you have a string 12 characters long. The first eight characters contain the tag address. The last four characters contain the tag status. To display only the status, specify a character offset of 10. The address part of the string will not be displayed. To display the address in a separate field, create another string display field, specify the same tag name and specify a field width of 10 and an offset of 0.

Now you will monitor both string display and string input on your display. Save the display and run it. Initially you will observe SCADA in both the fields because it is the initial text set by you in tag editor as shown in Fig. 10.45.

Fig. 10.45 Initial Text String

Update Field: Closes the on-screen keyboard and stores the new value on screen, for later download.

Download: For numeric and string input fields, closes the on-screen keyboard and downloads the value or text the operator typed.

Cancel: Discards the operator's changes and closes the on-screen keyboard.

Now click the display input to open an alphanumeric keyboard. Enter the desired text *e.g.* "RKDEMO" and click download button of keyboard. Then you will observe RKDEMO text in both fields as shown in Fig. 10.46.

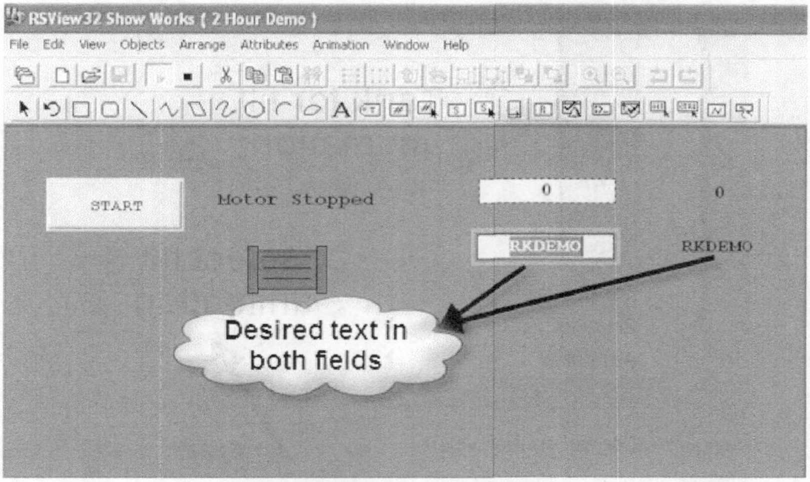

Fig. 10.46 Downloaded Text

10.8 FILL ANIMATION

Fill animation is used to check the level of liquid in an object that changes on the basis of tag's value in relation to its minimum and maximum values. The object's fill level is proportional to the value of the expression. For example, if the value of the expression is halfway between the minimum and maximum values, the object will be half full. In order to observe the effect of fill animation, select the tank library as shown in Fig. 10.47. Select a tank, copy it and paste it on the display.

Fig. 10.47 Tank Library Selection

Now right click the tank, select fill animation as shown in Fig. 10.48.

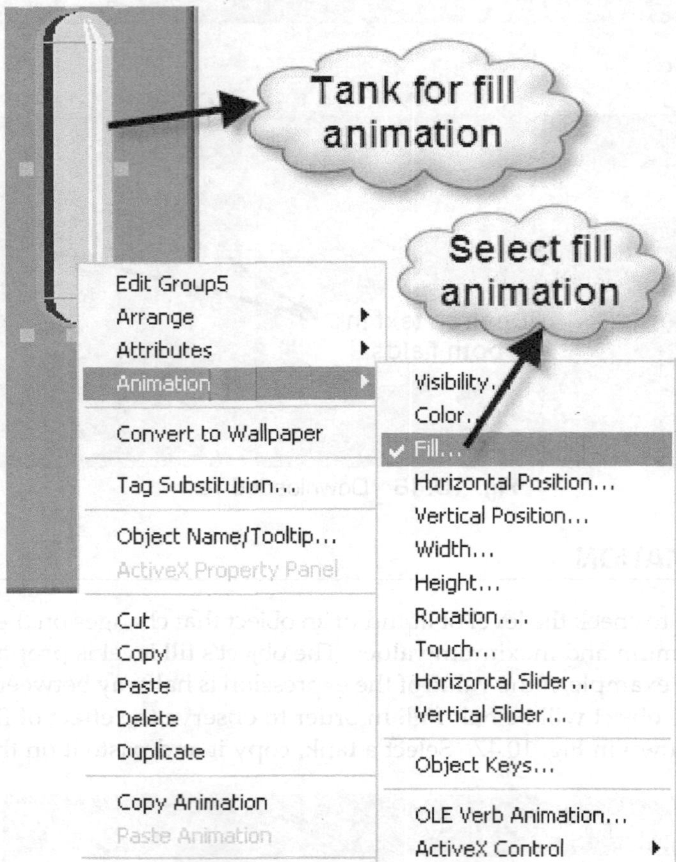

Fig. 10.48 Fill Animation Selection

Then a fill animation configuration window will appear. Assign analog tag and select fill direction as up as shown in Fig. 10.49.

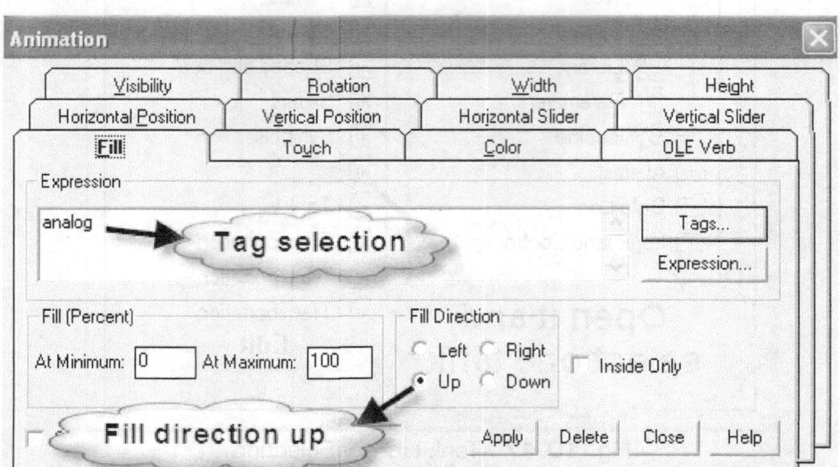

Fig. 10.49 Fill Animation Window

Fill animation configuration

1. Select the object.
2. Click the Fill tab in the Animation dialog box.
3. Create an expression. The expression can be a tag name or a more complex expression. If a tag name is used, the tag's value determines the fill levels.
4. Specify At Minimum and At Maximum values for the fill percentage levels.
5. Click a fill direction.
6. Select the Inside Only check box if you want the object's outline to remain constant so only the inside fill level varies.
7. To specify minimum and maximum values, select the Specify check box and type the values.
8. Click Apply

In order to select fill color, right click tank and select attributes and then select fill color as shown in Fig. 10.50.

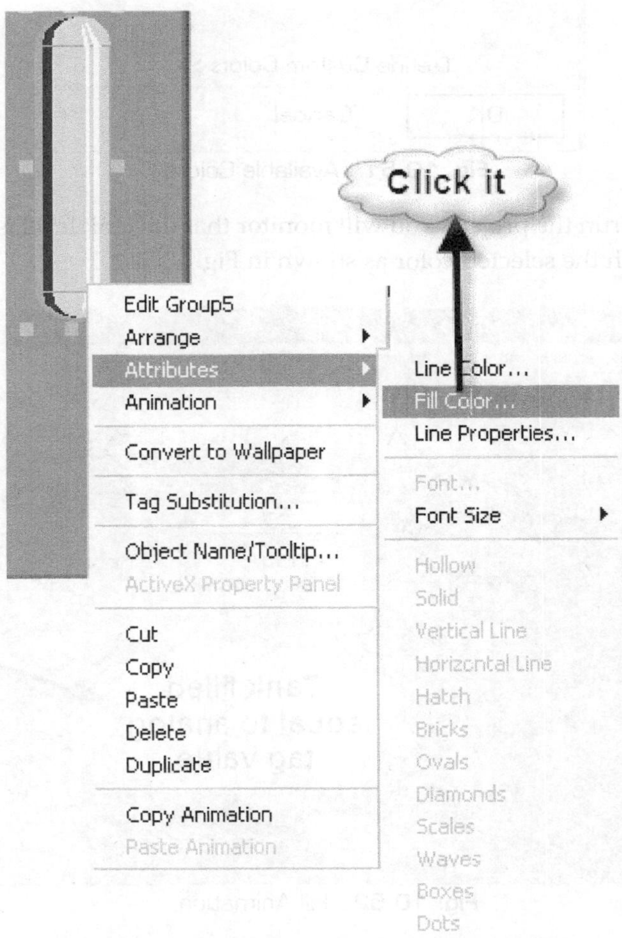

Fig. 10.50 Fill Color Selection

Select desired color from the window and click OK as shown in Fig. 10.51.

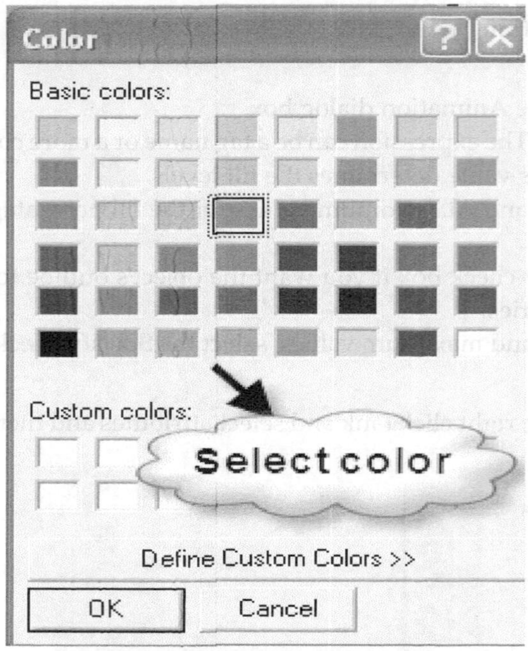

Fig. 10.51 Available Colors

Save the display and run the project. You will monitor that the tank level is equal to analog tag's value and it is filled with the selected color as shown in Fig. 10.52.

Fig. 10.52 Fill Animation

If you also want to view the empty portion of the tank then again go to fill animation configuration window and check the box "Inside only" as shown in Fig. 10.53.

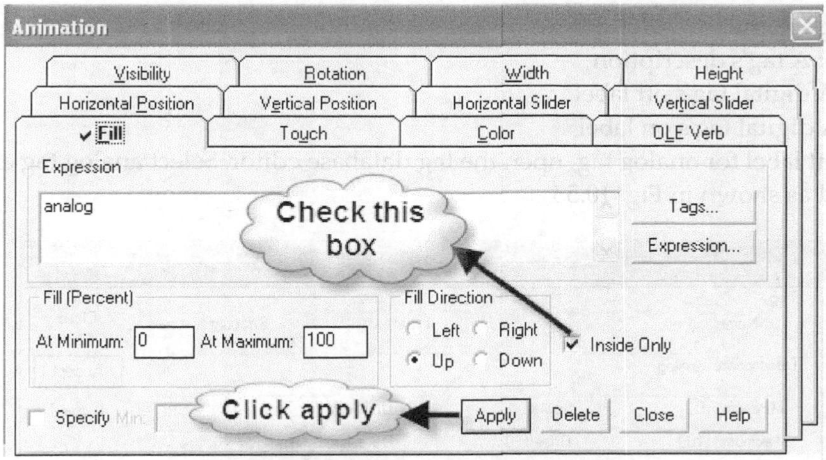

Fig. 10.53 Fill Animation Window

Again save the display and run it then you will also notice the empty potion of the tank as shown in Fig. 10.54.

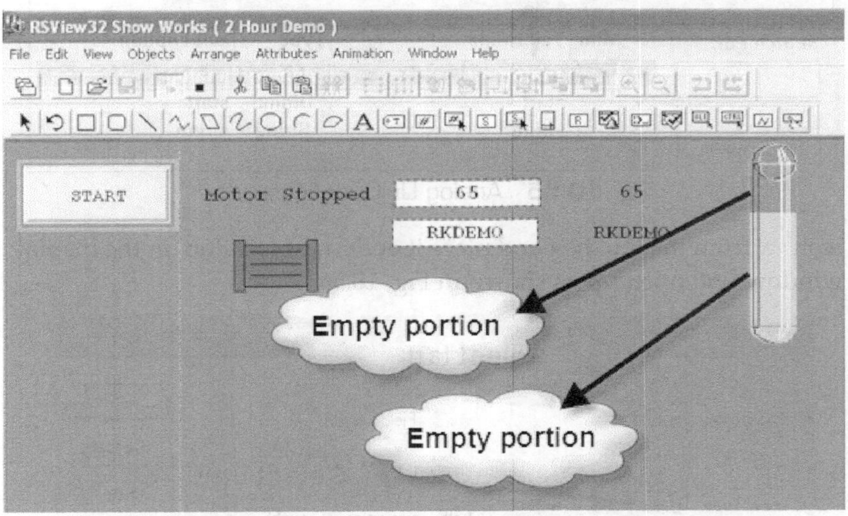

Fig. 10.54 Tank with Both Filled and Empty Portions

10.9 LABEL

Use the Label tool to create fields that display different types of tag information at runtime. All the information comes from fields you configure for the tag in the Tag Database editor. You can create the following types of labels:

Minimum: An analog tag's minimum value

Maximum: An analog tag's maximum value

Status: The current status of a digital tag. When the tag value is 1, "On Label" is displayed; when the tag value is 0, "Off Label" is displayed

Units: An analog tag's units label

Name: A tag's name
Description: A tag's description
Off Label: A digital tag's off label
On Label: A digital tag's on label

To create unit label for analog tag, open the tag database editor. Select analog tag and enter "%" in the unit field as shown in Fig. 10.55.

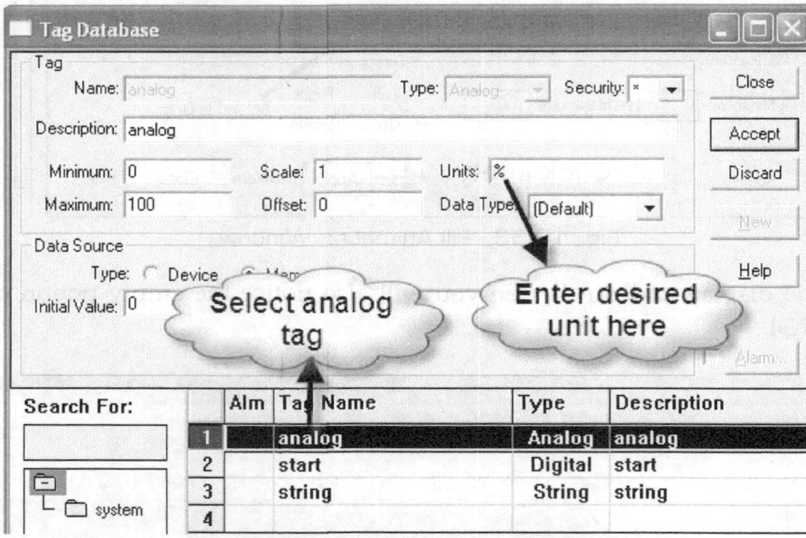

Fig. 10.55 Analog Unit Configuration

Select label icon from the tool box and place it at desired location on the display. Then a label configuration window will open up as shown in Fig. 10.56.

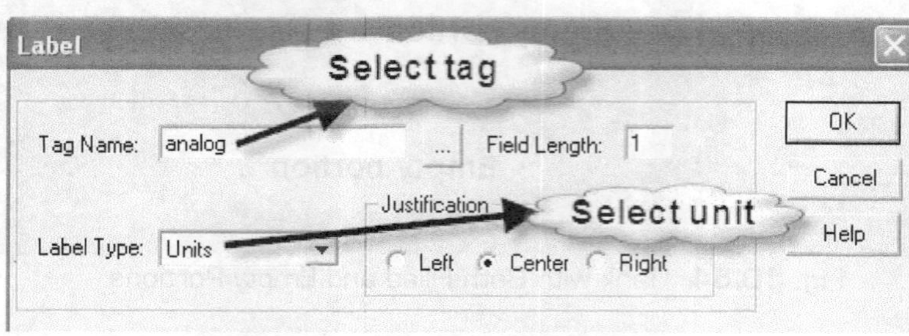

Fig. 10.56 Label Configuration Window

Fill in the fields as follows:

Tag Name: Specify a tag name or placeholder.

Field Length: Type a value to specify the maximum number of characters that will be displayed in this field. Be sure the field is long enough to accommodate the label. The maximum field width depends on the type of label you are creating.

Label Type: Select a type.

Justification: Select Left, Center, or Right to specify how the values will be aligned within the field.

Once the label configuration is complete then click OK. Now save the display and run it to view "%" as label in front of numeric display as shown in Fig. 10.57.

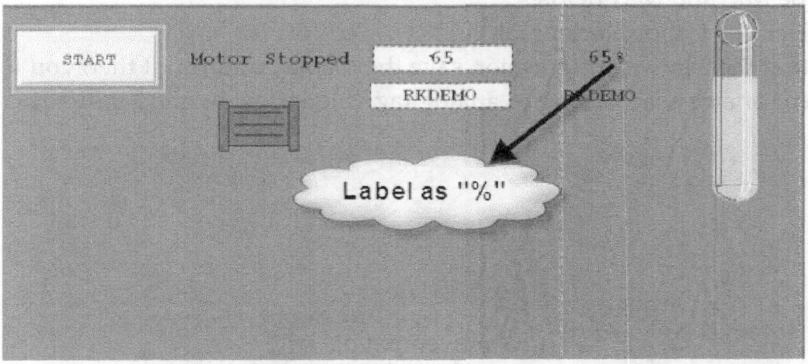

Fig. 10.57 Display Showing % as Label

10.10 ARROW

Use the Arrow tool to create arrows that move based on the results of an expression. Arrows can move vertically or horizontally. Vertical arrows move up or down in relation to a tag's minimum and maximum values. For vertical arrows, if the value is less than or equal to the minimum value, the arrow is at the bottom. If the value is equal to or greater than the maximum value, the arrow is at the top. Horizontal arrows move left and right in relation to the minimum and maximum values. For horizontal arrows, if the value is less than or equal to the minimum value, the arrow is at the left. If the value is equal to or greater than the maximum value, the arrow is at the right.

Select label icon from the tool box and place it at desired location on the display. Then an arrow configuration window will open up as shown in Fig. 10.58.

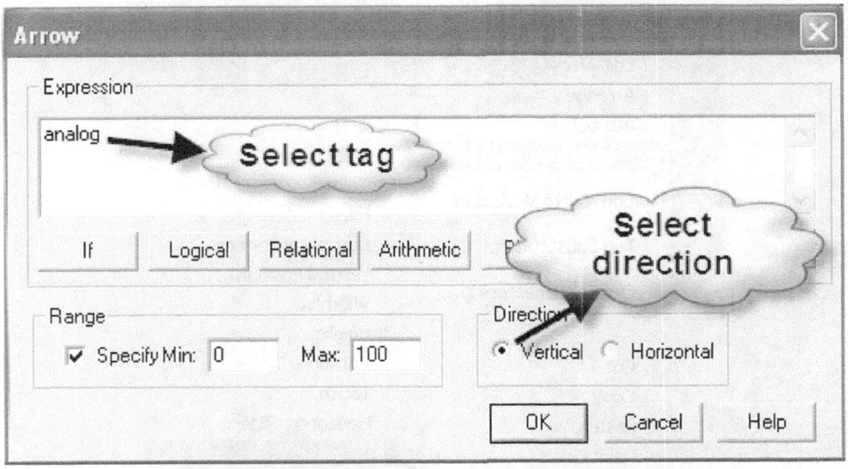

Fig. 10.58 Arrow Configuration Window

Fill in the fields as follows:

Expression: Create an expression. At runtime, the numeric result of the expression positions the arrow.

Range: To use the minimum and maximum values for the first tag in the expression, clear the Specify check box. To specify a minimum and maximum value that is different than the first tag's, Enter values in the Min and Max fields.

Direction: Select Vertical or Horizontal.

After completing the arrow configuration save the display and run it then you will observe the vertical movement of arrow according to analog tag value as shown in Fig. 10.59.

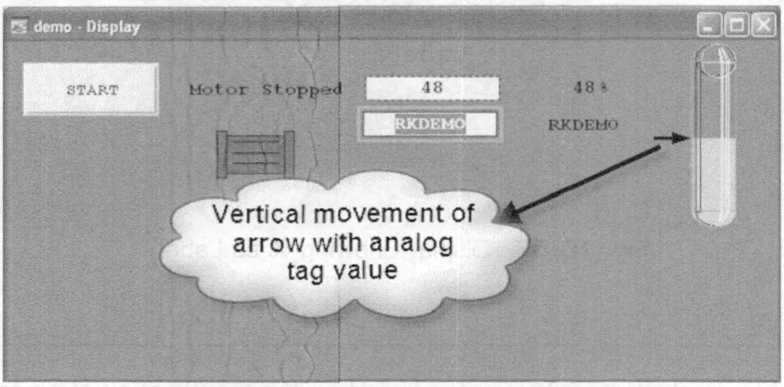

Fig. 10.59 Vertical Movement of Arrow

10.11 VERTICAL SLIDER

With vertical slider animation, you can create a graphic object that sets the value of a tag. To do this, define a path for the object and then use the mouse to move the object vertically. The pixel position of the object is translated into values that are written to the tag. Here we will attach vertical slider animation to arrow created in the previous section. In order to do so select the arrow and configure it for vertical slider animation as shown in Fig. 10.60.

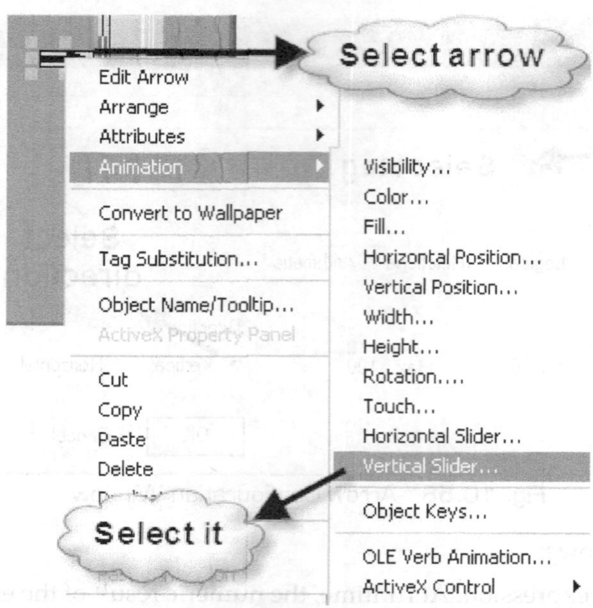

Fig. 10.60 Vertical Slider Animation Selection

The vertical slider animation window opens up for configuration. Complete it as shown in Fig. 10.61.

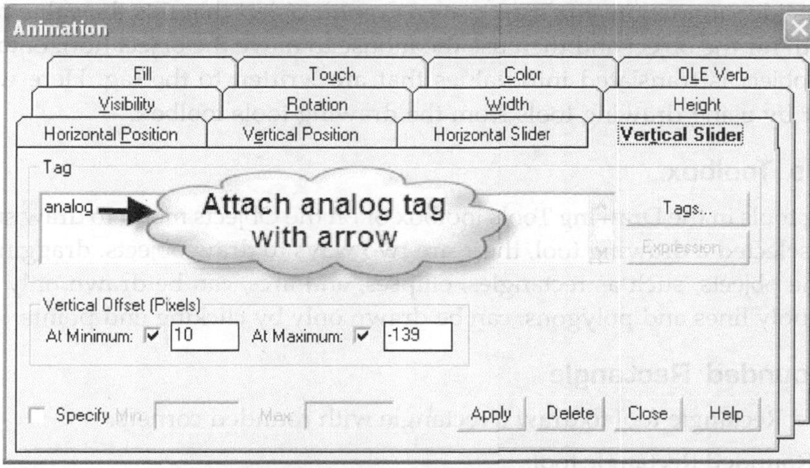

Fig. 10.61 Vertical Slider Configuration

To configure vertical slider animation:

1. Select the object.
2. Click the Vertical Slider tab in the Animation dialog box.
3. Type the name of the tag whose value will be determined by the slider.
4. Set the starting point for the slider object by dragging the object or by typing a value. Select the At Minimum check box.
5. Set the ending point for the slider object by dragging the object or by typing a value. Select the At Maximum check box.
6. To specify the minimum and maximum values that will be written to the tag, click the Specify check box and type the values.
7. Click Apply.

Now once again save the display and run it. In the display window select the arrow and drag it up and down then you will monitor change in numeric input, numeric display and tank fill level as shown in Fig. 10.62.

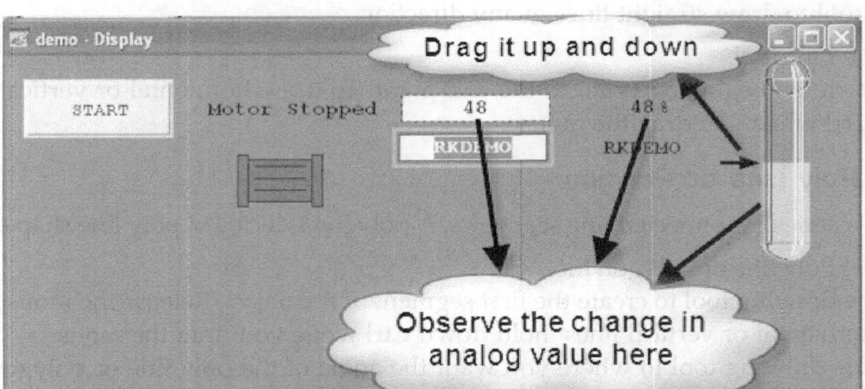

Fig. 10.62 Arrow as Vertical Slider

10.12 HORIZONTAL SLIDER

With horizontal slider animation, you can create a graphic object that sets the value of a tag. To do this, define a path for the object and then use the mouse to move the object horizontally. The pixel position of the object is translated into values that are written to the tag. Here we will create horizontal slider by using drawing tools from the drawing tools toolbox.

Drawing Tools Toolbox

Use the drawing tools in the Drawing Tools toolbox or on the Objects menu to draw simple objects. Once you have selected a drawing tool, there are two ways to draw objects: dragging, or clicking end points. Some objects, such as rectangles, ellipses, and arcs, can be drawn only by dragging. Others, such as poly lines and polygons, can be drawn only by clicking end points.

Drawing a Rounded Rectangle

Use the Rounded Rectangle tool to draw a rectangle with rounded corners.

1. Click the Rounded Rectangle tool.
2. Drag the drawing tool diagonally until the object is the size you want.

Drawing a Rectangle or Square

Use the Rectangle tool to draw a rectangle or square.

1. Click the Rectangle tool.
2. To draw a rectangle, drag the drawing tool diagonally until the object is the size you want. To draw a square, hold down Ctrl while you drag.

Drawing an Ellipse or Circle

Use the Ellipse tool to draw an ellipse or circle.

1. Click the Ellipse tool.
2. To draw an ellipse, drag the drawing tool diagonally until the object is the size you want. To draw a circle, hold down Ctrl while you drag.

Drawing a Line

Use the Line tool to draw straight lines in any direction.

1. Click the Line tool.
2. Drag from the beginning point to the end point. To draw horizontal or vertical lines, hold down Ctrl while you drag the mouse.

Drawing a Poly Line or Polygon

A poly line is a series of connected line segments. A polygon is a closed poly line shape.

1. Click the Poly line or Polygon tool.
2. Drag the drawing tool to create the first segment of the object. Release the mouse button. To draw horizontal or vertical lines hold down Ctrl while you drag the mouse.
3. Move the drawing tool to where you want the angle of the poly line or polygon to be and click. Repeat this step until you have completed the object.
4. To finish drawing, click the Select tool.

Drawing a Freehand Object

Drawing freehand is similar to drawing with a pen on paper.

1. Click the Freehand tool.
2. Drag the drawing tool to create the shape you want.

Drawing an Arc or Wedge

Arcs and wedges are drawn in two steps: first you create a circle, and then you reshape the circle. To draw the arc or wedge in 45-degree increments, hold down Ctrl while you draw.

1. Select the Arc or Wedge tool.
2. Drag to draw a circle. Release the mouse button. When you release the mouse button, a set of handles appears so you can reshape the circle.
3. Click a handle, and drag the mouse to 'cut out' part of the circle. You can also use the Arc and Wedge tools to reshape any arc, ellipse, or wedge.

Creating Text

1. Click the Text tool. The pointer becomes an I-beam ( ).
2. Type the text.

Once the Text tool is selected, you can create more than one text object. To do so, move to an empty spot in the drawing area, click, and then type the text. To edit text, select the text object and double-click. The I-beam appears. To delete text characters, use the Backspace and Delete keys. You cannot rotate text.

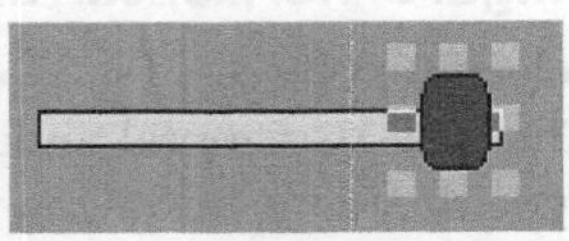

Fig. 10.63 Created Horizontal Slider

By using these drawing tools create one horizontal slider as shown in Fig. 10.63 and configure it for horizontal animation as shown in Fig. 10.64.

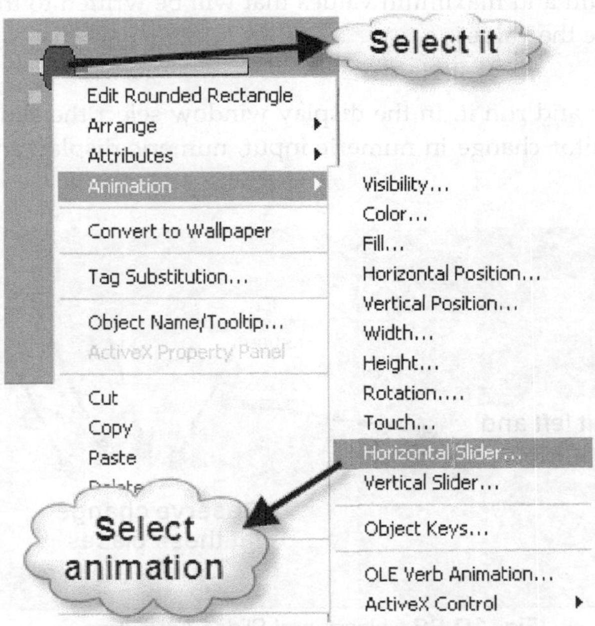

Fig. 10.64 Horizontal Animation Selection

The horizontal slider animation window opens up for configuration. Complete it as shown in Fig. 10.65.

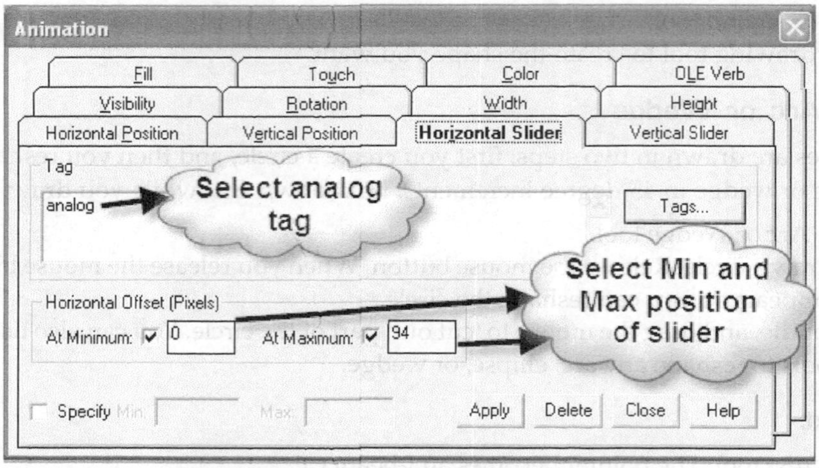

Fig. 10.65 Horizontal Slider Animation

To configure horizontal slider animation:

1. Select the object.
2. Click the Horizontal Slider tab in the Animation dialog box.
3. Type the name of the tag whose value will be determined by the slider.
4. Set the starting point for the slider object by dragging the object or by typing a value. Select the At Minimum check box.
5. Set the ending point for the slider object by dragging the object or by typing a value. Select the At Maximum check box.
6. To specify minimum and maximum values that will be written to the tag, click the Specify check box and type the values.
7. Click Apply.

Now save the display and run it. In the display window select the slider and drag it left and right then you will monitor change in numeric input, numeric display and tank fill quantity as shown in Fig. 10.66.

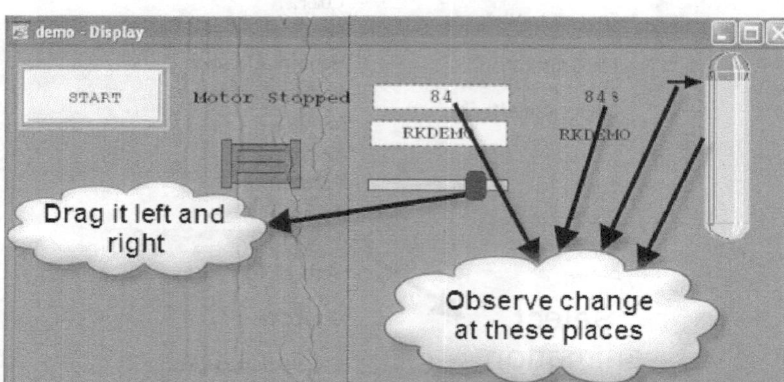

Fig. 10.66 Horizontal Slider Animation

10.13 HORIZONTAL POSITION ANIMATIONS

With horizontal position animation, an object moves horizontally based on the result of an expression in relation to its minimum and maximum values. The object's horizontal position is proportional to the value of the expression. For example, if the value of the expression is halfway between the minimum and maximum values, the object will be halfway between its minimum and maximum pixel offset. To view the effect of horizontal position animation, select the transportation library as shown in Fig. 10.67.

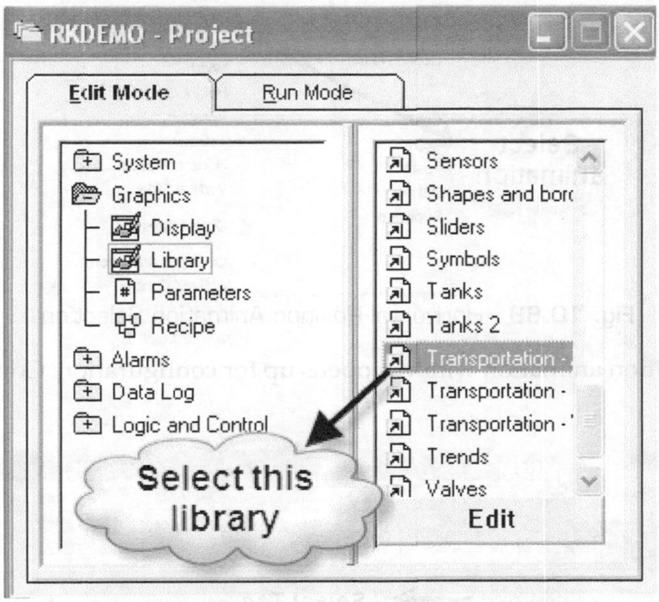

Fig. 10.67 Select Transportation Library

Select the desired object from transportation library, copy it and paste it on the display as shown in Fig. 10.68.

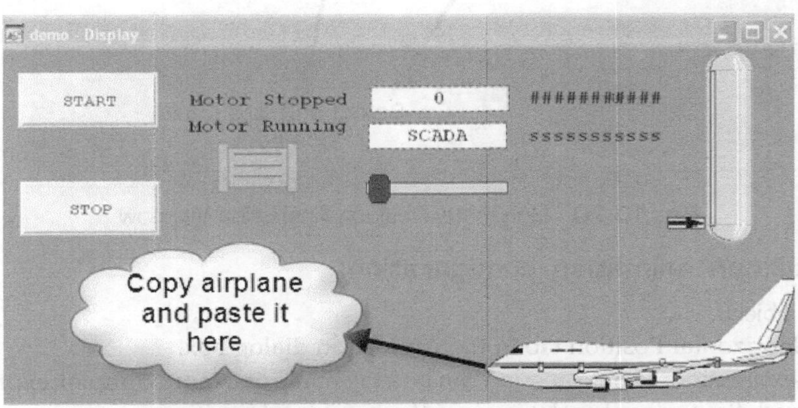

Fig. 10.68 Copy and Paste of an Airplane

Select this object and configure it for horizontal position animation as shown in Fig. 10.69.

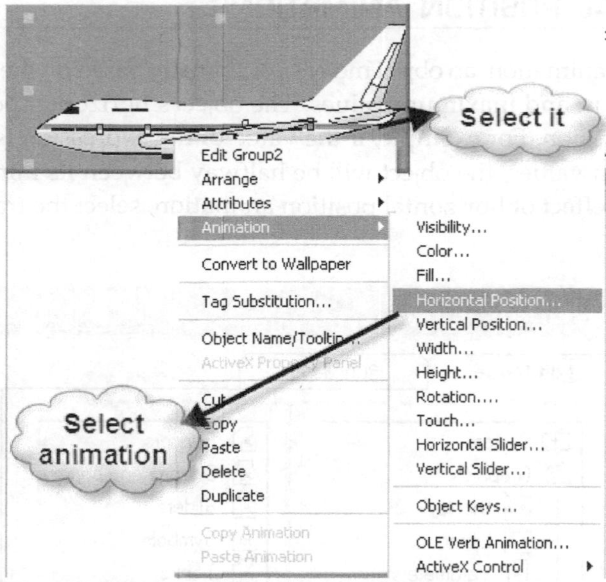

Fig. 10.69 Horizontal Position Animation Selection

The horizontal position animation window opens up for configuration. Complete it as shown in Fig. 10.70.

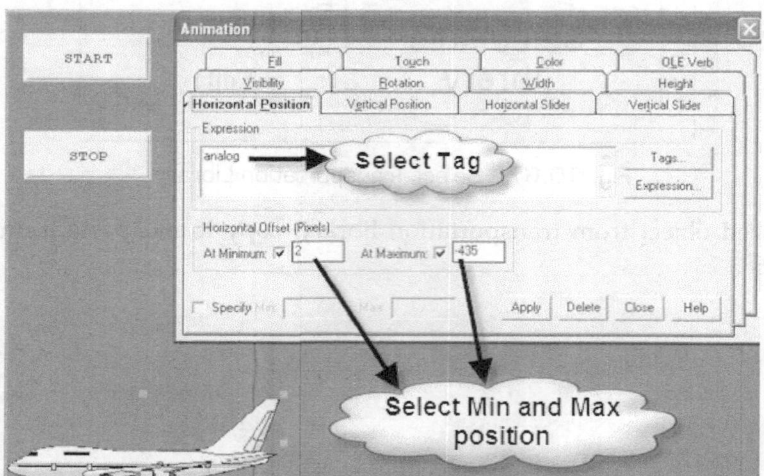

Fig. 10.70 Horizontal Position Animation Window

Horizontal position animation configuration:

1. Select the object.
2. Click the Horizontal Position tab in the Animation dialog box.
3. Create an expression. The expression can be a tag name or a more complex expression. If a tag name is used, the tag's value determines the horizontal position.
4. Set the starting point for the object by dragging the object or by typing a value. Click the At Minimum check box.

5. Set the ending point for the object by dragging the object or by typing a value. Click the At Maximum check box.
6. To specify minimum and maximum values, click the Specify check box and type the values.
7. Click Apply.

   Configure this same object for the following animations:
   (*a*) Vertical position animation
   (*b*) Width animation
   (*c*) Height animation

10.14 VERTICAL POSITION ANIMATION

With vertical position animation, an object moves vertically based on the result of an expression in relation to its minimum and maximum values. The object's vertical position is proportional to the value of the expression. For example, if the value of the expression is halfway between the minimum and maximum values, the object will be halfway between its minimum and maximum offset. Select vertical position animation in the animation configuration window as shown in Fig. 10.71.

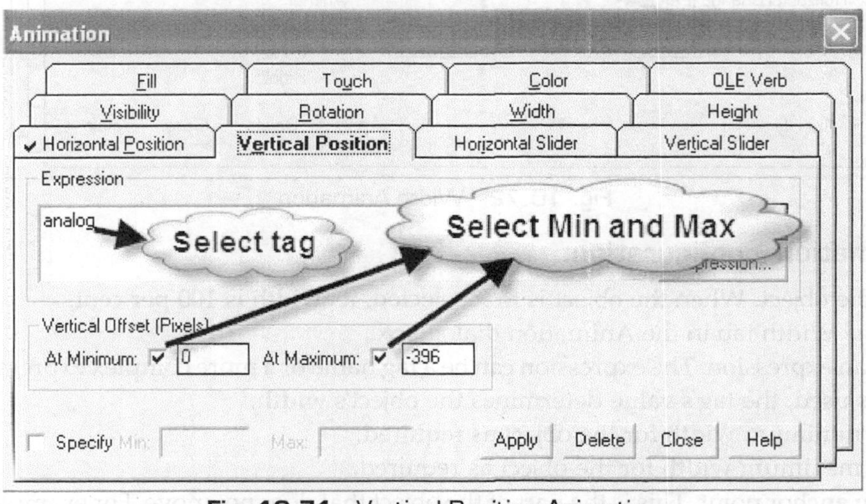

Fig. 10.71 Vertical Position Animation

Vertical position animation configuration:

1. Select the object.
2. Click the Vertical Position tab in the Animation dialog box.
3. Create an expression. The expression can be a tag name or a more complex expression. If a tag name is used, the tag's value determines the vertical position.
4. Set the starting point for the object by dragging the object or by typing a value. Click the At Minimum check box.
5. Set the ending point for the object by dragging the object or by typing a value. Click the At Maximum check box.
6. To specify minimum and maximum values, click the Specify check box and type the values.
7. Click Apply.

10.15 WIDTH ANIMATION

The width animation is used to change the width of an object based on a tag value or the result of an expression. The object's width is proportional to the value of the expression. For example, if the value of the expression is halfway between the minimum and maximum values, the object will be half the full width. Select width animation in the animation configuration window as shown in Fig. 10.72.

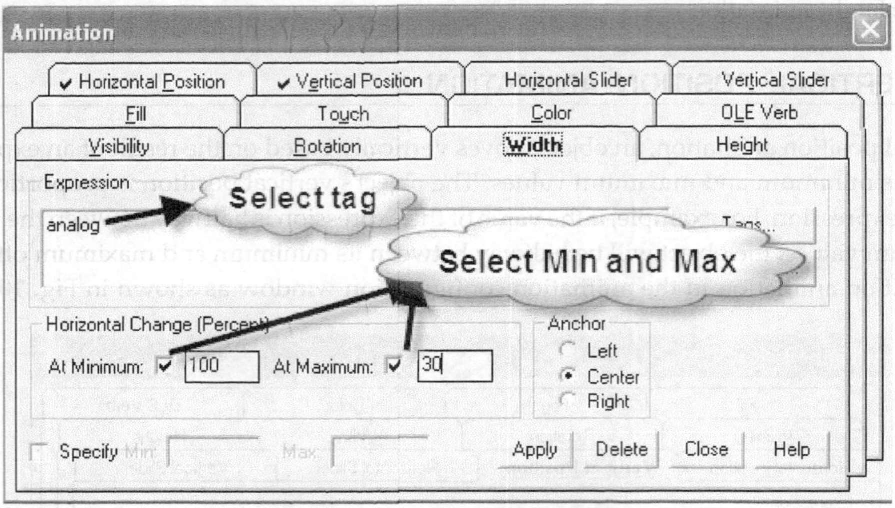

Fig. 10.72 Width Animation

Width animation configuration:

1. Select the object. When the object is first selected, its width is 100 per cent.
2. Click the Width tab in the Animation dialog box.
3. Create an expression. The expression can be a tag name or a more complex expression. If a tag name is used, the tag's value determines the object's width.
4. Set the minimum width for the object as required.
5. Set the maximum width for the object as required.
6. Click an anchor point. This is the part of the object that does not move. For example, click Left if you do not want the left side of the object to move. As the value of the expression changes, the object will grow to or shrink from the right.
7. To specify minimum and maximum values, click the Specify check box and type the values.
8. Click Apply.

10.16 HEIGHT ANIMATION

The height animation is used to change the height of an object based on a tag value or the result of an expression. The object's height is proportional to the value of the expression. For example, if the value of the expression is halfway between the minimum and maximum values, the object will be half the full height. Select height animation in the animation configuration window as shown in Fig. 10.73.

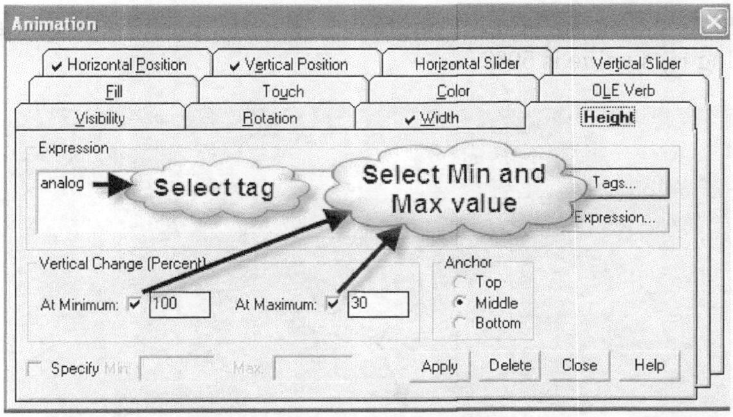

Fig. 10.73 Height Animation

Height Animation Configuration:

1. Select the object. When the object is first selected, its height is 100 per cent.
2. Click the Height tab in the Animation dialog box.
3. Create an expression. The expression can be a tag name or a more complex expression. If a tag name is used, the tag's value determines the object's height.
4. Set the minimum height for the object as required.
5. Set the maximum height for the object as required.
6. Click an anchor point. This is the part of the object that does not move. For example, click Top if you do not want the top of the object to move. As the value of the expression changes, the object will grow to or shrink from the bottom.
7. To specify minimum and maximum values, click the Specify check box and type the values.
8. Click Apply.

Once all the four animations are configured for airplane then save the display and run it for following cases:

Case 1: Figure 10.74 shows the vertical position, horizontal position, height and width of the airplane when analog tags value is 0%.

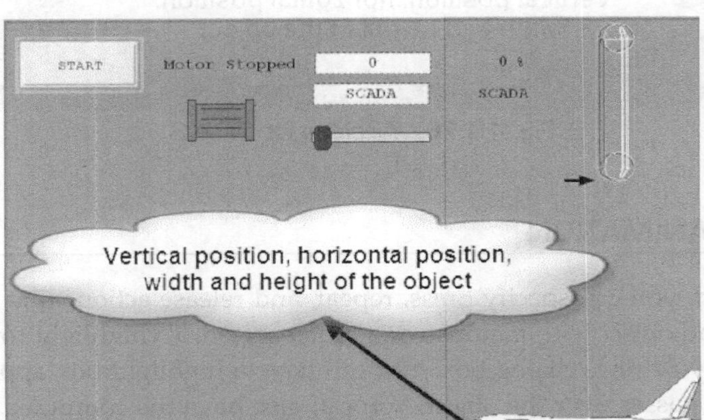

Fig. 10.74 Airplane for Case 1

Case 2: Figure 10.75 shows the vertical position, horizontal position, height and width of the airplane when analog tags value is 50%.

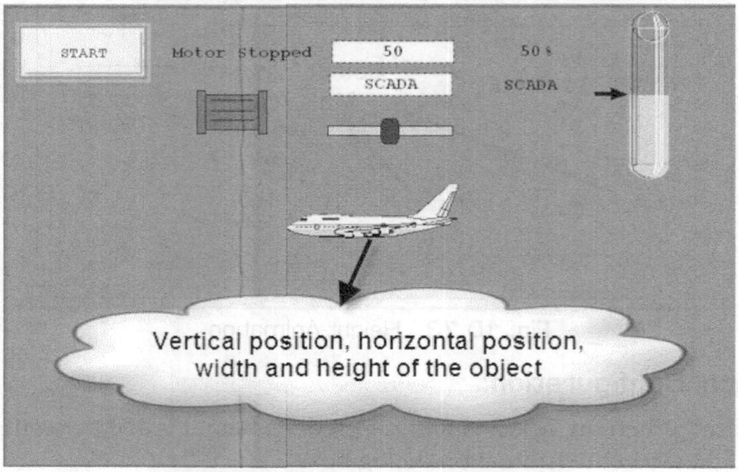

Fig. 10.75 Airplane for Case 2

Case 3: Figure 10.76 shows the vertical position, horizontal position, height and width of the airplane when analog tags value is 105%.

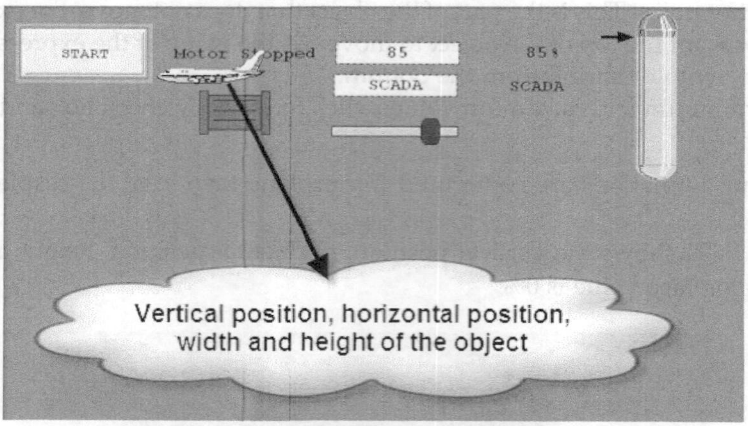

Fig. 10.76 Airplane for Case 3

10.17 TOUCH ANIMATION

With touch animation, you can specify press, repeat, and release actions that are triggered when an operator touches an object with a mouse or a touch screen. To highlight touch objects, use the options in the Display Settings dialog box. You can have a highlight box appear around a touch object when a cursor passes over the object. You can also have the computer beep when a touch object is selected.

To view the effect of touch animation, copy and paste one push button from button library on the display. Select the button and configure it for touch animation as shown in Fig. 10.77.

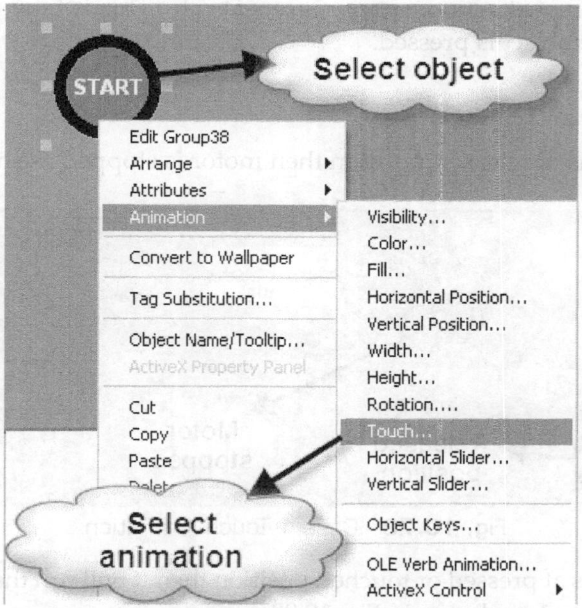

Fig. 10.77 Touch Animation Selection

The touch animation window opens up for configuration. Complete it as shown in Fig. 10.78.

Touch Animation Configuration:

1. Select the object.
2. Click the Touch tab in the Animation dialog box.

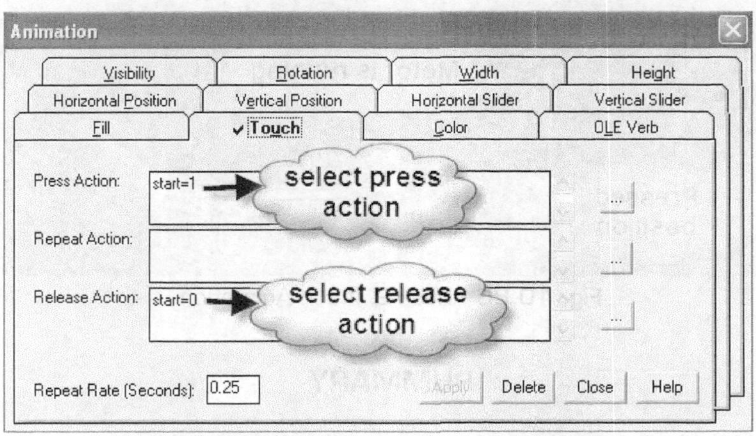

Fig. 10.78 Touch Animation Configuration

3. In the Press Action field, type an RSView32 command or a macro. Separate multiple commands or macros with a semi-colon (;) or place them on separate lines. Multiple commands are executed in the order they are listed. If you want repeat and release actions, type commands or macros in these fields as well. Press and release actions occur when the object is pressed or released. Repeat actions occur at the rate specified in the Repeat Rate field and do not wait for the press command to be completed.

4. If you specified a repeat action, specify a repeat rate. This is how often the action will be repeated while the object is pressed.
5. Click Apply.

Then save the display and run it.

Case 1: When button is at released position then motor is stopped as shown in Fig. 10.79.

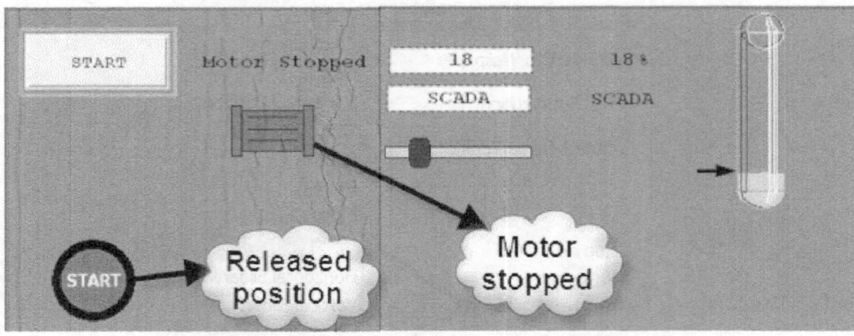

Fig, 10.79 Case 1 Touch Animation

Case 2: When button is at pressed or touched position then it will run the motor by changing the value of start tag from 0 to 1 as shown in Fig. 10.80.

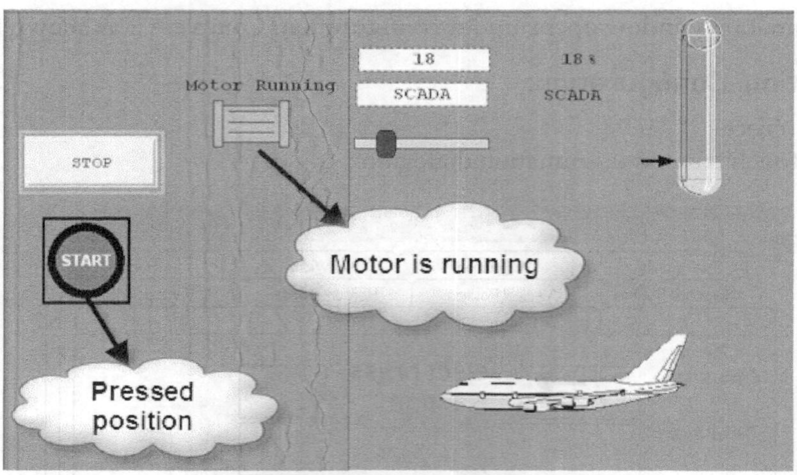

Fig. 10.80 Case 2 Touch Animation

SUMMARY

After creating graphic objects, we can attach animation that links an object to a tag so that the object's appearance changes to reflect changes in the tag's value. The Animation dialog box is a floating dialog box, which means you can have it open all the time and can move it around the screen. The range of motion is related to the minimum and maximum values for the expression. When configuring animation for objects, you are linking objects to tags, so you have to specify a tag name or tag placeholder. You can use tag names that you have already added to the tag database or you can use a new tag name. The tag database is used to define the data that we want RSView32 to monitor. With visibility animation, an object becomes visible or invisible based on a tag value or the result of a logical expression. If an object is invisible, no other animation attached to the object is evaluated to prevent unnecessary processing. Fill animation is used to check the level of liquid in an object that changes on the basis of tag's value in relation to its minimum and maximum values. The object's fill level is proportional to the value of the expression. Numeric Display tool is used to create fields that operators can use to display the current value of an analog or digital tag or a value based on an expression. The width animation is used to change the width of an object based on a tag value or the result of an expression. Label tool is used to create fields that display different types of tag information at runtime. With vertical position animation, an object moves vertically based on the result of an expression in relation to its minimum and maximum values.

REVIEW QUESTIONS

1. What is an animation dialog box? Name its different sections and explain them.
2. Explain the function and configuration of the following with the help of suitable examples:
   (*a*) Button (*b*) Numeric Input (*c*) Numeric Display
   (*d*) String Input (*e*) String Display (*f*) Label
   (*g*) Arrow
3. Explain the working and configuration of the following animations with the help of examples:
   (*a*) Color animation (*b*) Visibility animation
   (*c*) Fill animation (*d*) Vertical slider animation
   (*e*) Horizontal slider animation (*f*) Vertical position animation
   (*g*) Horizontal position animation (*h*) Width animation
   (*i*) Height animation (*j*) Touch animation

11

ALARMING AND DATA LOGGING

LEARNING OBJECTIVES

After reading this chapter, you will be able to explain :

1. Use of Alarm, Alarm configurations
2. Procedure for Alarm set up
3. Alarm start up and operation
4. Concept of Alarm summary and Data logging

11.1 ALARMING

Alarms are important part of most plant control applications so as to identify any accident in plant operation. It is often equally important to have a record of the alarm and whether the alarm was acknowledged. The following type of alarms can be configured:

1. Analog Alarms
2. Digital Alarms

The alarm operation basically consists of three steps:

(*i*) Alarm Configuration
(*ii*) Alarm setup
(*iii*) Alarm startup

11.2 ALARM CONFIGURATION

11.2.1 Digital Alarm

Digital tags are either on or off. Therefore, alarms for digital tags do not have thresholds. They have alarm states which are listed below:

(*a*) **On:** Triggers an alarm when a tag has a value of 1.
(*b*) **Off:** Triggers an alarm when a tag has a value of 0.
(*c*) **Any Change:** Triggers an alarm when a tag value changes from 0 to 1 or from 1 to 0.
(*d*) **Changes to On:** Triggers an alarm only when a tag value changes from 0 to 1.
(*e*) **Changes to Off:** Triggers an alarm only when a tag value changes from 1 to 0.

To configure a digital alarm, open the tag database editor. Select the digital tag and check alarm box for that digital tag as shown in Fig. 11.1.

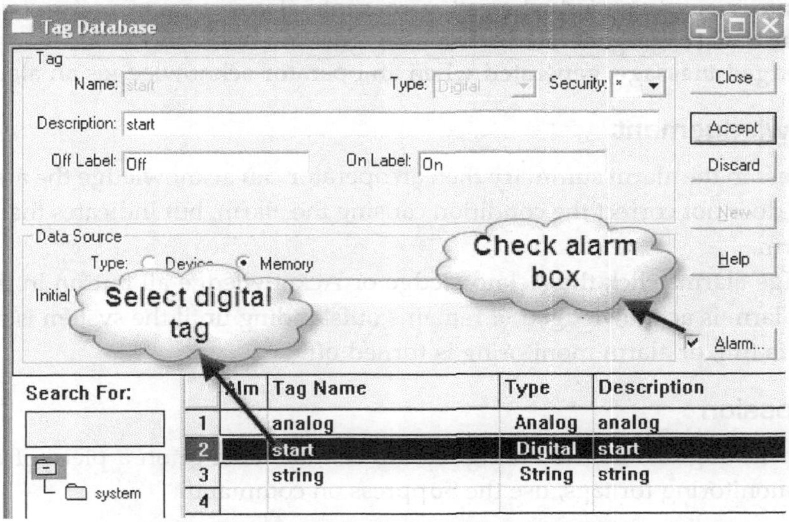

Fig. 11.1 Digital tag selection for alarm

When you check the alarm box then a new window will open up for digital alarm configuration as shown in Fig. 11.2.

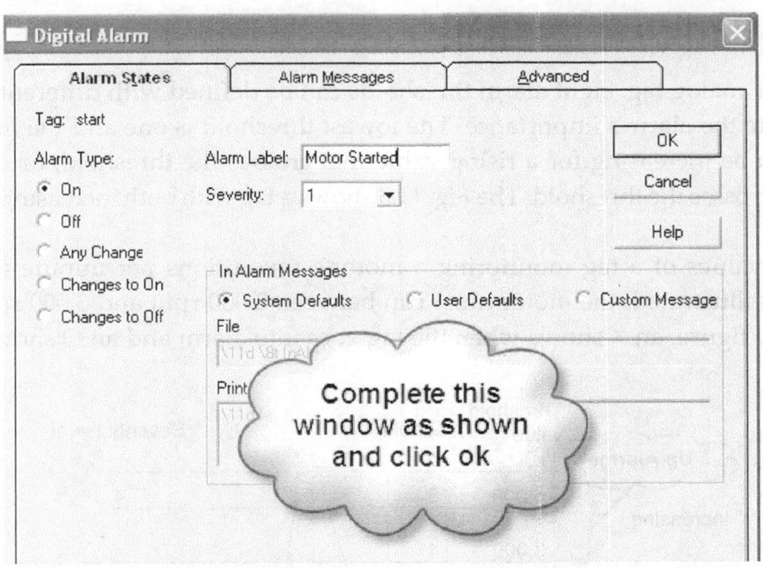

Fig. 11.2 Digital alarm configuration

Alarm Severity

Alarms can range in severity from one to eight to indicate different levels of importance. One is most severe, eight is least severe. For example, a level-four alarm might warn that a vat is half full of liquid while a level-one alarm indicates that the vat is about to overflow. Both alarms monitor the same tag but have different severity levels.

Alarm Messages

Alarm messages report information about alarms. There are three types of messages:

(*a*) **In Alarm message:** generated when a tag goes into alarm.
(*b*) **Out of Alarm message:** generated when a tag goes out of alarm.
(*c*) **Acknowledged message:** generated when an operator acknowledges an alarm.

Alarm Acknowledgement

If an alarm appears in the alarm summary then an operator can acknowledge the alarm. Acknowledging an alarm does not correct the condition causing the alarm, but indicates that an operator is aware of the alarm.

To acknowledge alarms click the Acknowledge or Acknowledge all button in the alarm summary unless an alarm is acknowledged, it remains outstanding until the system is shut down, the alarm summary is full, or alarm monitoring is turned off.

Alarm Suppression

This is useful for testing or for performing repairs or maintenance on a piece of equipment. To suppress alarm monitoring for tags, use the Suppress on command.

11.2.2 Analog Alarm

An analog tag can trigger a number of alarms when it crosses various predefined threshold levels whereas a digital tag can be either on or off.

Thresholds

When defining an analog tag, eight alarm thresholds can be defined with different levels of alarm severity to indicate the alarm's importance. The lowest threshold is one and the highest is eight.

Thresholds can be increasing for a rising value that crosses the threshold, or decreasing for a falling value that crosses the threshold. The Fig. 11.3 shows a tag with both increasing and decreasing thresholds.

The changing values of a tag monitoring a motor's revolutions per minute (rpm). With the threshold settings illustrated, the motor must run between 2,000 rpm and 3,000 rpm, or an alarm is triggered. In the figure, an X shows when the tag goes into alarm and an O shows when the tag goes out of alarm.

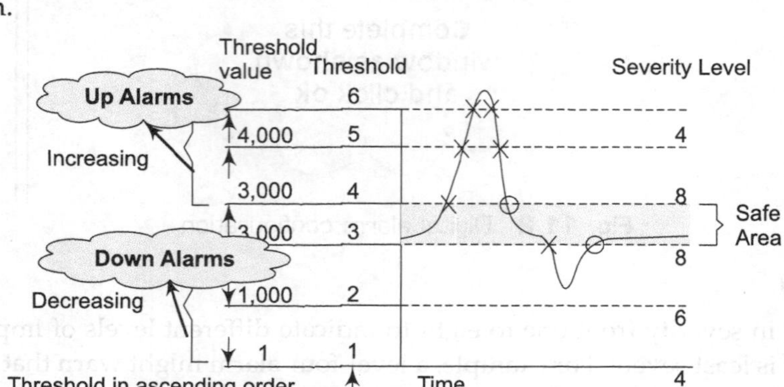

Fig. 11.3 Analog alarming example

To configure an analog alarm, open the tag database editor again. Select the analog tag and check alarm box for that tag as shown in Fig. 11.4.

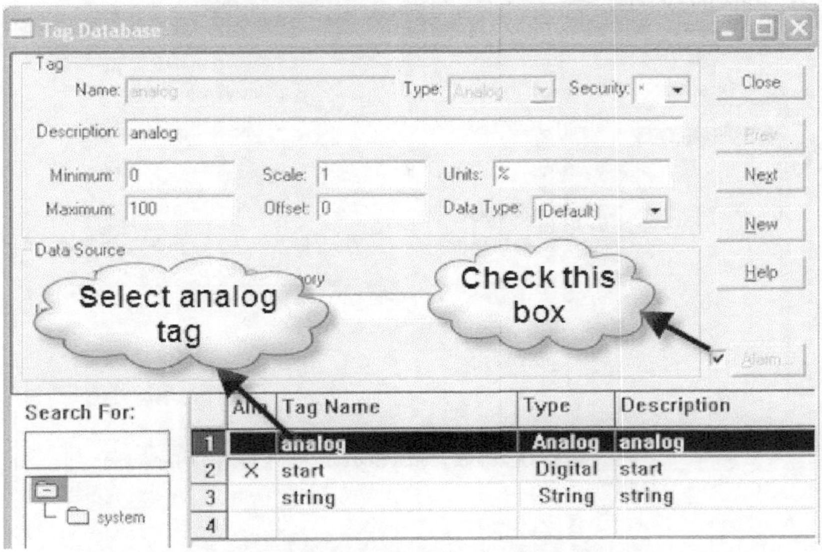

Fig. 11.4 Analog tag configuration

When you check the alarm box then a new window will appear for analog alarm configuration as shown in Fig. 11.5. Configure four thresholds for this tag as shown in Fig. 11.5 to Fig. 11.8.

Threshold 1: Configure it as shown in Fig. 11.5.

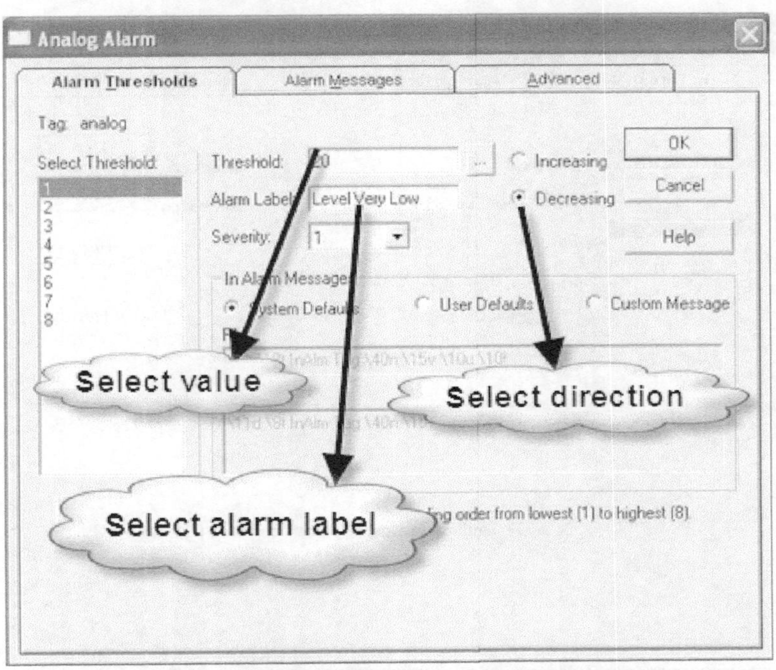

Fig. 11.5 Threshold 1 for analog tag

Threshold 2: Configure it as shown in Fig. 11.6.

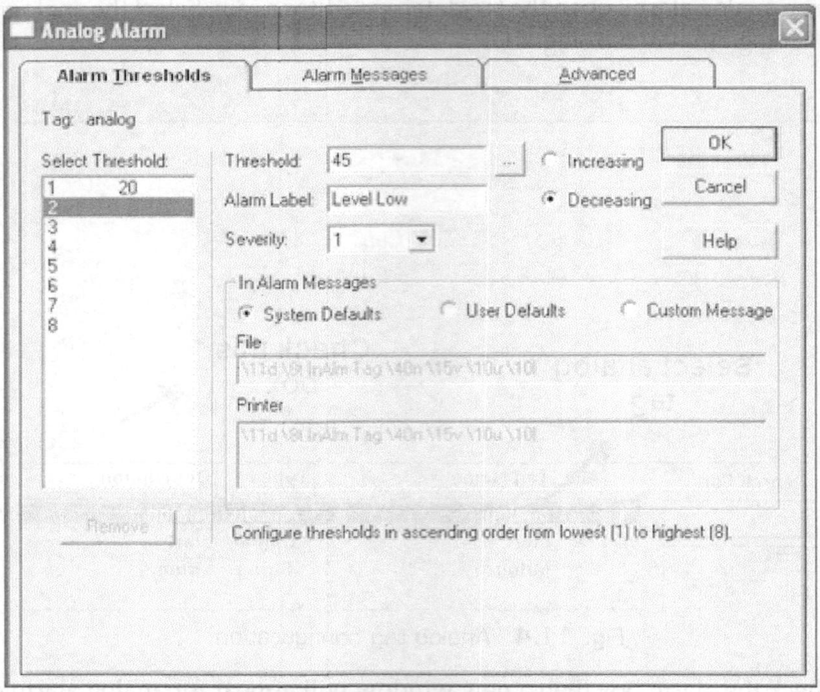

Fig. 11.6 Threshold 2 for analog tag

Threshold 3: Configure it as shown in Fig. 11.7.

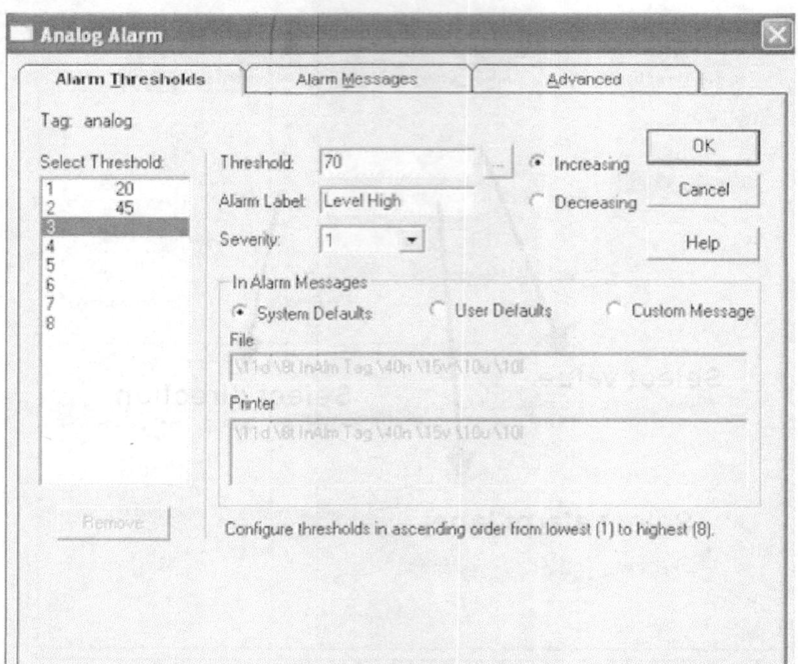

Fig. 11.7 Threshold 3 for analog tag

Threshold 4: Configure it as shown in Fig. 11.8.

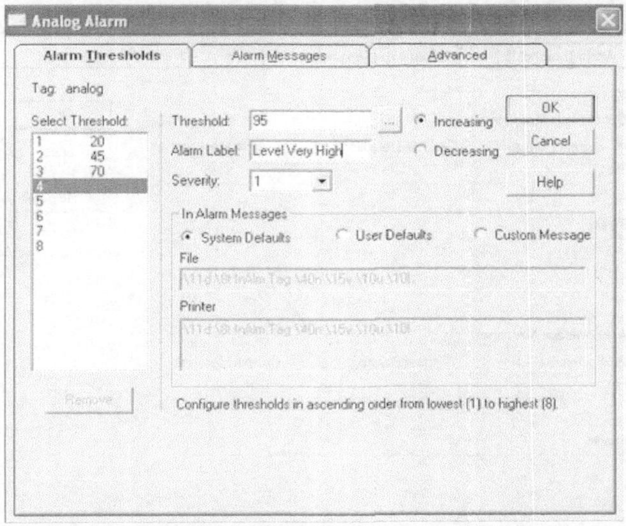

Fig. 11.8 Threshold 4 for analog tag

11.2.2(a) Alarm Threshold Tab

Fill in the fields as follows:

Threshold

To define a constant threshold, type a number. To define a variable threshold, specify a tag name.

Increasing/Decreasing

To monitor for a rising value that passes the threshold, click Increasing. To monitor for a falling value that passes the threshold, click Decreasing.

Alarm Label

Type a description of the alarm up to 21 characters long. This description is displayed in the alarm summary and alarm banner.

Severity

Select a severity level. Severity 1 is most severe, Severity 8 is least severe. Severities are configured in the Alarm Setup editor.

In Alarm Messages

Select the message you want when the tag goes into alarm. If you click Custom Message, type the message in the File and/or Printer fields. To configure additional thresholds, select another threshold and fill in each field. When you select another threshold, the information you configured for the previous threshold is saved.

11.2.2(b) Alarm Messages Tab

Alarm messages tab is shown in Fig. 11.9. Select the file and printer messages for when a tag goes out of alarm and when an operator acknowledges an alarm.

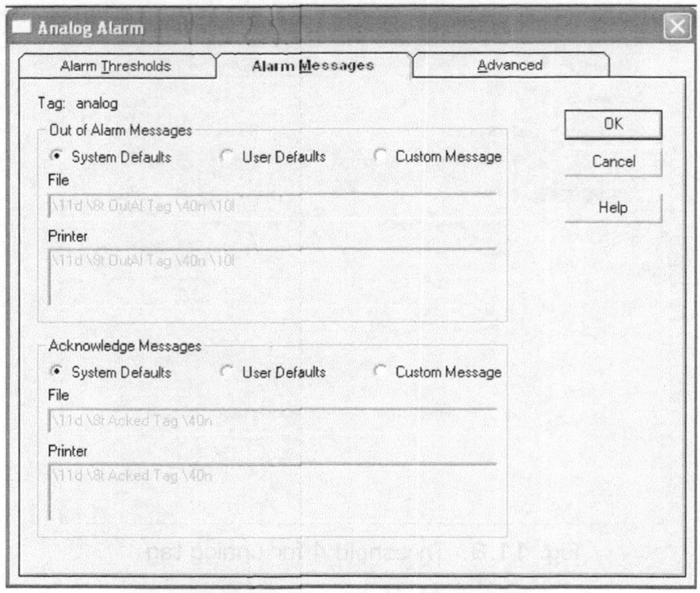

Fig. 11.9 Alarm Message tab

11.2.2(c) Advanced Tab

Click the Advanced tab as shown in Fig. 11.10

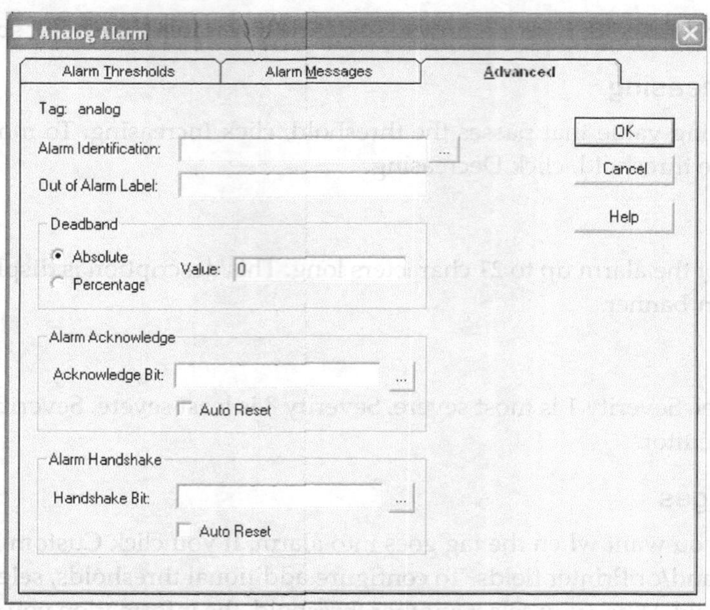

Fig. 11.10 Advanced Tab

Fill in the fields as follows:

Alarm Identification

Type a command or macro file name, or click the Selection button to open the Command Wizard.

Out of Alarm Label

Type a message up to 21 characters long that will be displayed in the alarm summary when the tag is no longer in alarm.

Dead Band

Click Absolute or Percentage and then type a value. The dead band is a buffer zone on the edge of an alarm threshold.

Alarm Acknowledge

In the Acknowledge bit field, type the name of the tag that refers to the acknowledge bit's address. When the operator acknowledges the alarm, the acknowledge bit is set to 1 and

When another RSView32 station detects the acknowledge bit changing from 0 to 1, the RSView32 station acknowledges all tags associated with this acknowledge bit and RSView32 logs a remote acknowledgment to the alarm log file. The bit is set only once per acknowledgment. To have the Acknowledge bit automatically reset when the tag goes back into alarm, check the Auto Reset box.

Alarm Handshake

In the Handshake Bit field, type the name of the tag that refers to the handshake bit's address. When a tag goes into alarm, the handshake bit is set to 1. To have the handshake bit automatically reset when the tag goes out of alarm, check the Auto Reset box. Otherwise the programmable controller must reset the bit.

Once analog tag configuration is complete, Click OK. Then you will notice a cross mark against both digital and analog tags as shown in Fig. 11.11.

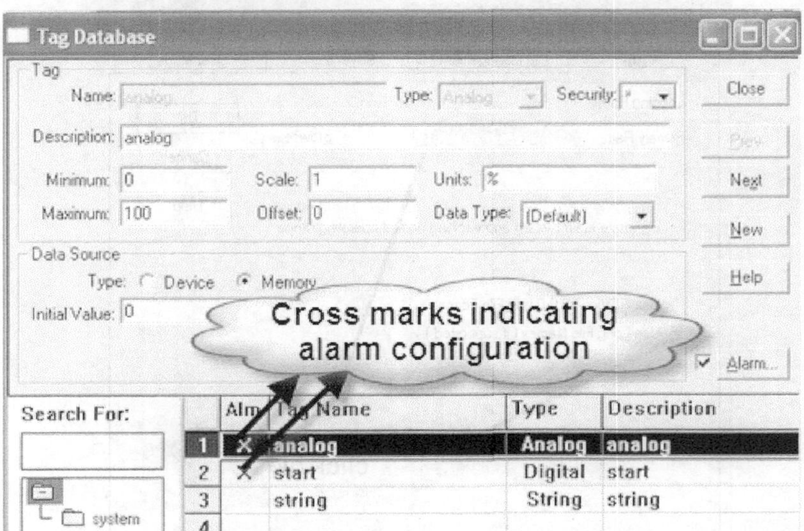

Fig. 11.11 Alarm indication

11.3 ALARM SETUP

To open the Alarm Setup Editor in the Project Manager, open the Alarms folder as shown in Fig. 11.12.

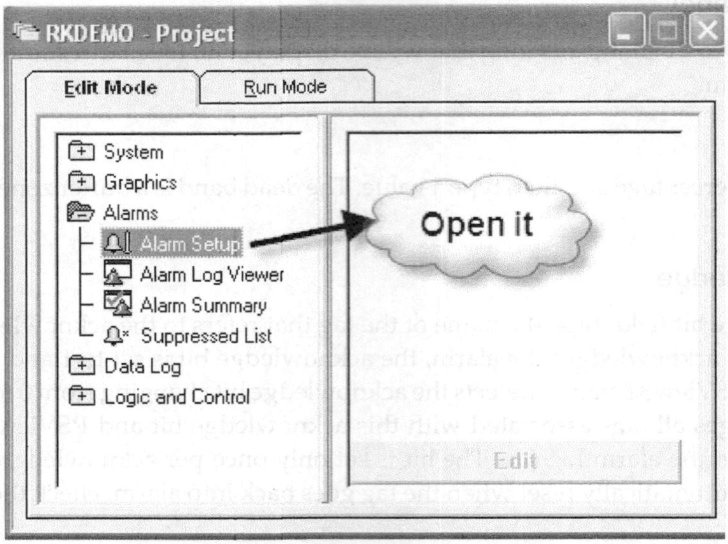

Fig. 11.12 Open alarm setup

Now open the alarm set by double clicking it to view a new alarm setup editor window as shown in Fig. 11.13. Complete this alarm setup and click OK.

11.3.1 Setup Tab

Open the setup tab as shown in Fig. 11.13 and configure it as required.

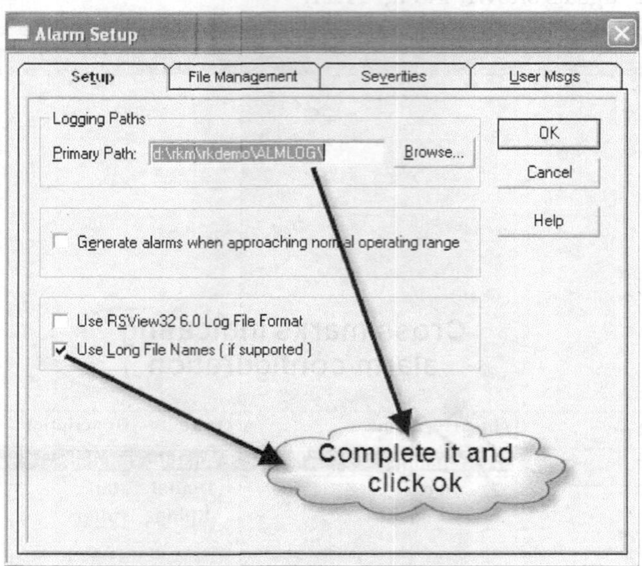

Fig. 11.13 Setup tab

Specify a new path if required. When log files are created, they will be stored in the directory specified here, Click OK. The new log files can be created:

(*a*) periodically
(*b*) at specified times
(*c*) when a particular event occurs
(*d*) never

11.3.2 File Management Tab

In the Alarm Setup editor, click the File Management then a new window will appear as shown in Fig. 11.14.

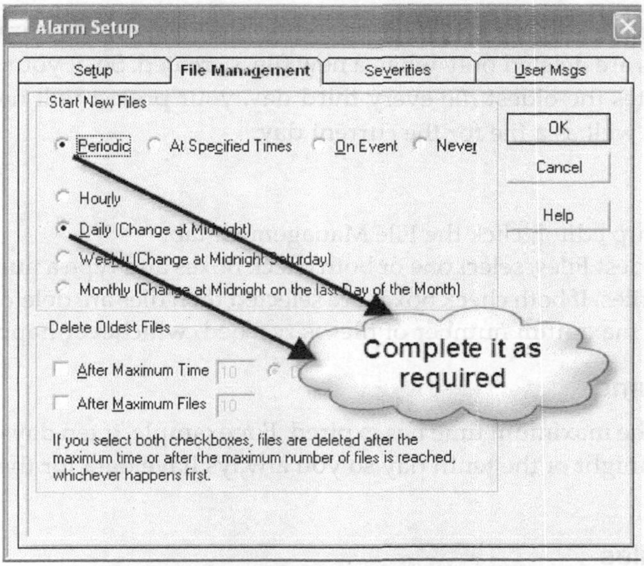

Fig. 11.14 File Management Tab

Under Start New Files, click a button to specify when you want to have new files created.

Periodic

Click Periodic, and then click a time period. A new file is created after the specified interval. For this period the new file is created

(*a*) **Hourly:** Approximately on the hour
(*b*) **Daily:** Each day approximately at midnight
(*c*) **Weekly:** Each Sunday approximately at midnight
(*d*) **Monthly:** On the first day of each month approximately at midnight

Specific Times

Click at Specified Times, and then type a time or list of times when you want data to be logged to a new file. Start a new line for each new time and allow at least five minutes between start times. Use any of the following to specify time.

Time Format

(*a*) **Day:** Sun, Mon, Tue, Wed, Thu, Fri, or Sat
(*b*) **Date:** 1 – 31 Specifying 31 means those months that do not have 31 days will not have new log files.
(*c*) **Hour:** 00: to 23:
(*d*) **Minute:** :00 to :511

On Event

1. Click On Event.
2. In the Expression field, create the expression that will trigger creation of a new file.

Never Creating New Files

Click Never. Logged data is added to a single file.

Deleting Log Files

The log files can be deleted after a specified period or once a specified number of files has been created. Alarm log files are deleted only when a new file is created. So, if your project creates a new file each day and deletes the oldest file every third day, your project will have files for the three previous days' data as well as a file for the current day.

File Deletion

1. In the Alarm Setup editor, click the File Management tab.
2. Under Delete Oldest Files, select one or both check boxes and type a number to specify when to delete the log files. If both check boxes are selected then files are deleted after the maximum time or after the maximum number of files is reached, whichever happens first.

After Maximum Time

Files are deleted after the maximum time has expired. For example, if ten days is the specified time, files are deleted at midnight of the tenth day so you always have data for the current day and the nine previous days.

After Maximum Files

The oldest log file is deleted after the specified maximum has been reached. The files currently being logged to are not included in this number. For example, if you specify 20, you will have a maximum of 21 alarm log files at any time—20 old ones and the current one. When a new set is started, the oldest file is deleted.

11.3.3 Severities Tab

To configure alarm severity in the Alarm Setup editor, click the Severities tab as shown in Fig. 11.15.

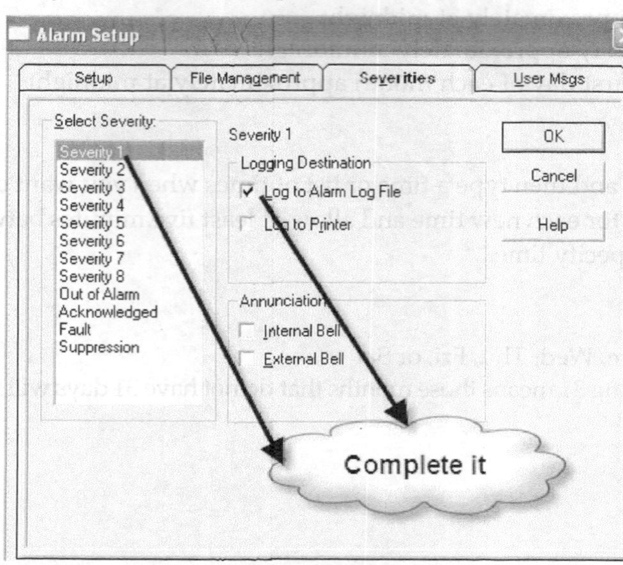

Fig. 11.15 Severities Tab

1. Select an alarm severity or an incident.
2. Specify the logging destination.
3. For an alarm severity, specify how you want the alarm to be annunciated.

Internal Bell

The internal bell is a sound produced by the computer running RSView32. To use the internal bell, select the Internal Bell check box. The sound for the internal bell can come from your computer's speaker or from a sound card. To set up the sound, open the Windows Control Panel and then double-click the Sounds icon to open the Sounds Properties dialog box. In the Events field, click Exclamation, and then assign a. wav file to the Exclamation event. When an alarm occurs, the assigned sound is repeated once per second. When the alarm is acknowledged or silenced, the sound is stopped. Because the sound is played repeatedly, assign only short sounds to the Exclamation event.

External Bell

The external bell is a sound produced by a horn or some other audio device. To set up the external bell, select the check box and specify the name of a tag that is linked to an audio device.

The specified tag must refer to a bit in the programmable controller. When an alarm occurs, this bit is set and the logic in the programmable controller turns on the audio device. When the alarm is acknowledged or silenced, the bit in the programmable controller is reset and the logic in the programmable controller turns off the audio device.

11.3.4 User Messages Tab

To configure alarm user messages in the Alarm Setup editor, click the user messages tab as shown in Fig. 11.16. Alarm messages report information about alarms. There are several types of messages which are:

(*a*) **In alarm message:** Generated when an analog tag's alarm threshold has been crossed or when a digital tag has changed state.

(*b*) **Out of alarm message:** Generated when a tag is no longer in alarm.

(*c*) **Acknowledged message:** Generated when an alarm has been acknowledged.

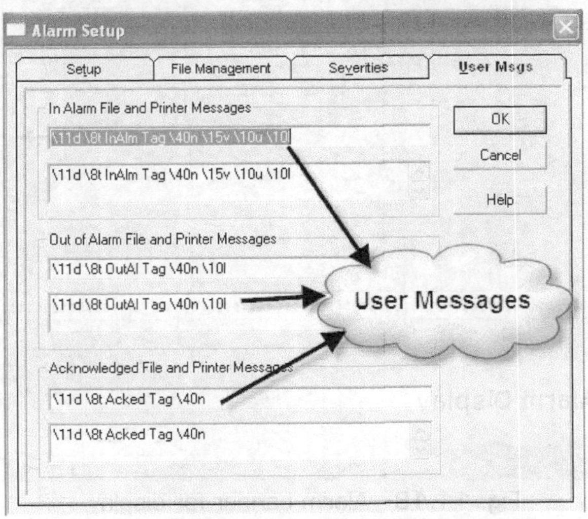

Fig. 11.16 User messages tab

Once alarm setup configuration is complete click OK.

11.4 ALARM STARTUP AND DISPLAY

11.4.1 Alarm Display

To display an alarm open the alarm information library as shown in Fig. 11.17.

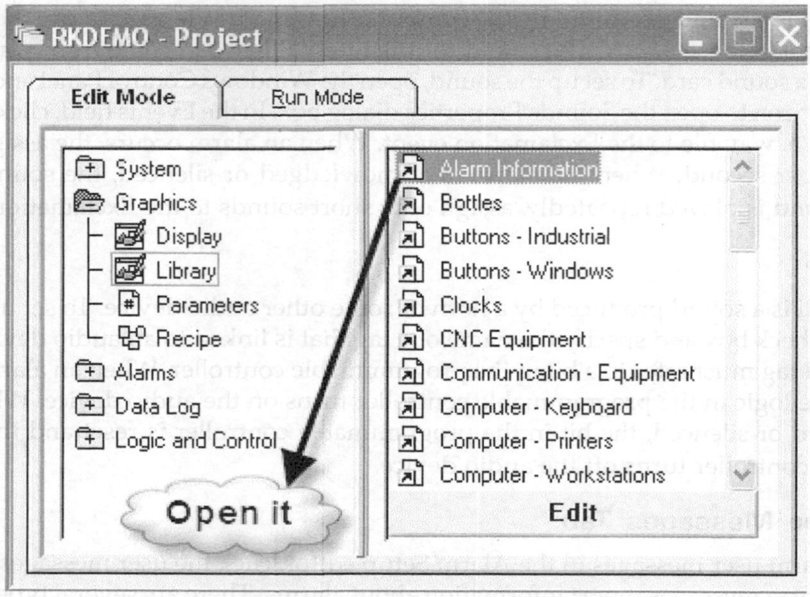

Fig. 11.17 Alarm information library

Copy alarm banner form this library and paste it on the display as shown in Fig. 11.18.

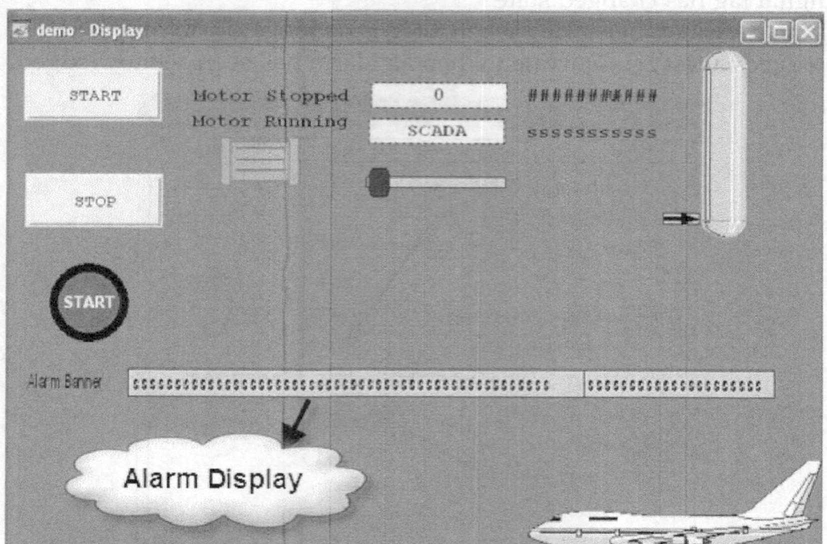

Fig. 11.18 Alarm banner for display

11.4.2 Alarm Startup

In project manager window, select system folder and open the startup as shown in Fig. 11.19.

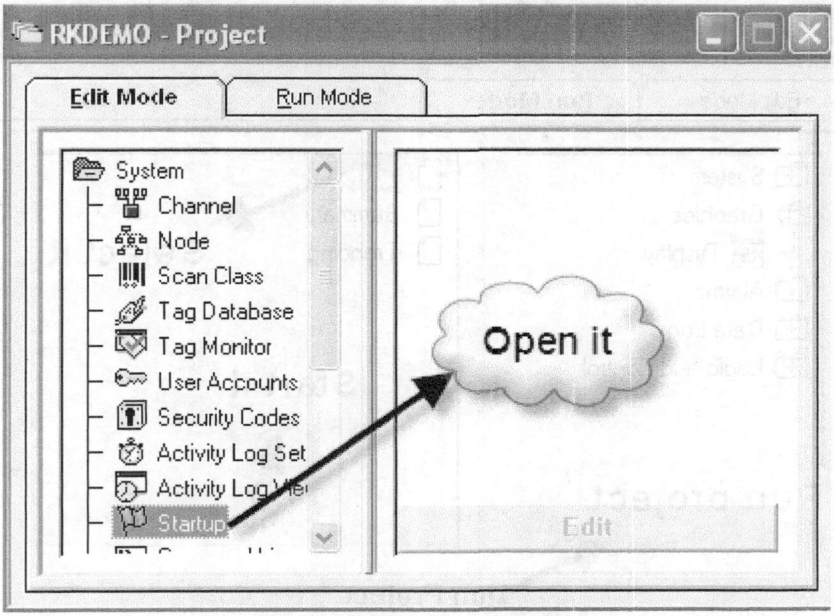

Fig. 11.19 Open startup

In startup window select "Alarming" as shown in Fig. 11.20 and then click OK.

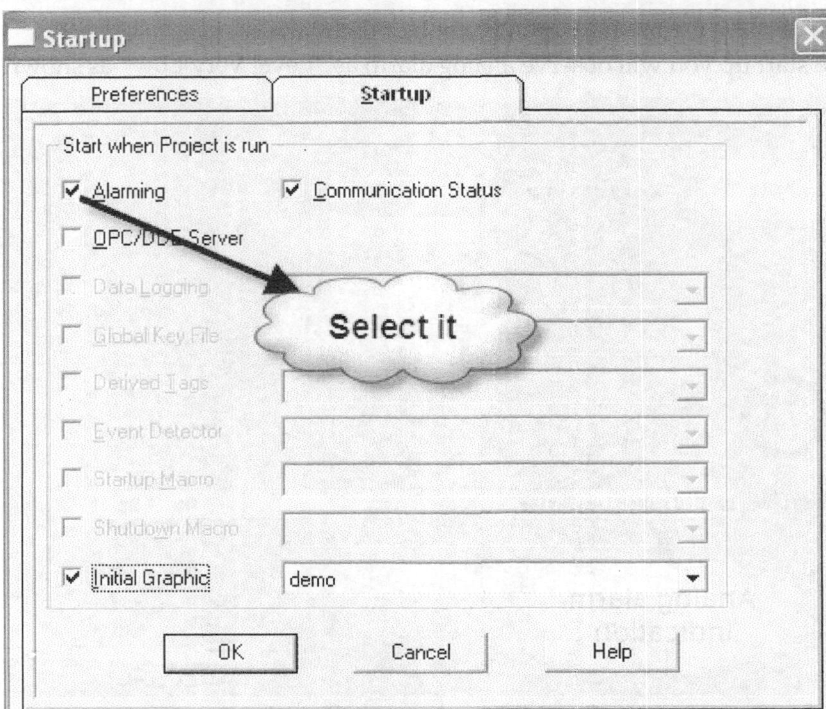

Fig. 11.20 Startup alarming

Now save the display and run it. Start it as shown in Fig. 11.21.

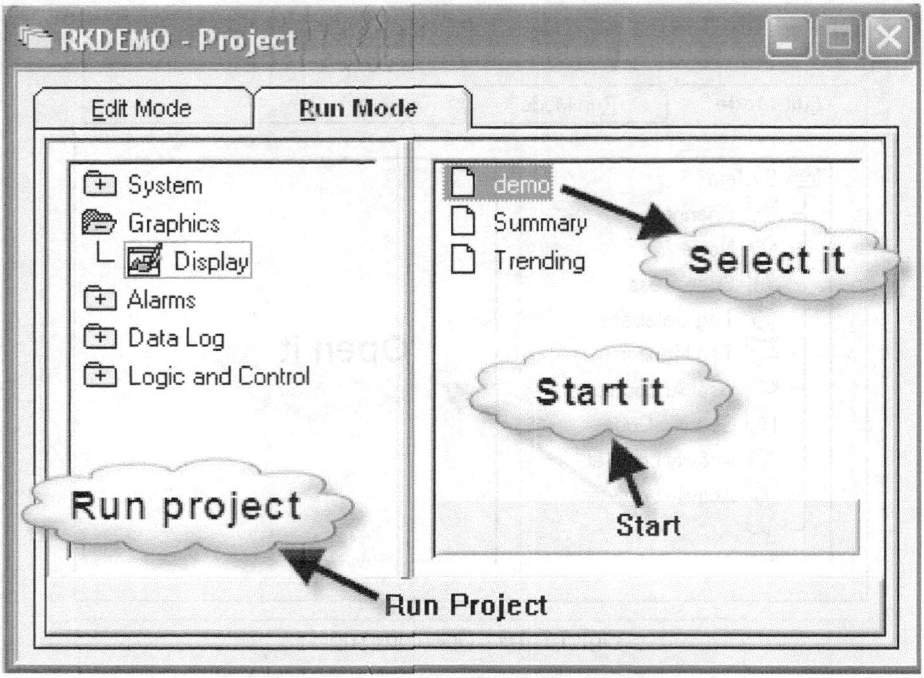

Fig. 11.21 Running your Project

Then you will observe different cases and some of them are:

Case 1: At the start up you will observe analog alarm as "Level Very Low" as shown in Fig. 11.22.

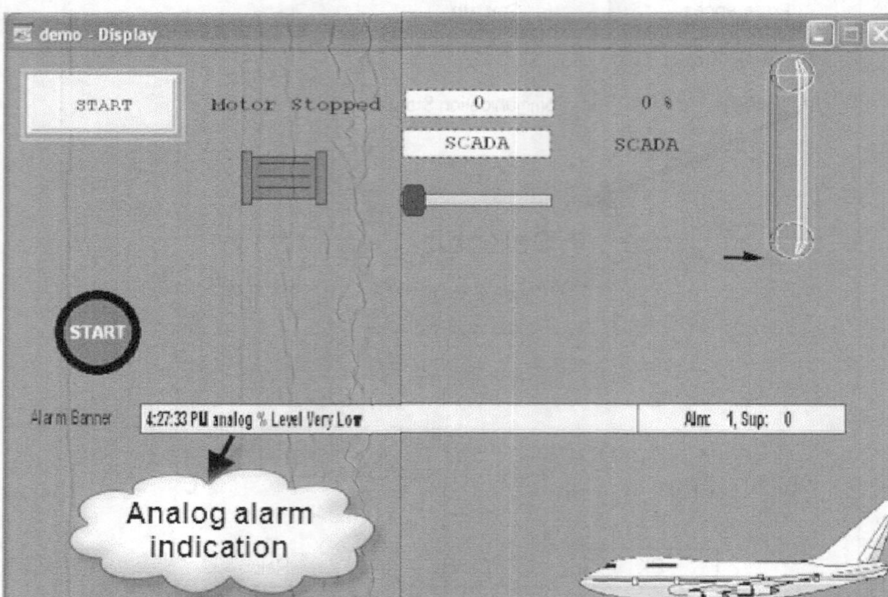

Fig. 11.22 Alarming Case 1

Case 2: When you press start button then you will observe digital alarm as "Motor Started" as shown in Fig. 11.23.

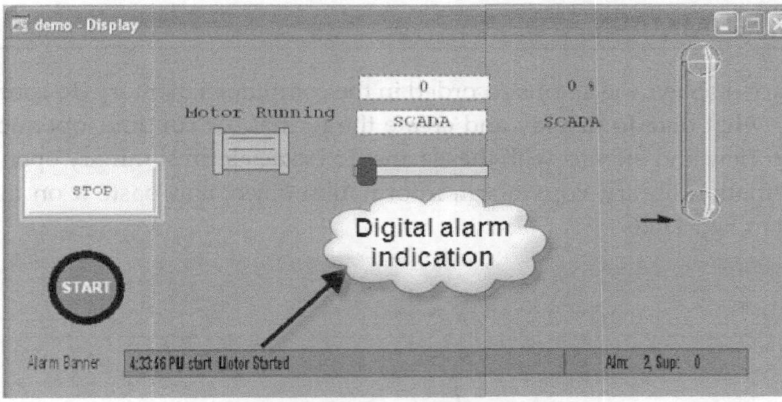

Fig. 11.23 Alarming Case 2

Case 3: When you drag the arrow upwards then you will observe analog alarm as "Level High" when it crosses level high threshold as shown in Fig. 11.24.

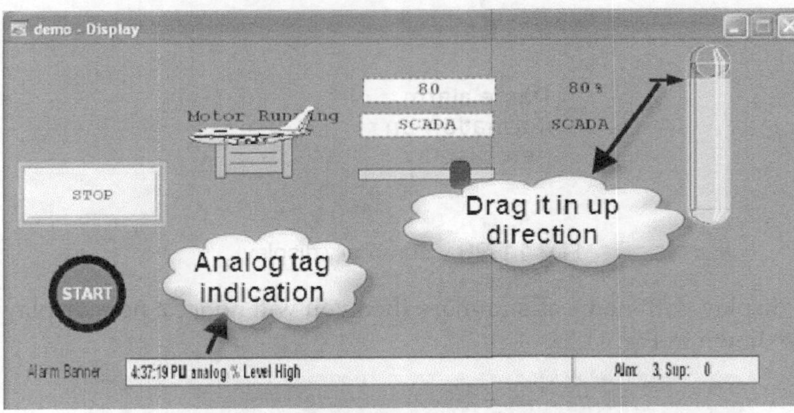

Fig. 11.24 Alarming Case 3

Case 4: When you drag the arrow downwards then you will observe analog alarm as "Level Low" when it goes below low level threshold as shown in Fig. 11.25.

Fig. 11.25 Alarming Case 4

11.5 ALARM SUMMARY

The alarm summary displays the alarms recorded in the computer's memory. To configure an alarm summary, specify what data to display and create the format. At runtime, operators can use the alarm summary to view and interact with the alarms. To create alarm summary open a new display. Open alarm information library, copy alarm information object and paste it on the new created display as shown in Fig. 11.26.

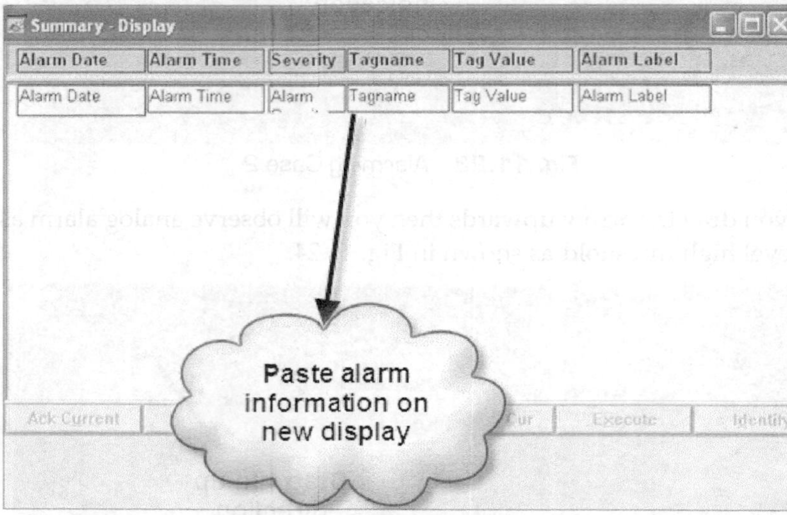

Fig. 11.26 Summary display

Save this new display with name as summary then you will notice a new display on right pane of the window as shown in Fig. 11.27.

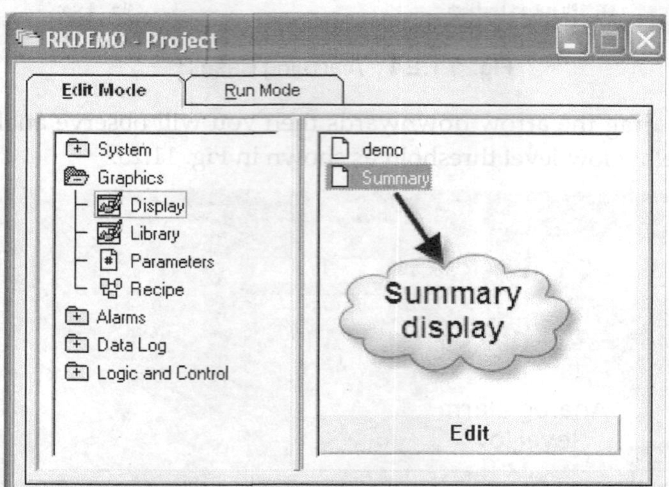

Fig. 11.27 Window with Summary display

To open up this summary display while running demo display just create one button with up appearance as "Summary". In action tab for this button enter "Display Summary" in command field as shown in Fig. 11.28.

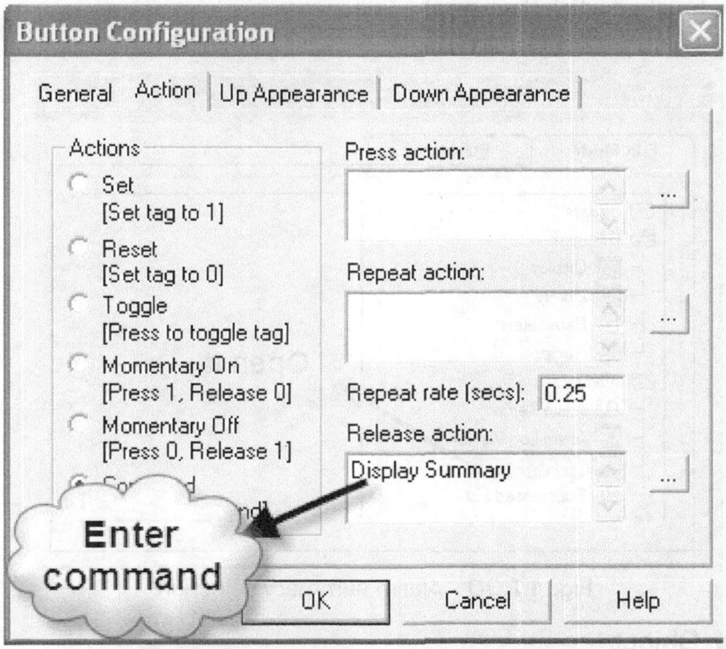

Fig. 11.28 Summary button configuration

Now save the demo display and run it. Do some changes in digital and analog tags of demo display then click the new created summary button to open the summary display in new window as shown in Fig. 11.29.

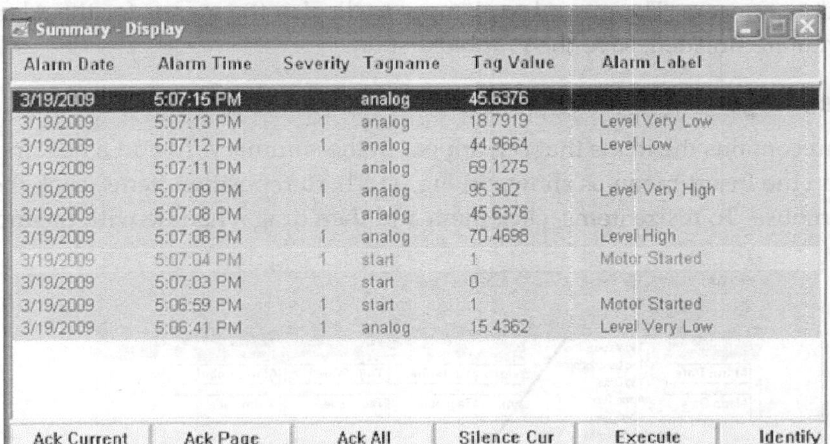

Fig. 11.29 Summary display

Project Manager for Alarm Summary

Alarm summary can also be configured by selecting alarm summary in project manager window as shown in Fig. 11.30.

1. In the Project Manager, open the Alarms folder.
2. Open the Alarm Summary editor.
3. Configure the appearance of the alarm summary.

4. When you are finished, click Save on the File menu.
5. Type a file name.

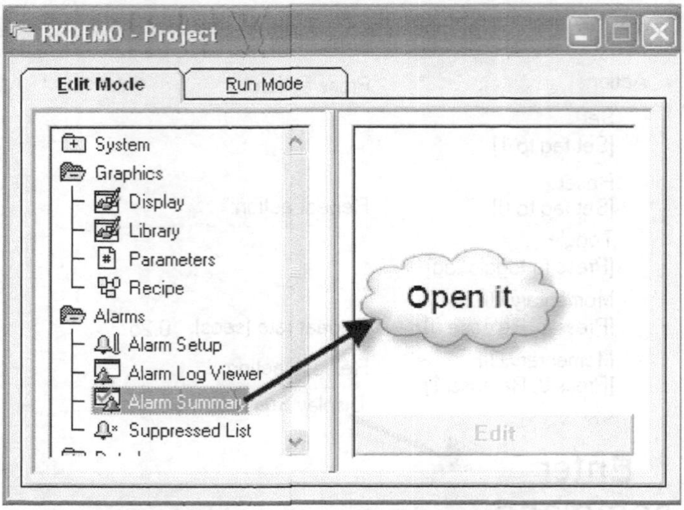

Fig. 11.30 Alarm summary selection

Alarm Summary Object

An alarm summary object can be part of a graphic display or it can be the entire graphic display. To use the pre-built alarm summary, called Alarm Information, in the graphics objects library. It can also be created by using the alarm summary icon. Drag the mouse to create a box approximately the size you want for the alarm summary. When you release the mouse, an empty alarm summary window appears. The menu bar changes to display the menus for the Alarm Summary editor. When you are finished, save the graphic display.

Inserting Headings

The Insert menu contains the items that can appear in the summary. To add a heading for an item, click the item on the Insert menu as shown in Fig. 11.31. To reposition items, click them and drag them with the mouse. To resize items, click them and then drag a handle with the mouse.

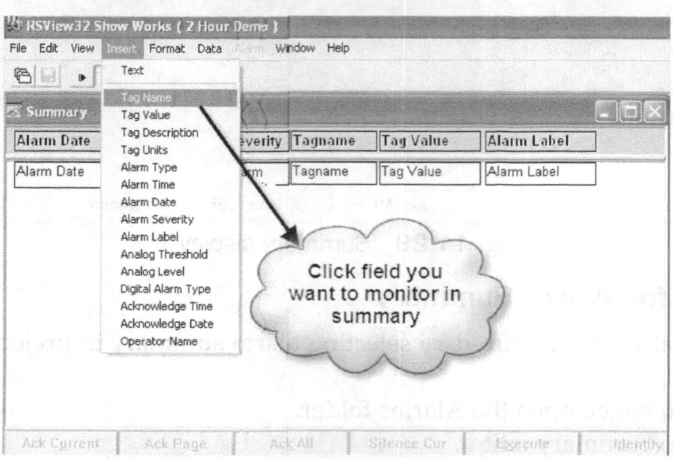

Fig. 11.31 Field selection for alarm summary

Suppressing Alarm Monitoring

This is useful for testing or performing repairs or maintenance on a piece of equipment. To suppress alarm monitoring for tags, use the SuppressOn command. To suppress all alarms for the specified tags, issue the SuppressOn command before the AlarmOn command. To view a list of the tags not being monitored, use the Suppressed List. To suppress any tag use SuppressOn <*tag name*> command in the command line. To view the suppressed tags open the suppressed list in project manager as shown in Fig. 11.32.

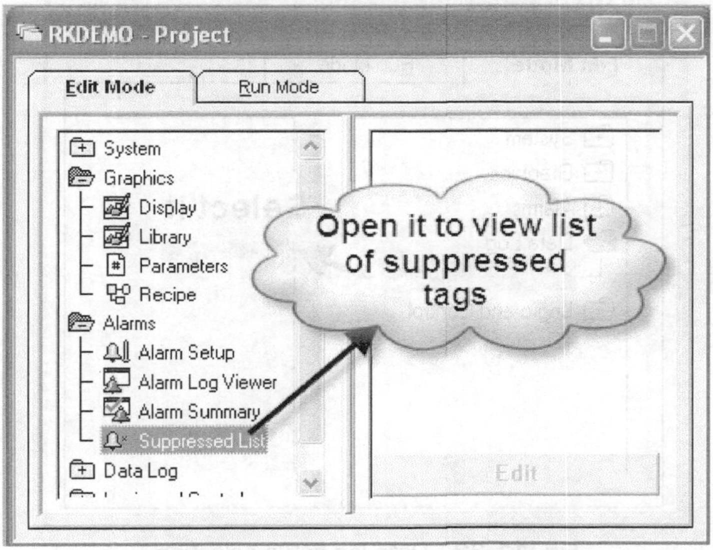

Fig. 11.32 Suppressed tag monitoring

11.6 DATA LOGGING

The logged data is stored in either dBASE IV (.dbf) format or using the ODBC connectivity standard. Data logged in the .dbf format can be stored in narrow or wide format.

Narrow .dbf File Format

The narrow .dbf format stores one date, one time, and one tag value per line. This format has the following features:

(*a*) Limit of 10,000 tags in the model.
(*b*) Most useful for storing data that changes infrequently.
(*c*) Data is harder to use for sums and averages because it crosses rows.
(*d*) Does not create a new log file set if tags are added to or deleted from the model.
(*e*) Stores numeric and string values in separate files.
(*f*) Preferred for queries because you can join tables.

Wide .dbf File Format

The wide .dbf format stores one date, one time, and multiple tag values per line. This format has the following features

(*a*) Limit of 100 tags in the model.
(*b*) Most useful for storing frequently changing data.

(*c*) Particularly useful for storing data from batch applications.
(*d*) Data is more easily used for sums and averages because it is all in one column.
(*e*) Creates a new log file set if tags are added to or deleted from the model.
(*f*) Stores numeric and string values in the same file.
(*g*) Can't join tables for queries.

In project manager window, select datalog folder and open data log setup as shown in Fig. 11.33.

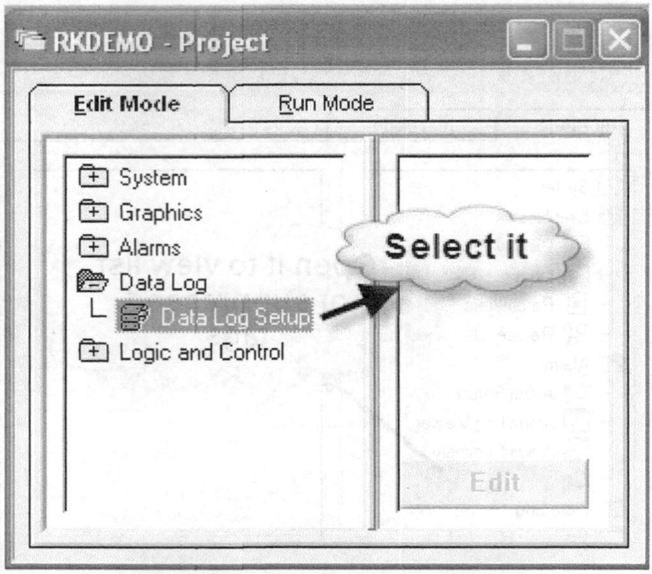

Fig. 11.33 Data log setup selection

A new data log editor window will appear. In setup tab, Fill in the information as shown in Fig. 11.34.

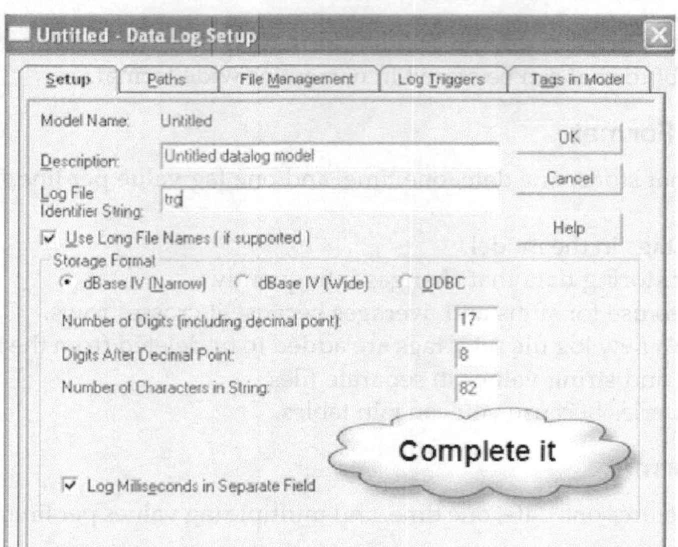

Fig. 11.34 Setup tag configuration

11.6.1 Setup Tag Configuration

1. In the Data Log Setup editor, click the Setup tab.
2. Type a description of the model. This is for your information only.
3. Specify whether to use long file names for the data log files. If you choose to use long file names, type a log file identifier string, up to 20 characters. This string of text forms part of the file name for the data log files.
4. Click a file storage format.
   (*a*) Narrow dBASE IV format
   (*b*) Wide dBASE IV format
   (*c*) ODBC format

Number of Digits (Including Decimal Point)

Type the number of digits, including the decimal point, to be stored for floating point numbers.

Digits after Decimal Point

Type the number of digits after the decimal points that are to be stored for floating point numbers.

Number of Characters in String

Type the number of characters to be logged for all string tags in the model.

Log Milliseconds in Separate Field

Specify whether to log the millisecond portion of the time in a separate field in the .dbf file.

5. If the ODBC storage format is selected then specify the ODBC data source. Type the path and file name of an existing ODBC data source.

11.6.2 Path Tab Configuration

1. In the Data Log Setup editor, click the Paths tab as shown in Fig. 11.35.

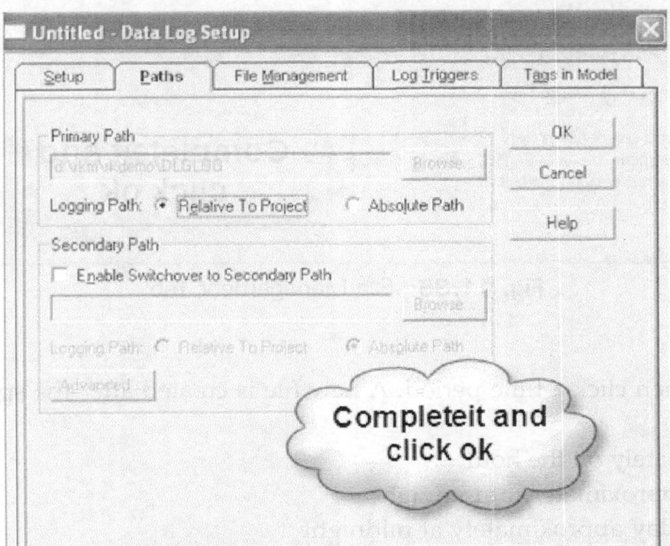

Fig. 11.35 Path tab configuration

2. For the dBASE IV storage format, specify the primary path where you want to store the .dbf data log files.

Relative To Project

Click this to store the log files in the project directory. RSView32 creates a directory called Dlglog and a subdirectory with the same name as the model name and stores the files there.

Absolute Path

Click this to specify a particular path. If the path you type does not exist, RSView32 will create it. The first time data logging runs, RSView32 creates a subdirectory with the same name as the model name and stores the files there.

11.6.3 File Management Tab Configuration

1. In the Data Log Setup editor, click the File Management tab as shown in Fig. 11.36.
2. Under Start New Files, click a button to specify when you want to have new files created.

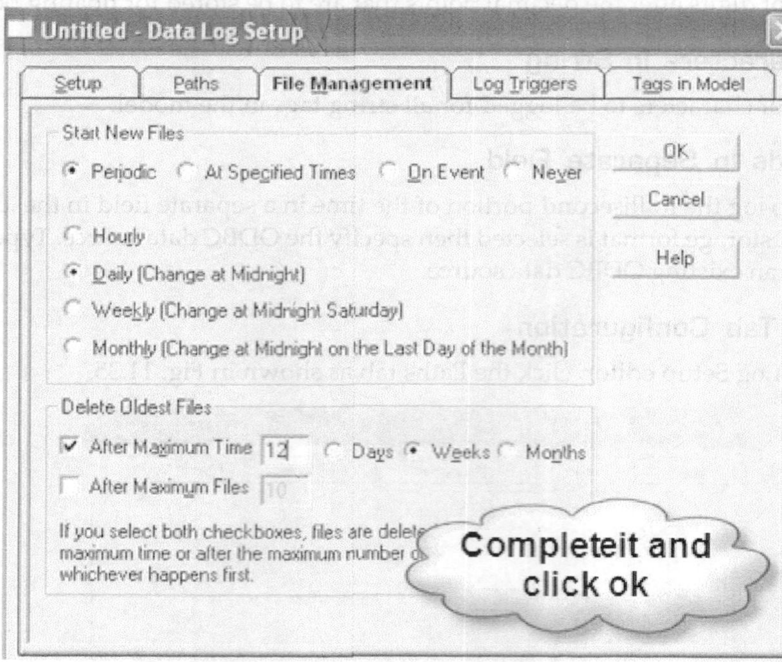

Fig. 11.36 File Management Tab

Periodic

Click Periodic, and then click a time period. A new file is created after the specified interval has elapsed.

Hourly: Approximately on the hour
Daily: Each day approximately at midnight
Weekly: Each Sunday approximately at midnight
Monthly: On the first day of each month approximately at midnight

Specific Times

Click at Specified Times, and then type a time or list of times when you want data to be logged to a new file. Start a new line for each new time and allow at least five minutes between start times.

On Event

1. Click On Event.
2. In the Expression field, create the expression that will trigger creation of a new file.

Never

Click Never. Logged data is added to a single file.

To manage disk space if a log file grows too large, stop data logging, move or delete the data log file, and then restart data logging. You cannot delete a log file while data is being logged to it.

Deleting ODBC Database Records and .dbf Log Files

To specify when to delete .dbf files or ODBC database data:

1. In the Data Log Setup editor, click the File Management tab.
2. For ODBC storage format, under Purge Oldest Records in ODBC Database, select the After Maximum Time check box. Type a number to specify when to delete the records, and select whether the number is measured in days, weeks, or months.

11.6.4 Log Trigger Setup

To specify what should trigger logging:

1. In the Data Log Setup editor, click the Log Triggers tab as shown in Fig. 11.37.
2. Click a button to specify a log trigger and fill in any additional information.

Logging Periodically

Periodic logging is used to take a snapshot of all tag values at a particular point in time. For periodic logging to dBASE IV files, the dBASE IV (wide) file format is most efficient for storing the data.

1. Click Periodic.
2. In the Interval field, type a time and click a time unit to specify how often tag values will be logged. All tags in the model will be logged each time this interval expires. Do not set the interval faster than the scan class rate, OPC update rate, or DDE poll rate, or you will log redundant data.

Logging on Change

On-change logging is used to log only tags whose values have changed, when the change occurs.

1. Click On Change.
2. In the Change Percentage field, type the per cent the tag value has to change in order to trigger logging. Zero means all changes are logged.
3. In the Heartbeat field, type a time, and then click a time unit to specify how often tag values will be logged even if no change has occurred. If you do not want to use the heartbeat, type 0. The heartbeat ensures that the data in the log file is current. The heartbeat is also a good way to ensure that data logging is working and acquiring valid data.

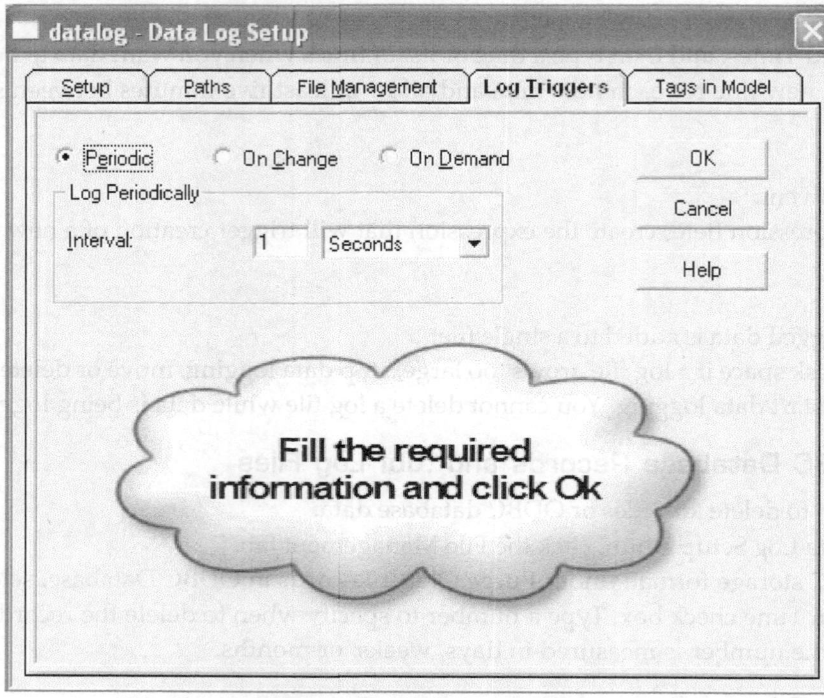

Fig. 11.37 Log trigger setup

Logging on Demand

Logging on demand means that data is logged for a model only when the DataLogSnapshot command is issued. When the DataLogSnapshot command is executed, tag values for all tags in the specified model or models are logged.

1. Click On Demand.
2. Run the model or models for which data will be logged
3. Issue the DataLogSnapshot command.

11.6.5 Tags in Model Tab

1. In the Data Log Setup editor, click the Tags in Model tab as shown in Fig. 11.38
2. Add the required tags, you want to configure for data logging.

Editing the data log model

1. In the Project Manager, open the data log model you want to change.
2. Make the required changes.
3. Save the changes.

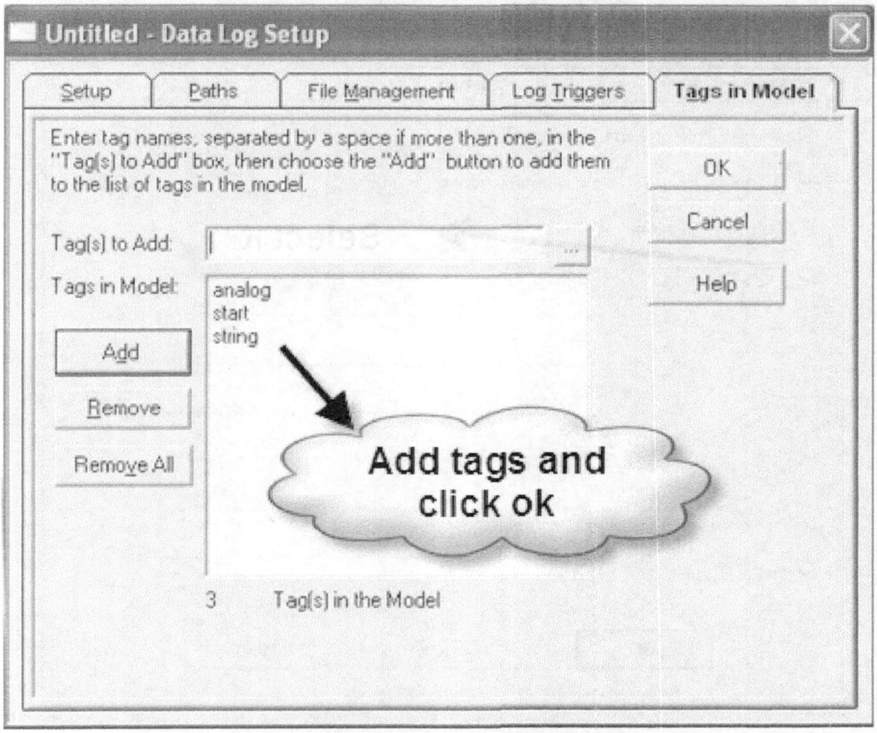

Fig. 11.38 Tag in models Tab

Once the configuration is complete then save it with some name *e.g.* "datalog" as shown in Fig. 11.39 and close it.

Fig. 11.39 Save window

Now go to startup window in display settings and select data loggining as shown in Fig. 11.40.

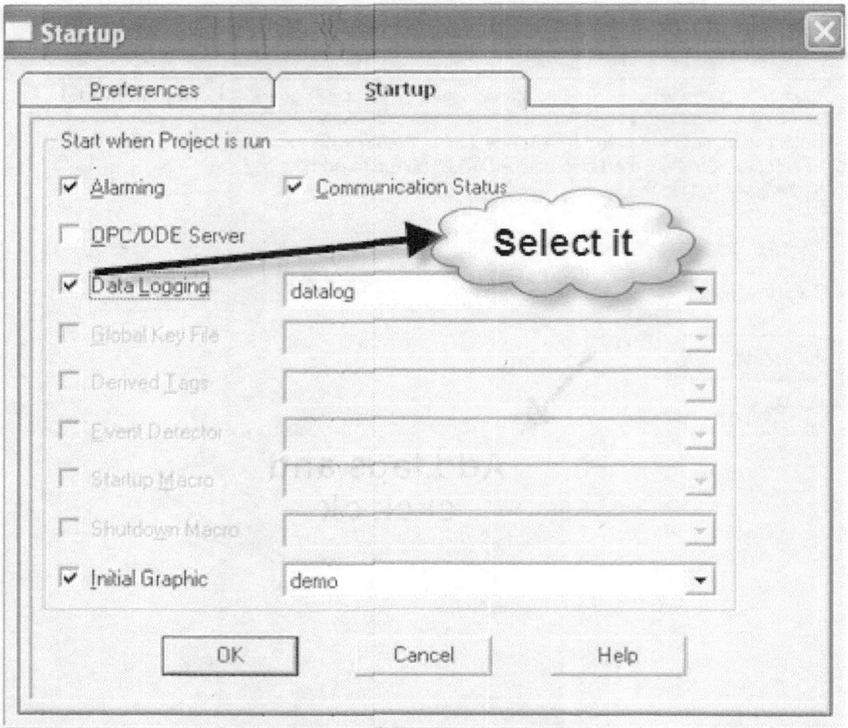

Fig. 11.40 Data Logging in startup

Run the project and change the tag values that you want to view in data logging file. Then open the MS Office Excel and select the exact path where the required file is saved as shown in Figs 11.41 to 11.43.

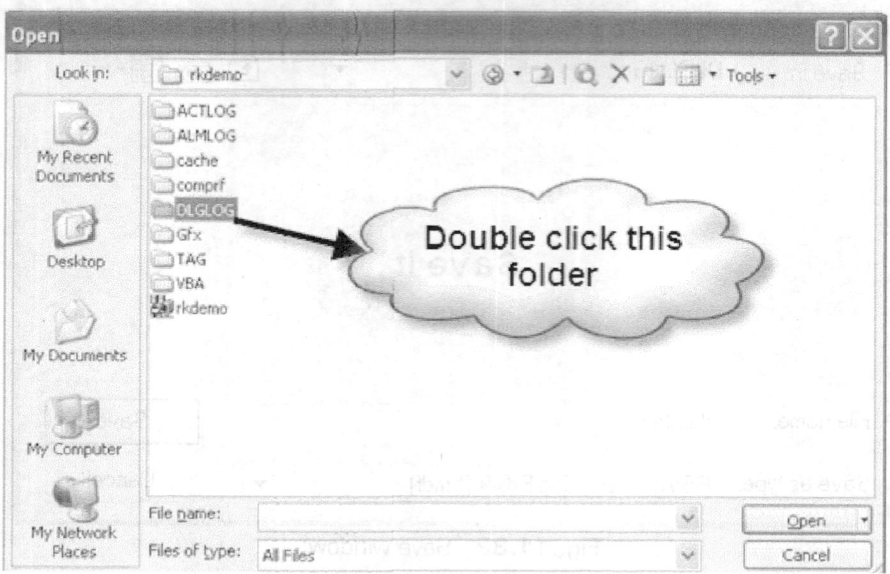

Fig. 11.41 DLGLOG folder

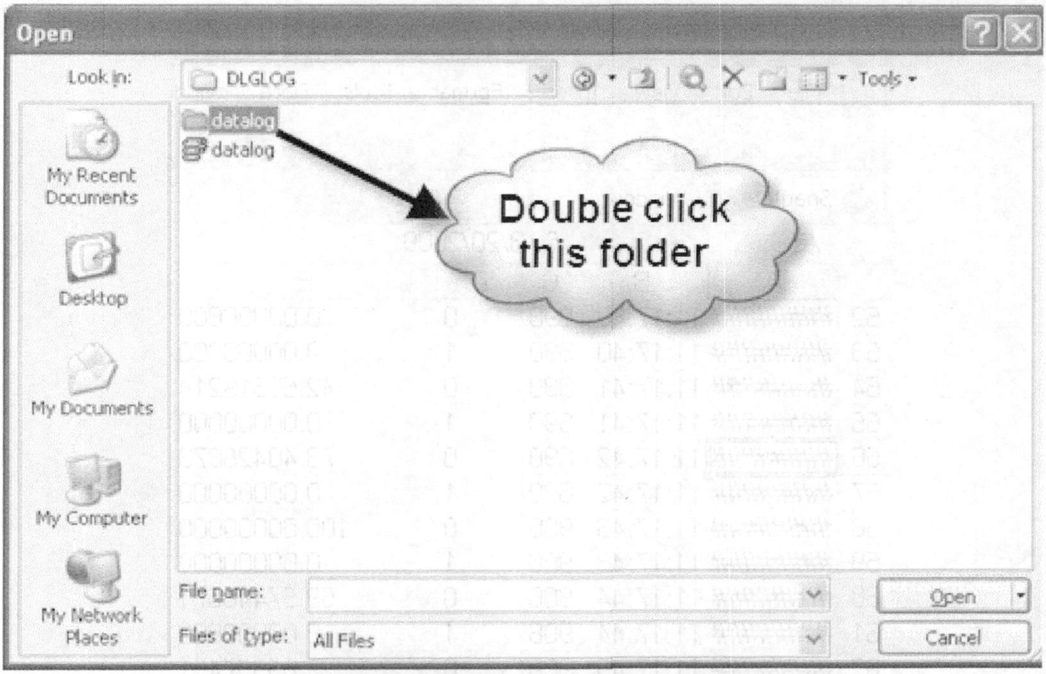

Fig. 11.42 Open datalog folder

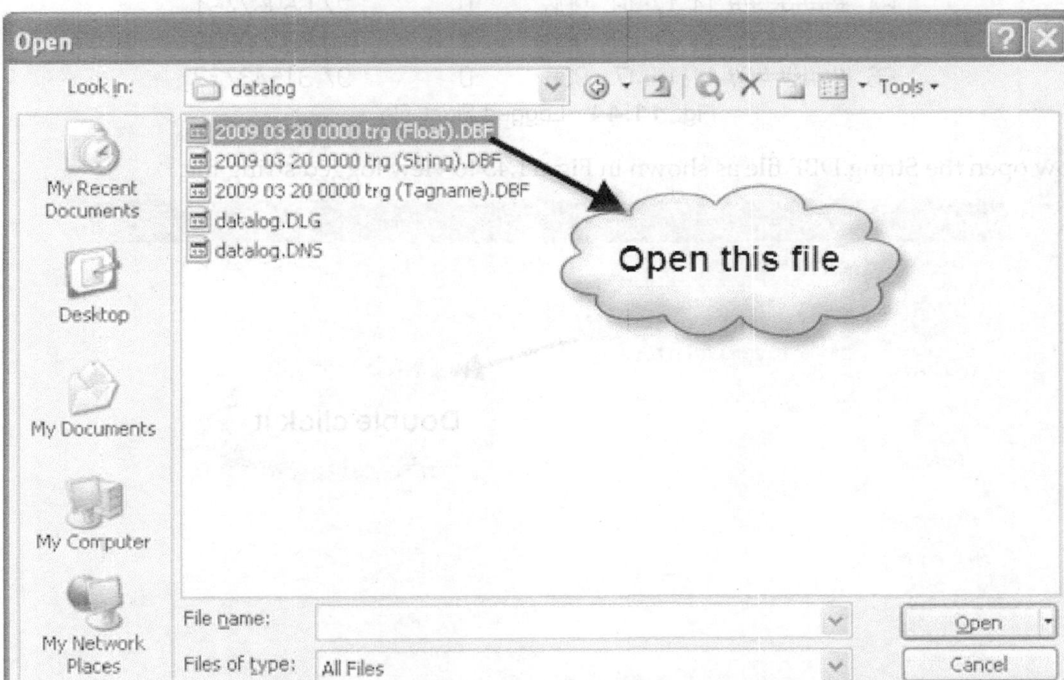

Fig. 11.43 Float.DBF File

Then data logged float file will open in excel format as shown in Fig. 11.44.

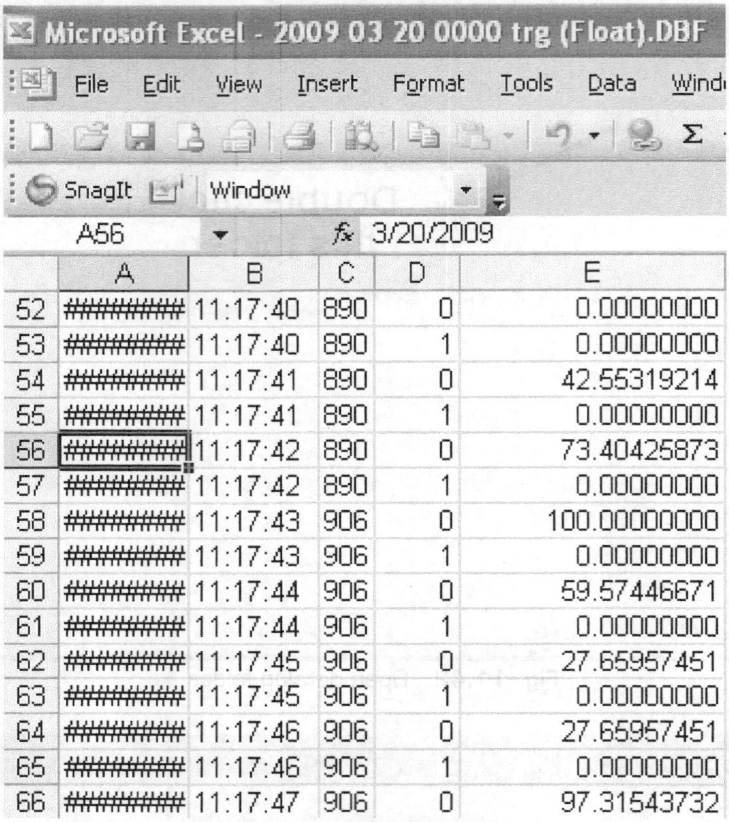

	A	B	C	D	E
52	########	11:17:40	890	0	0.00000000
53	########	11:17:40	890	1	0.00000000
54	########	11:17:41	890	0	42.55319214
55	########	11:17:41	890	1	0.00000000
56	########	11:17:42	890	0	73.40425873
57	########	11:17:42	890	1	0.00000000
58	########	11:17:43	906	0	100.00000000
59	########	11:17:43	906	1	0.00000000
60	########	11:17:44	906	0	59.57446671
61	########	11:17:44	906	1	0.00000000
62	########	11:17:45	906	0	27.65957451
63	########	11:17:45	906	1	0.00000000
64	########	11:17:46	906	0	27.65957451
65	########	11:17:46	906	1	0.00000000
66	########	11:17:47	906	0	97.31543732

Fig. 11.44 Logged Float file

Now open the String.DBF file as shown in Fig. 11.45 to view logged string file.

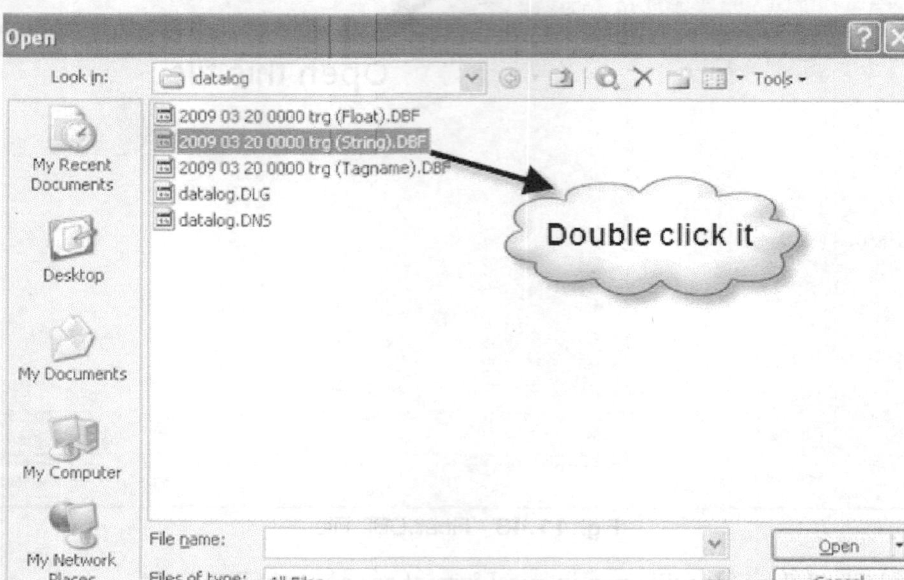

Fig. 11.45 String.DBF file

Then data logged string file will open in excel format as shown in Fig. 11.46.

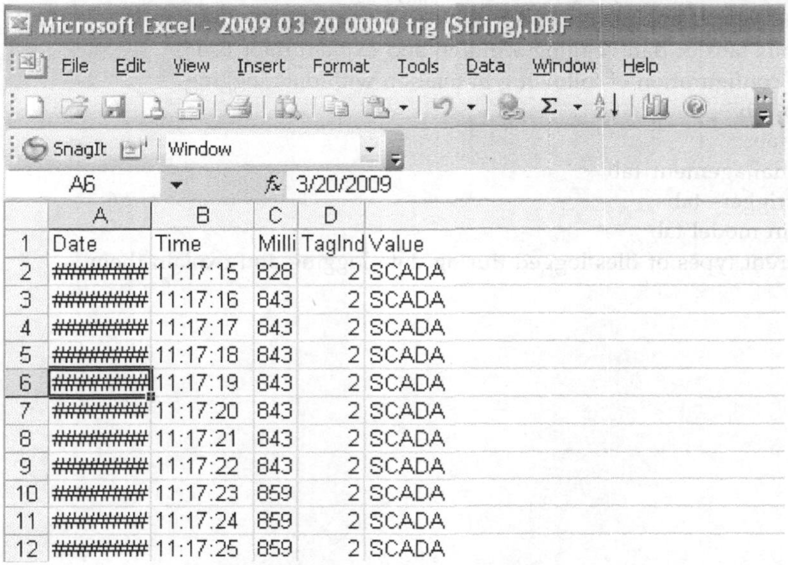

	A	B	C	D	
1	Date	Time	Milli	TagInd	Value
2	########	11:17:15	828	2	SCADA
3	########	11:17:16	843	2	SCADA
4	########	11:17:17	843	2	SCADA
5	########	11:17:18	843	2	SCADA
6	########	11:17:19	843	2	SCADA
7	########	11:17:20	843	2	SCADA
8	########	11:17:21	843	2	SCADA
9	########	11:17:22	843	2	SCADA
10	########	11:17:23	859	2	SCADA
11	########	11:17:24	859	2	SCADA
12	########	11:17:25	859	2	SCADA

Fig. 11.46 Logged string file

SUMMARY

Alarms are important part of most plant control applications so as to identify any accident in plant operation. The alarm operation basically consists of three steps: Alarm Configuration, Alarm setup and Alarm startup. Alarms for digital tags do not have thresholds. Alarms can range in severity from one to eight to indicate different levels of importance. One is most severe, eight is least severe.

If an alarm appears in the alarm summary then an operator can acknowledge the alarm. Acknowledging an alarm does not correct the condition causing the alarm, but indicates that an operator is aware of the alarm. The internal bell is a sound produced by the computer running RSView32. To use the internal bell, select the Internal Bell check box. The external bell is a sound produced by a horn or some other audio device. To set up the external bell, select the check box and specify the name of a tag that is linked to an audio device. The alarm summary displays the alarms recorded in the computer's memory. To configure an alarm summary, specify what data to display and create the format. The logged data is stored in either dBASE IV (.dbf) format or using the ODBC connectivity standard. Data logged in the .dbf format can be stored in narrow or wide format. The narrow .dbf format stores one date, one time, and one tag value per line.

REVIEW QUESTIONS

1. What do we need alarms? Name different type of alarms and explain their use.
2. Explain the configuration and working of:
   (*a*) Digital Alarms
   (*b*) Analog Alarms
3. Compare digital and analog alarms.
4. Name three major steps in alarming and explain them.
5. Explain the steps involved in setting up an alarm.

6. Explain the steps involved in display and start up of alarms.
7. What is an alarm summary? Explain its configuration.
8. What is data logging? Explain two formats used for data logging.
9. Name various tabs of data logging window and explain their use.
10. Explain the configuration of following in relation with data logging.
    (*a*) Setup tab
    (*b*) Path tab
    (*c*) File management tab
    (*d*) Log triggers tab
    (*e*) Tags in model tab
11. Name different types of files logged during data logging and explain them.

12

SCADA SUPPLEMENTS

LEARNING OBJECTIVES

After reading this chapter, you will be able to explain :

Various SCADA supplement like Event detection, Derived tags, Tens, OLE, Security and its significance in PLC Control.

12.1 INTRODUCTION

In this chapter we will discuss the following SCADA supplements:

1. Event Detection
2. Derived Tags
3. Macros
4. Keys Definitions
5. Trends
6. OLE
7. Security

12.2 EVENT DETECTION

Events are expressions that trigger actions. Expressions are equations containing tag values, mathematical operations, if-then-else logic, and other built-in RSView32 functions. Actions are RSView32 commands, symbols, or macros. At runtime, up to 20 event files can run simultaneously.

To configure it in the Project Manager, open the Logic and Control folder as shown in Fig. 12.1 and double click event.

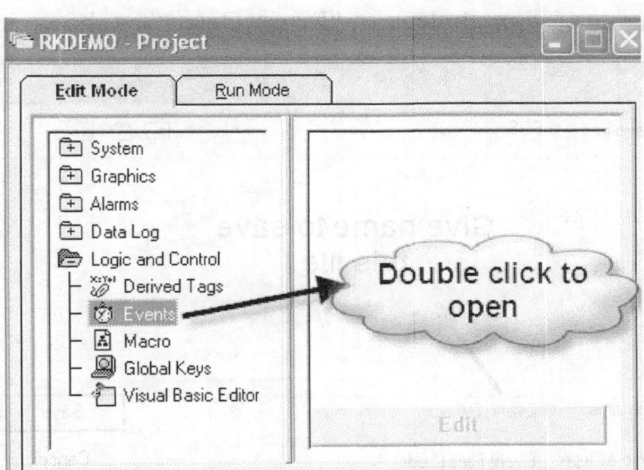

Fig. 12.1 Event selection

Then a new event editor window will open as shown in Fig. 12.2. In the window configure event action as "Display summary" when start tag changes its value from 0 to 1.

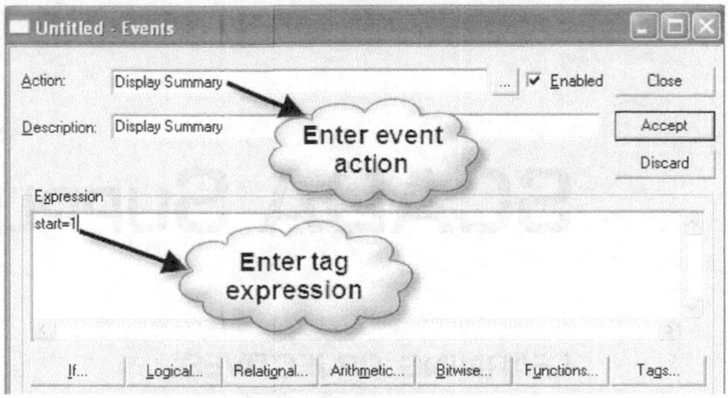

Fig. 12.2 Event editor

Using the Accept and Discard Buttons

When you enter information in the editor, the previous and next buttons change to accept and discard. Click Accept to save information. Click Discard to discard information.

Events Creation

An event is made up of an expression and an action. When the expression changes from false on the previous evaluation to true on the current evaluation, the action is triggered. The following illustration shows an event file.

1. In the Action field, type an RSView32 command, a macro, or a symbol that will run when the expression goes from false to true.
2. To disable the event, clear the Enabled check box. When the event file runs, this disabled event will not be evaluated.
3. In the Description field, type a brief description to document the event's function.
4. In the Expression area, create an expression to specify the conditions that will trigger the action.
5. Click Accept.
6. Repeat Steps 1 through 5 to create more events. Use the Next button to move to a new record in the spreadsheet.

Save this file with a name of your choice *e.g.*, "Event_Detect" as shown in Fig. 12.3.

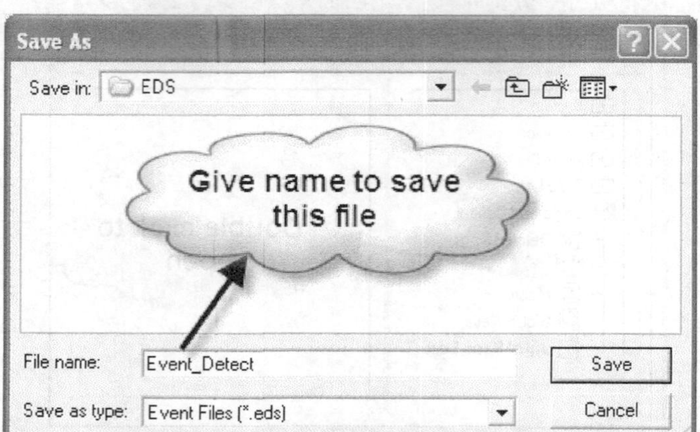

Fig. 12.3 Save event file

In setup, select event setup as shown in Fig. 12.4 and in event setup select continuous as shown in Fig. 12.5.

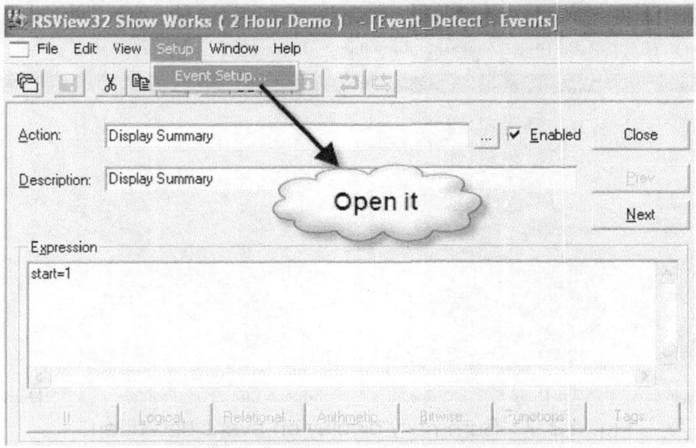

Fig. 12.4 Event setup opening

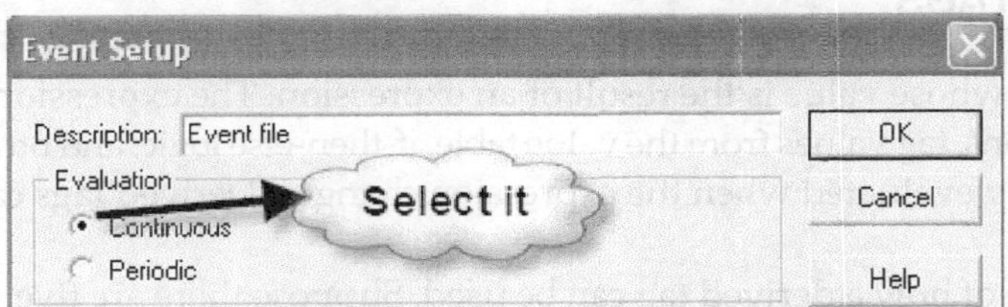

Fig. 12.5 Event setup configuration

To start event processing, open Startup editor. Then check the Event Detector box and specify an event file name as shown in Fig. 12.6.

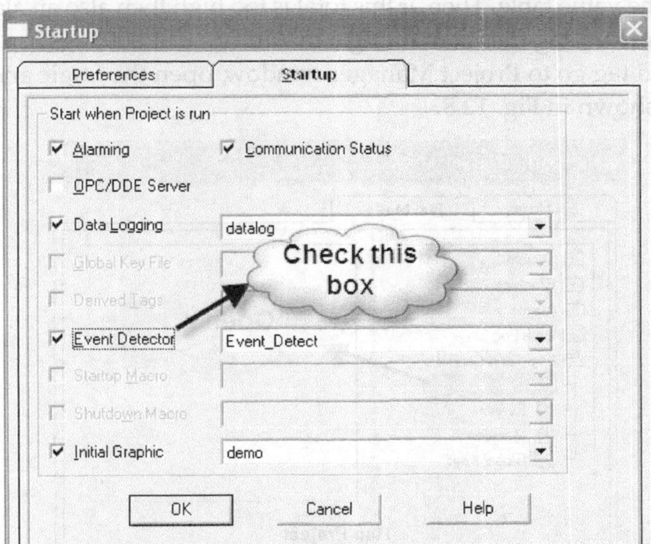

Fig. 12.6 Event detection startup

Now save the project and run it. You will observed that when start tag's value is 1 then an event will be detected which will open the display summary window as shown in Fig. 12.7.

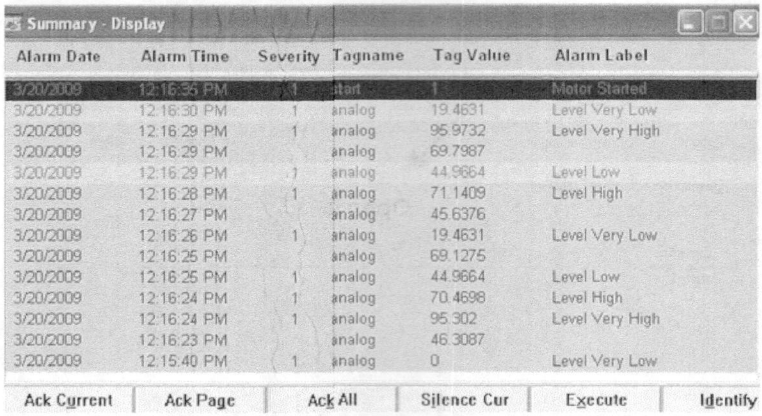
Summary - Display

Alarm Date	Alarm Time	Severity	Tagname	Tag Value	Alarm Label
3/20/2009	12:16:35 PM	1	start	1	Motor Started
3/20/2009	12:16:30 PM	1	analog	19.4631	Level Very Low
3/20/2009	12:16:29 PM	1	analog	95.9732	Level Very High
3/20/2009	12:16:29 PM		analog	69.7987	
3/20/2009	12:16:29 PM	1	analog	44.9664	Level Low
3/20/2009	12:16:28 PM	1	analog	71.1409	Level High
3/20/2009	12:16:27 PM		analog	45.6376	
3/20/2009	12:16:26 PM	1	analog	19.4631	Level Very Low
3/20/2009	12:16:25 PM		analog	69.1275	
3/20/2009	12:16:25 PM	1	analog	44.9664	Level Low
3/20/2009	12:16:24 PM	1	analog	70.4698	Level High
3/20/2009	12:16:24 PM	1	analog	95.302	Level Very High
3/20/2009	12:16:23 PM		analog	46.3087	
3/20/2009	12:15:40 PM	1	analog	0	Level Very Low

Fig. 12.7 Summary display on event detection

12.3 DERIVED TAGS

A derived tag is a tag whose value is the result of an expression. The expression can be made up of mathematical operations, tag values from the value table, if-then-else logic, and other special functions. The derived tag is only evaluated when the expression changes. Derived tags can be device tags or memory tags.

Here is an example of how a derived tag can be used. Suppose there are five weight sensors on a conveyor belt. The tag database contains one tag for each sensor, so the weight at each point on the conveyor belt is monitored. If the weight at any point is excessive then an alarm will be activated. However, what happens if no individual sensor detects an excessive weight, but the total of all five sensors is too high? In this case, you could set up a derived tag to sum the weights of all five sensors and store the result in the value table. Then, if this total is too high then also an alarm will be activated. At runtime, up to 20 derived tag files containing a maximum of 1,000 tags can run simultaneously.

To create the derived tag go to Project Manager window, open the Logic and Control folder and select derived tags as shown in Fig. 12.8.

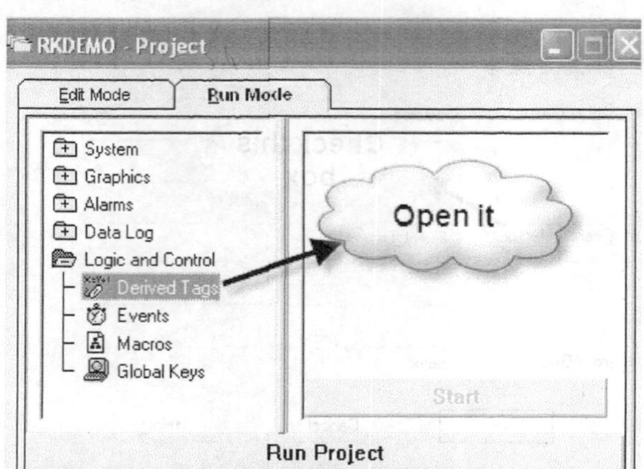

Fig. 12.8 Derived tag selection

A new derived tag editor window will appear edit it as shown in Fig. 12.9.

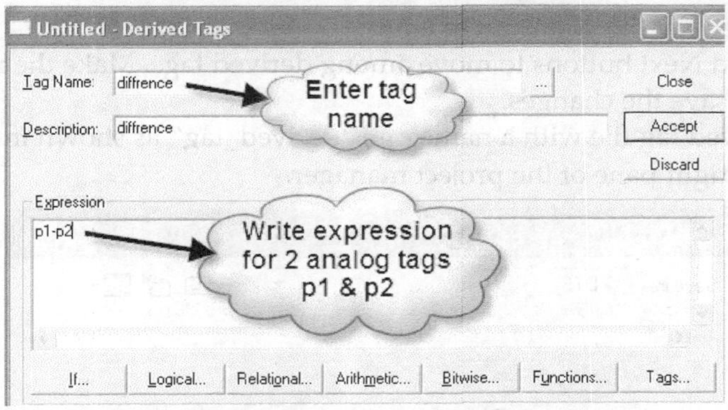

Fig. 12.9 Derived tag configuration

After editing derived tag click accept. Then you will notice a message asking to create tag with name "difference". Click yes and create an analog tag with name as difference in tag editor window. Likewise you will also get messages for another two tags p1 and p2. Create these tags also in the tag editor window as analog tags.

Then you will observe edited derived tag in the down window as shown in Fig. 12.10.

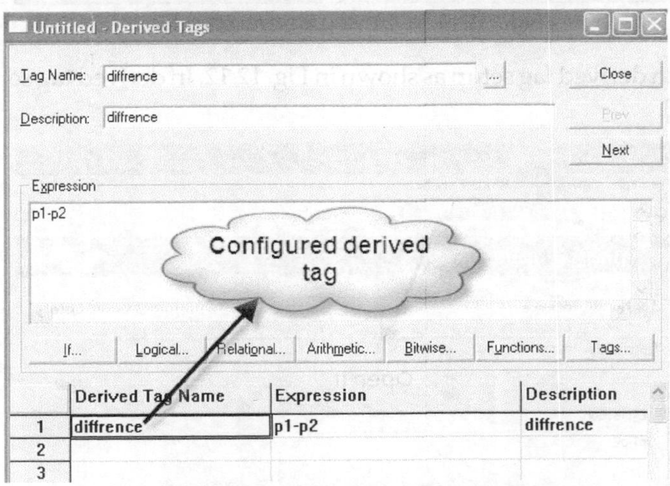

Fig. 12.10 Configured derived tag

Using the Accept and Discard Buttons

When information is entered in the editor, the Prev and Next buttons change to Accept and Discard. Click Accept to save information. Click Discard to discard information. In the Tag Name field, specify the name of the tag that will hold the derived tag value. This tag can be analog, digital, or string and can have device or memory as the data source. The tag name is also the name of the derived tag file. In the Description field, type a brief description of the derived tag's function. This description is for your information only and does not appear anywhere else. In the Expression area, create the expression that will determine the derived tag value. Click Accept and Repeat Steps 1 through 4 to create more derived tags. Use the Next button to move to a new record in the spreadsheet.

Editing Derived Tags

1. Open the derived tag file you want to edit.
2. Use the Prev and Next buttons to move among derived tags. Make the required changes.
3. Click Accept to save the changes.

Then save the derived tag file with a name *e.g.*, "derived_tag" as shown in Fig. 12.11. Then this file can be viewed on right pane of the project manager.

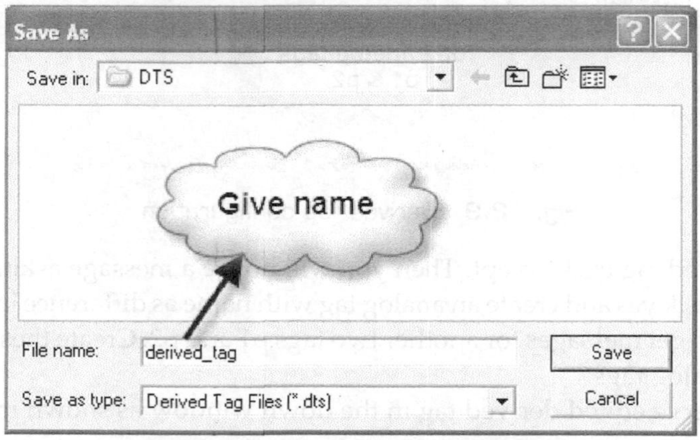

Fig. 12.11 Saving derived tag file

Go to setup and open derived tag setup as shown in Fig. 12.12. In derived tag setup select continuous as shown in Fig. 12.13.

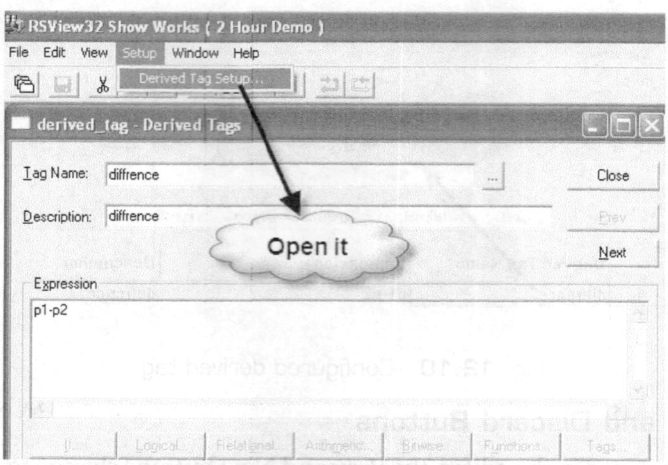

Fig. 12.12 Open derived tag setup

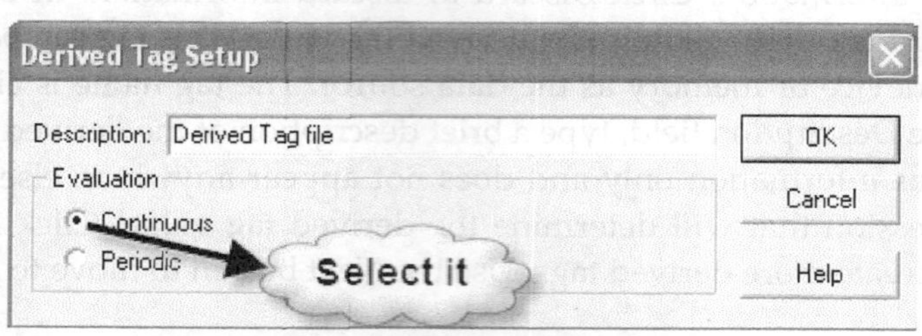

Fig. 12.13 Configure derived tag setup

To start derived tag processing open the Startup editor, check the Derived Tags box and specify a derived tag file as shown in Fig. 12.14.

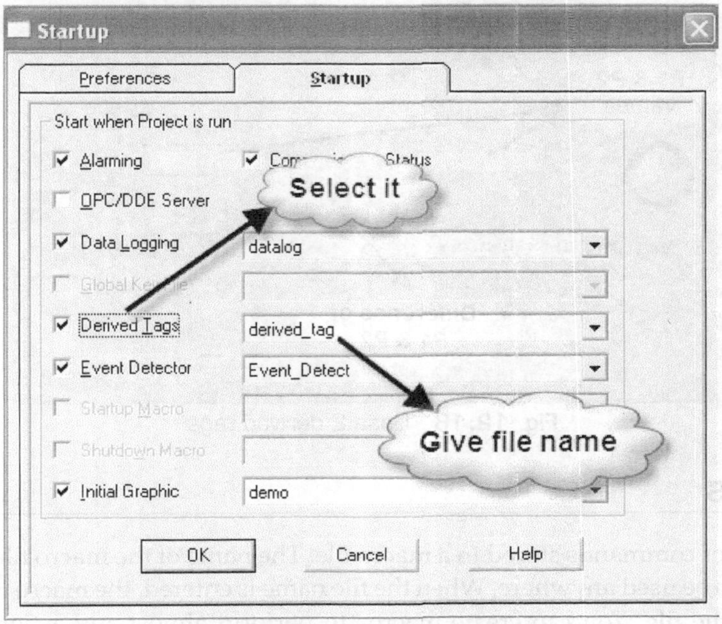

Fig. 12.14 Derived tag startup

In demo display create two numeric inputs and attach tags p1 and p2 respectively with these numeric inputs. Then create one numeric display and attach difference tab with this display. Save this display and run it. Then enter the values in p1 and p2 input fields with help of onscreen keyboard and observe the output of difference display.

Case 1: When p1 = 25 and p2 = 0 then Difference is 25 as shown in Fig. 12.15

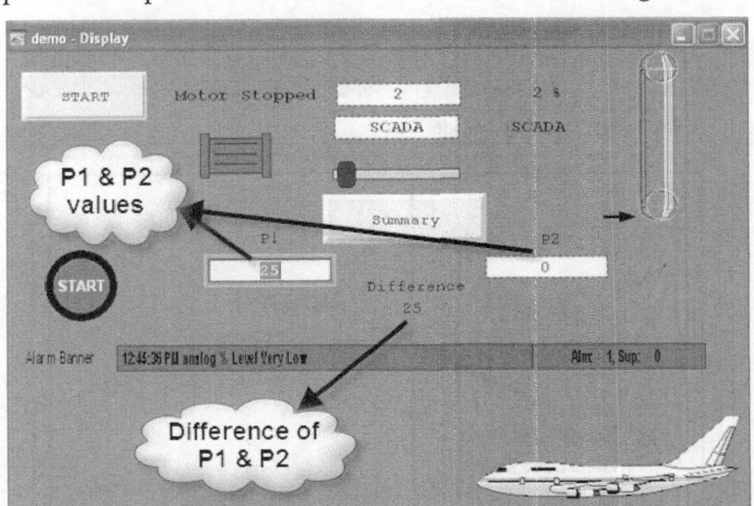

Fig. 12.15 Case 1 derived tags

Case 2: When p1 = 25 and p2 = 15 then Difference is 10 as shown in Fig. 12.16

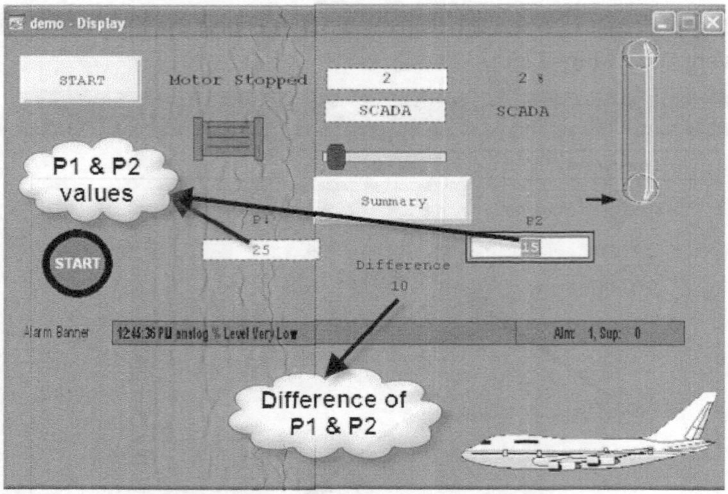

Fig. 12.16 Case 2 derived tags

12.4 MACROS

A macro is a series of commands stored in a macro file. The name of the macro file is then used like a command and can be used anywhere. When the file name is entered, the macro runs, executing all the commands in the file. You can create macros to perform almost any action. For example, a macro can:

(*a*) Open a group of windows and define their initial positions
(*b*) Define temporary key definitions
(*c*) Close any open windows
(*d*) Set tag values

To configure the macro go to Project Manager, open the Logic and Control folder as shown in Fig. 12.17.

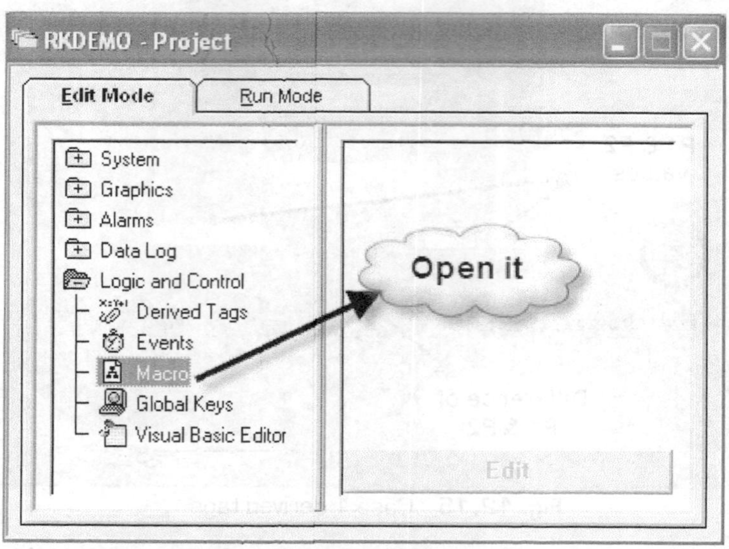

Fig. 12.17 Macro editor opening

Then a new macro editor window will open as shown in Fig. 12.19. Write the text in it as shown in Fig. 12.18.

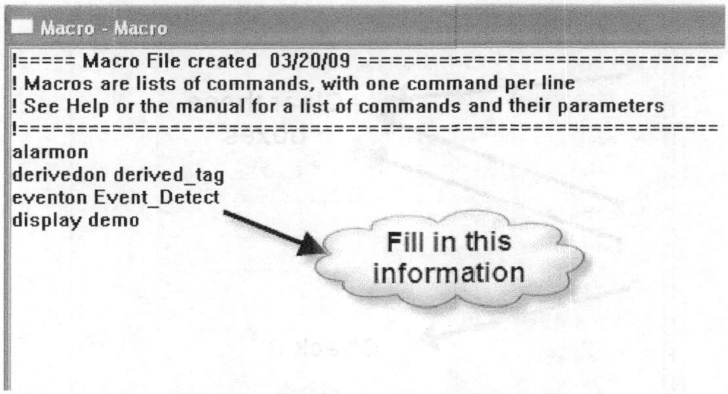

Fig. 12.18 Macro editor window

Type your macro commands and, if required, add command parameters. When entering commands, follow these guidelines:

(*a*) Separate each identifier, specification, or string with a space or a tab.

(*b*) Start each command on a new line, or separate commands on the same line with a semicolon (;).

(*c*) Precede comments with an exclamation mark (!). The comment lasts until the next semicolon (;) or line break.

(*d*) To indicate a per cent in a macro, use two per cent signs (%%) because a single per cent sign indicates a parameter.

Save the macro with a name *e.g.*, "Macro" as shown in Fig. 12.19.

Fig. 12.19 Macro save window

Using Parameters

Macros can accept parameters. To specify a parameter in a macro, type a percent (%) sign followed by a number. Up to nine parameters are allowed. To run the macro and parameters, specify the macro name followed by the parameters. Separate multiple parameters with spaces.

To startup macros in startup window, select the macros box and clear box against alarming, derived tag, event detector and initial display as shown in Fig. 12.20.

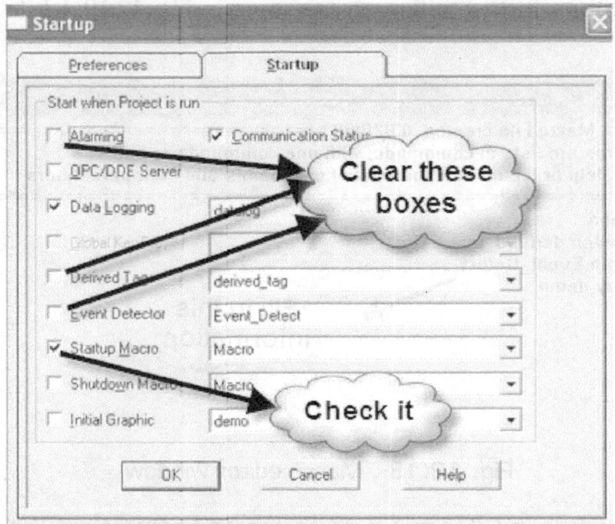

Fig. 12.20 Macro startup window

Then run the demo display to view the working of alarms, event detection, and derived tag. You will notice that all objects of display are working normally as shown in Fig. 12.21. So with the help of macros we can startup alarming, derived tags event detection etc.

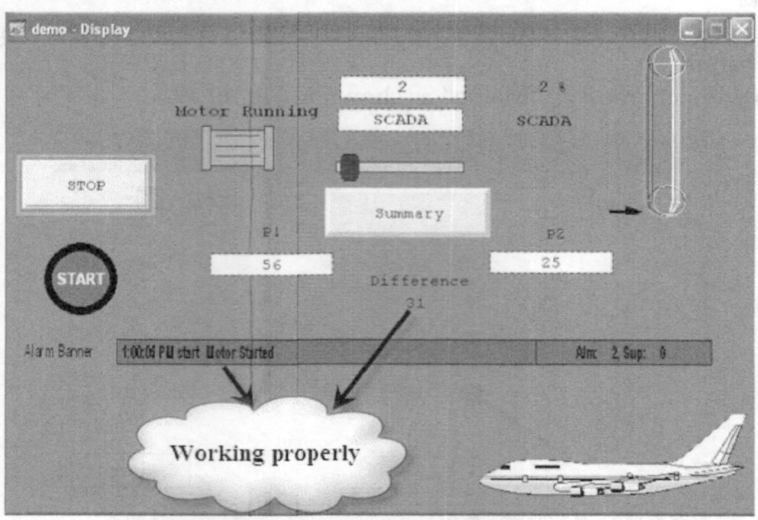

Fig. 12.21 Display working with macros

12.5 KEY DEFINITIONS

The three different types of keys with RSView 32 are:

1. Object Keys
2. Display Keys
3. Global Keys

Object keys and display keys are different from global keys. Object keys and display keys are active only while a particular object or display has input focus. Global keys are active at all times.

12.5.1 Object Keys

Example: Using object keys

Imagine you have a display showing a tank with two valves. The two valves control the flow in and out of the tank. Both valves have been configured with object key animation as follows:

Valve 1	F2 = Open
	F3 = Close
Valve 2	F2 = Open
	F3 = Close

At runtime, operators can select either valve, press F2 to open it and press F3 to close it.

To view the working of object keys, copy one valve from the library and paste it on demo display. Give color animation to valve by creating and assigning one new tag named "open" to this object as shown in Fig. 12.22.

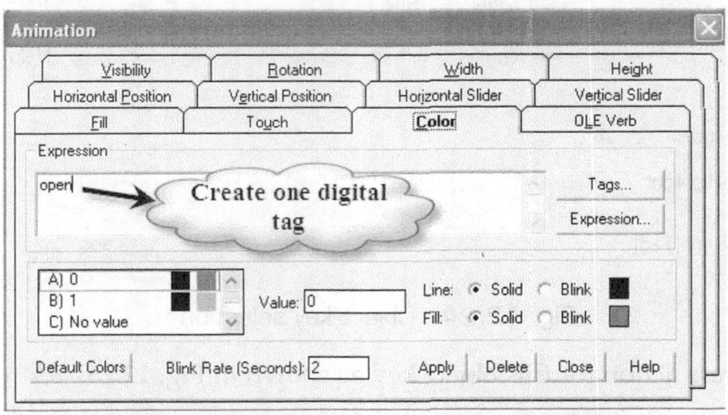

Fig. 12.22 Color animation for valve

Select object then right click to open animation. Select and open object keys as shown in Fig. 12.23.

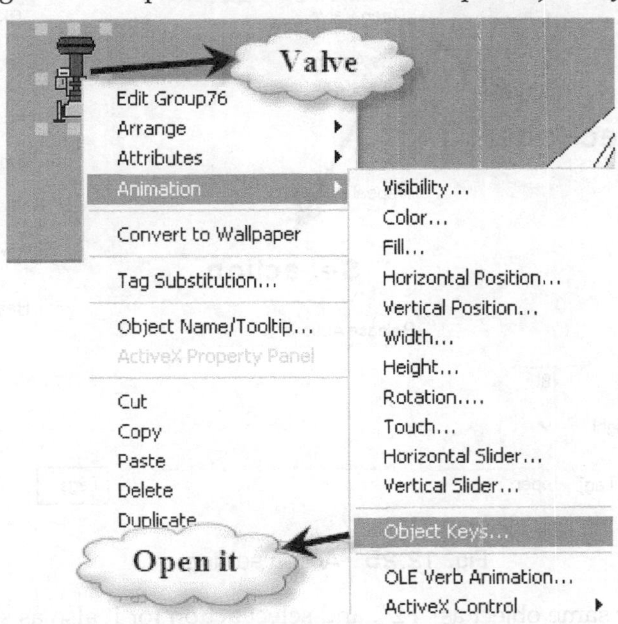

Fig. 12.23 Opening object keys

Then a new object key configuration window will appear as shown in Fig. 12.24. In this new window select add and then enter a key *e.g.*, "F1" and click OK.

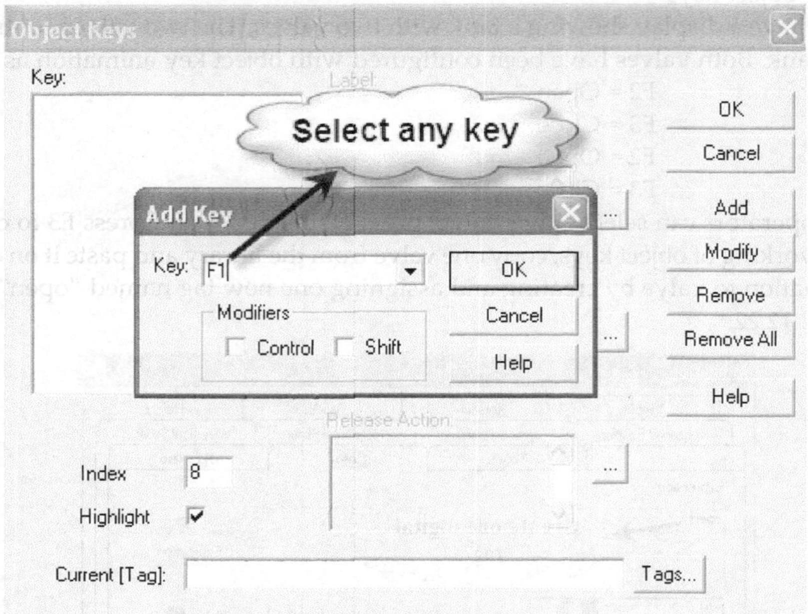

Fig. 12.24 Object key selection

Select label and press action for this object key as shown in Fig. 12.25.

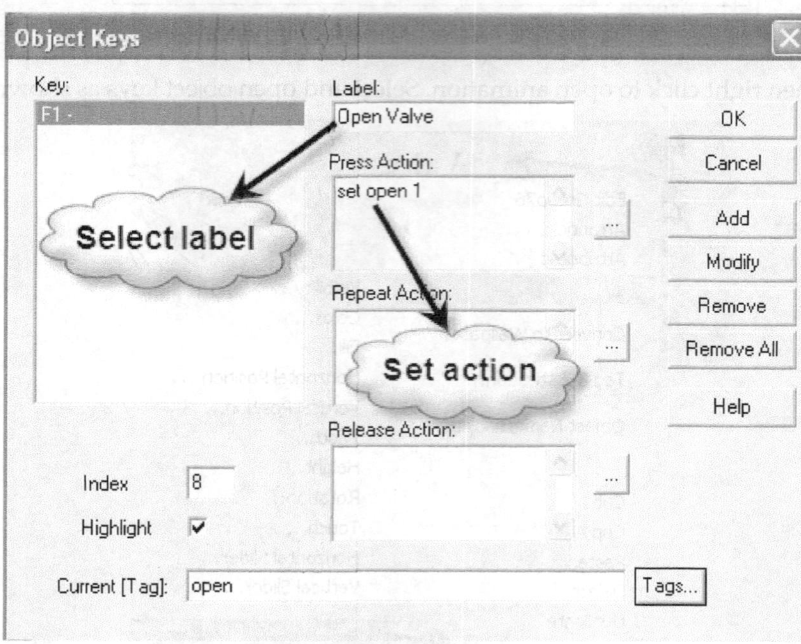

Fig. 12.25 Action setting

Create another key for same object as "F2", and select action for it also as shown in Fig. 12.26.

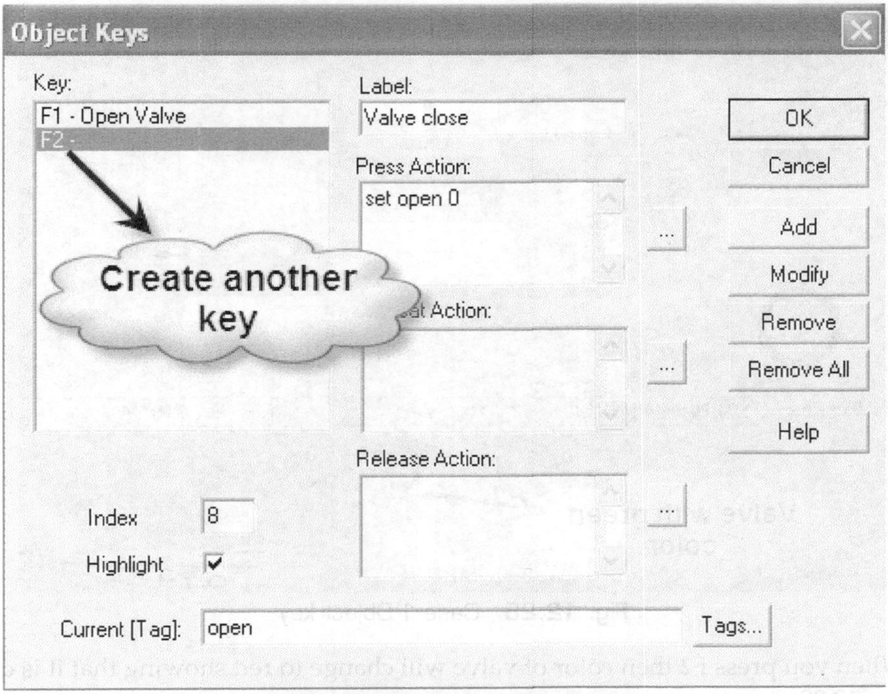

Fig. 12.26 F2 as Object key

Save the display and run it and select the valve. Then you will observe the created object keys with it as shown in Fig. 12.27.

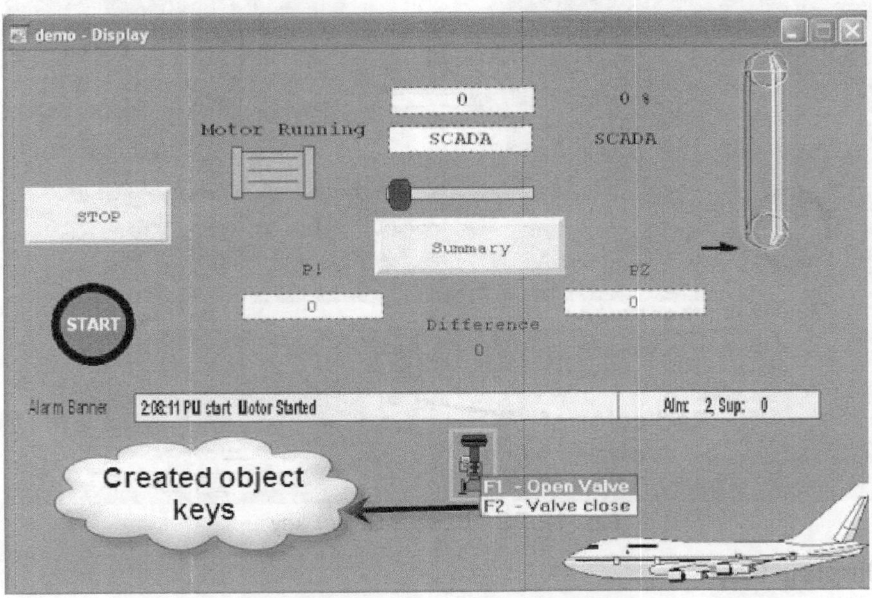

Fig. 12.27 Valve with object keys

Case 1: When you press F1 then color of valve will change to green showing that it is open. It can be seen in Fig. 12.28.

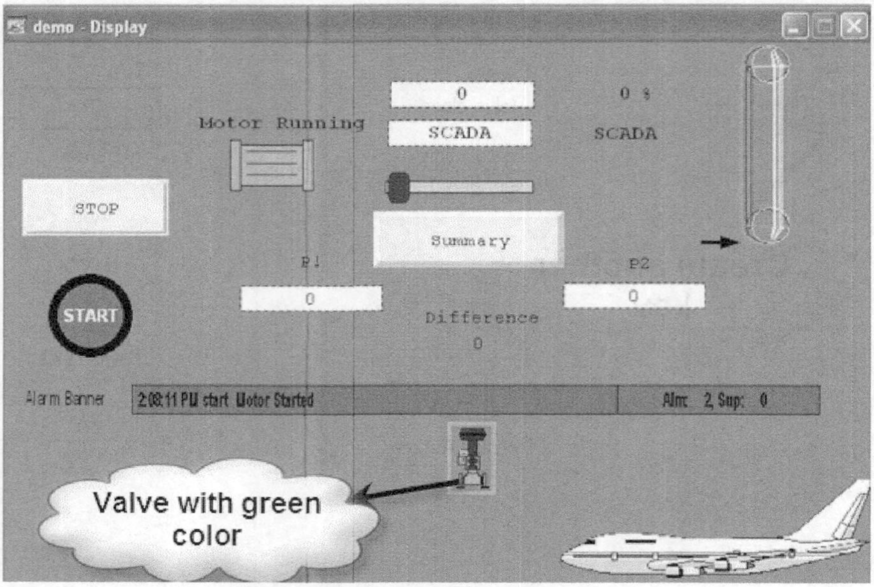

Fig. 12.28 Case 1 Object key

Case 2: When you press F2 then color of valve will change to red showing that it is close. It can be seen in Fig. 12.29.

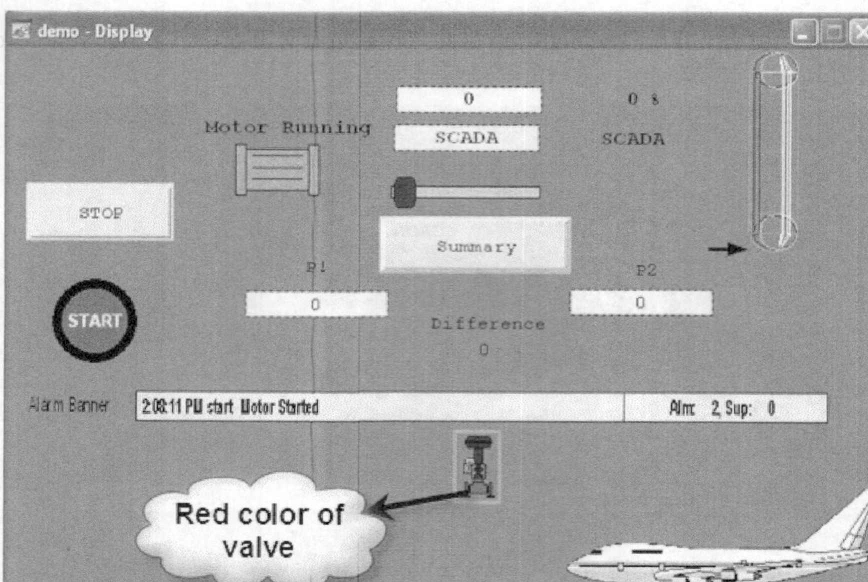

Fig. 12.29 Case 2 Object key

12.5.2 Display Key

Display keys are used to attach function keys with displays. These function keys works with that particular display in which these are created. To create display key, select animation and open display key as shown in Fig. 12.30.

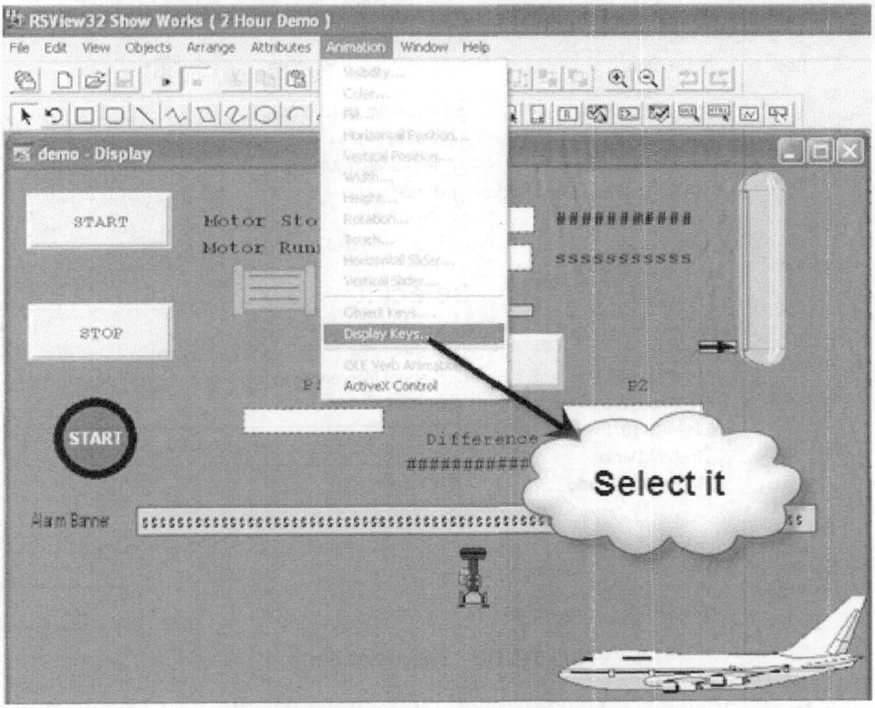

Fig. 12.30 Display key selection

Then a new window will appear as shown in Fig. 12.31. Select required command in this window and select OK.

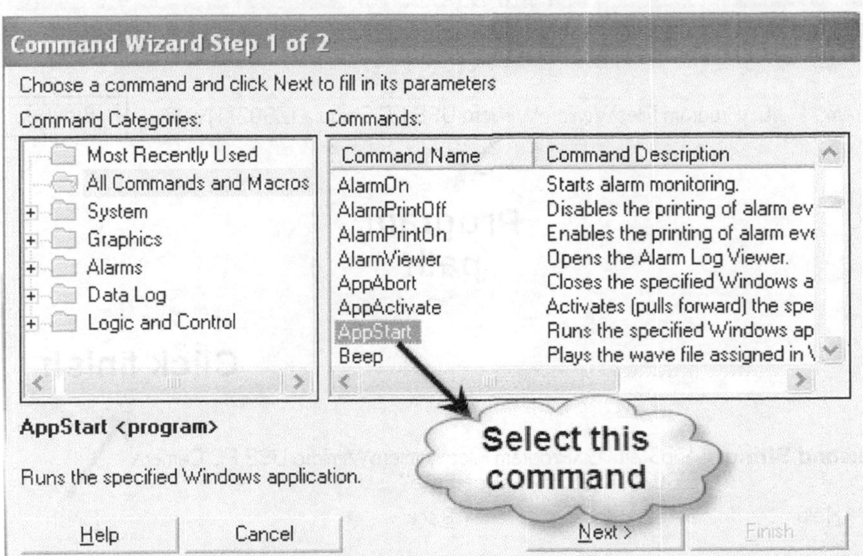

Fig. 12.31 Command selection

Then in new window select the required application that you want to open with display key. For example to open webcam select the required file as shown in Fig. 12.32.

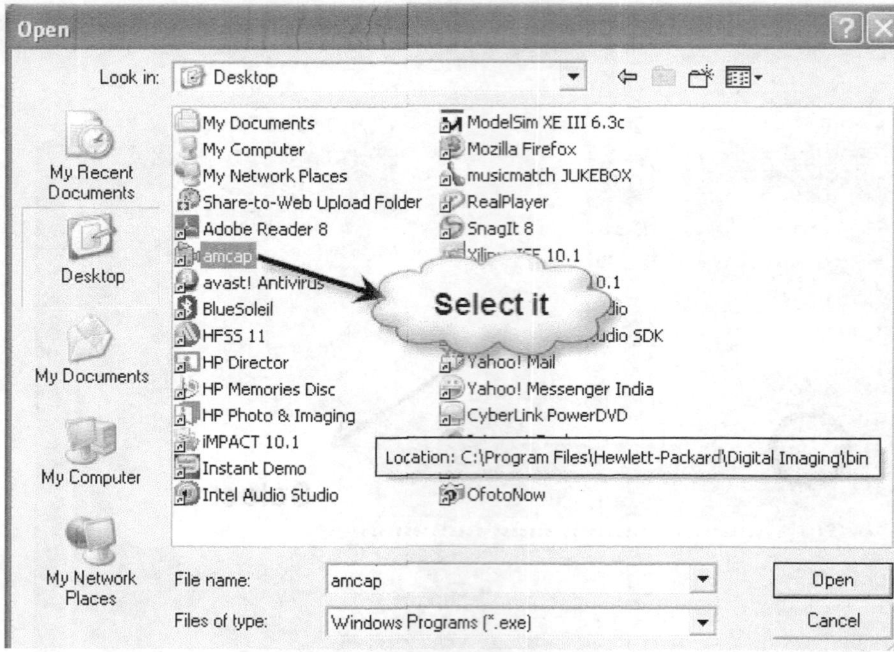

Fig. 12.32 File selection

Then new window will appear showing complete path of the file as shown in Fig. 12.33. Click finish in this window.

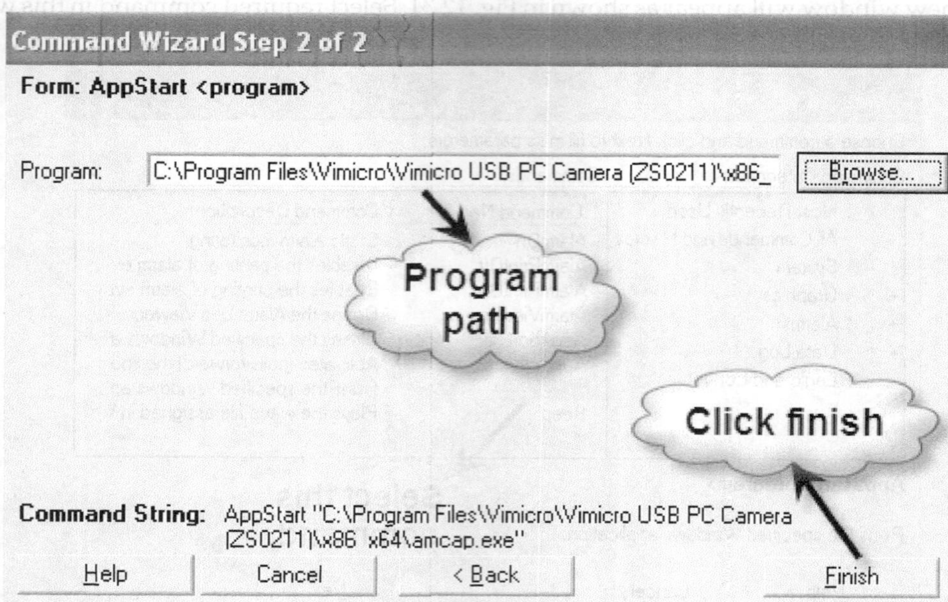

Fig. 12.33 Program path

In new window select the function key that you want to configure as display *e.g.*, "F3" as shown in Fig. 12.34. Give label to the key and click OK.

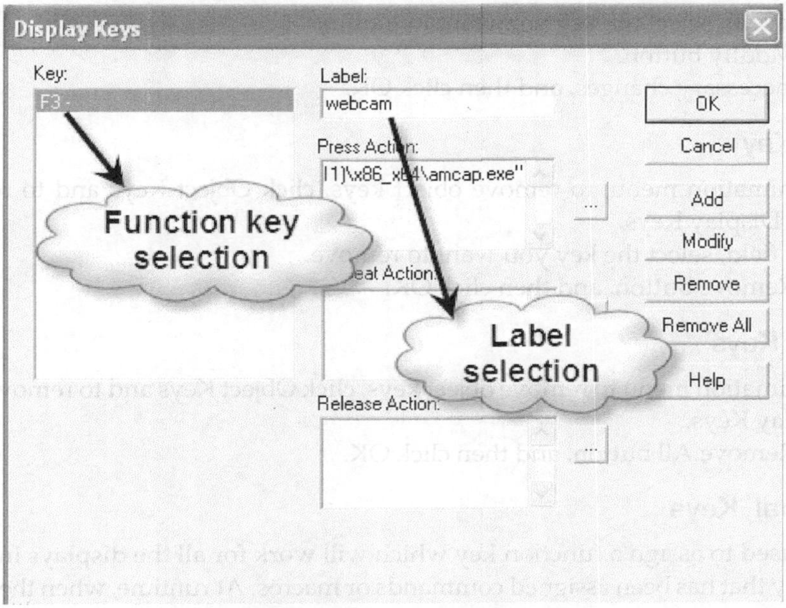

Fig. 12.34 Function key selection

Run the display and click F3 then webcam window will be opened on your display as shown in Fig. 12.35.

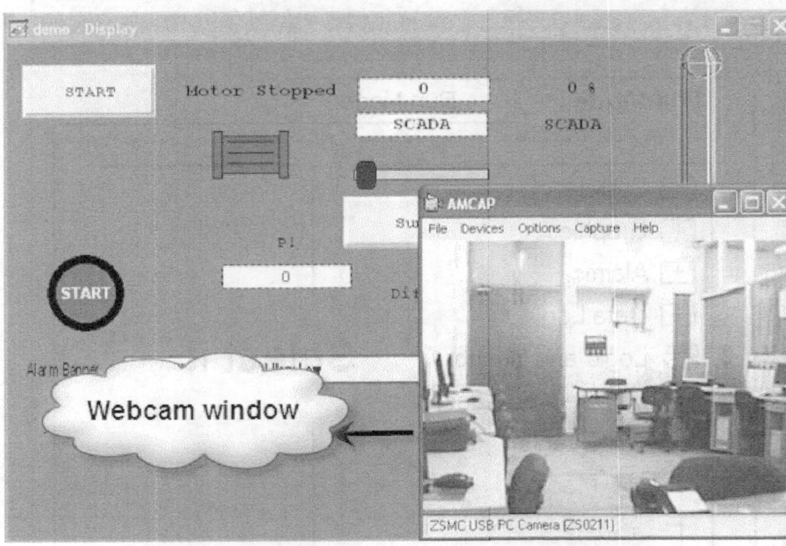

Fig. 12.35 Webcam window on display

Editing Display and Object Keys

Editing display and object keys is the same except that for object keys you must first select an object.

Modifying a Key

1. On the Animation menu, to modify object keys click Object Keys and to modify display keys, click Display Keys

2. In the Key field, select the key you want to modify.
3. Click the Modify button.
4. Make the necessary changes, and then click OK.

Removing a Key

1. On the Animation menu, to remove object keys, click Object Keys and to remove display keys, click Display Keys.
2. In the Key field, select the key you want to remove.
3. Click the Remove button, and then click OK.

Removing all Keys

1. On the Animation menu to remove object keys, click Object Keys and to remove display keys, click Display Keys.
2. Click the Remove All button, and then click OK.

12.5.3 Global Keys

A global key is used to assign a function key which will work for all the displays in one project. A global key is a key that has been assigned commands or macros. At runtime, when the key is pressed, the assigned action is triggered. A global key is active at all times because it is not associated with a particular graphic object or display. In the Project Manager, open the Logic and Control folder as shown in Fig. 12.36.

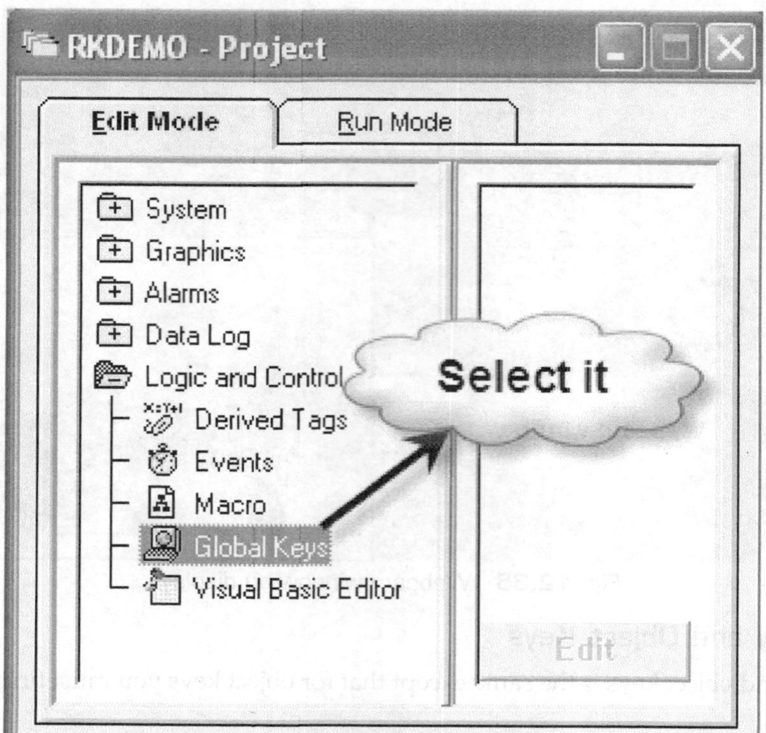

Fig. 12.36 Global key selection

In the new window select F4 as global key and click OK as shown in Fig. 12.37.

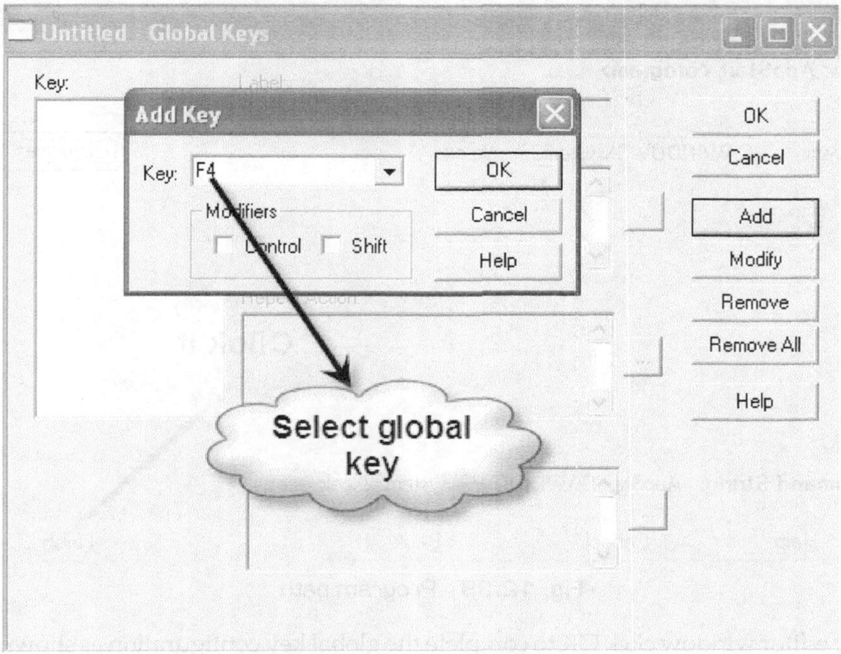

Fig. 12.37 F4 as global key

Select the file that you want to open with this global key as shown in Fig. 12.38.

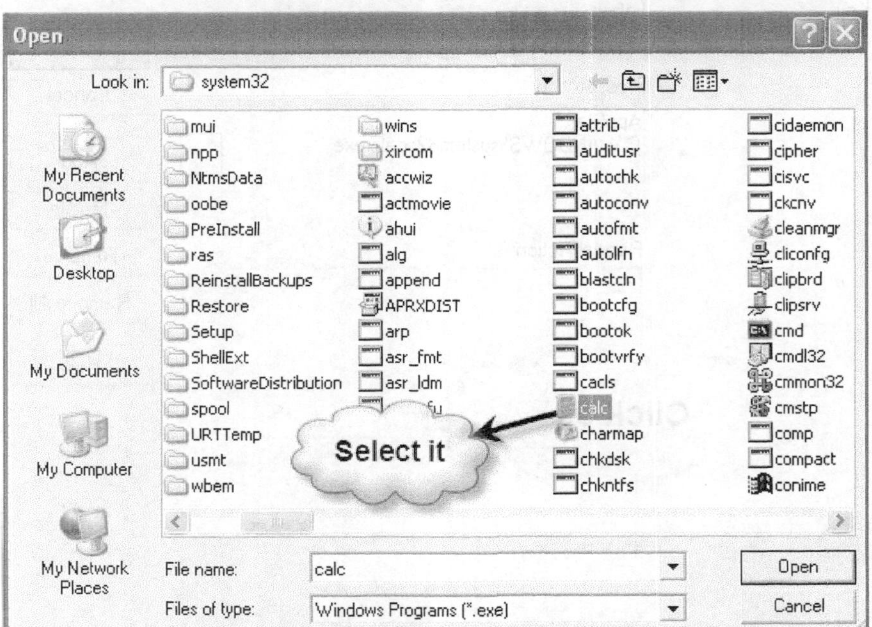

Fig. 12.38 Calculator selection

Then you will notice a new window that will show the complete path of the file as shown in Fig. 12.39. Click finish in this window.

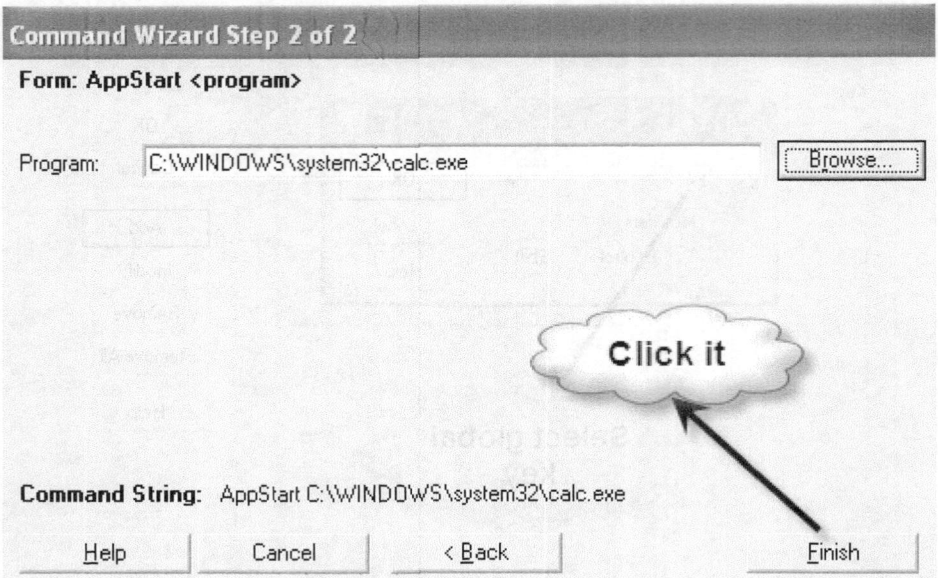

Fig. 12.39 Program path

In global key editor window click OK to complete the global key configuration as shown in Fig. 12.40.

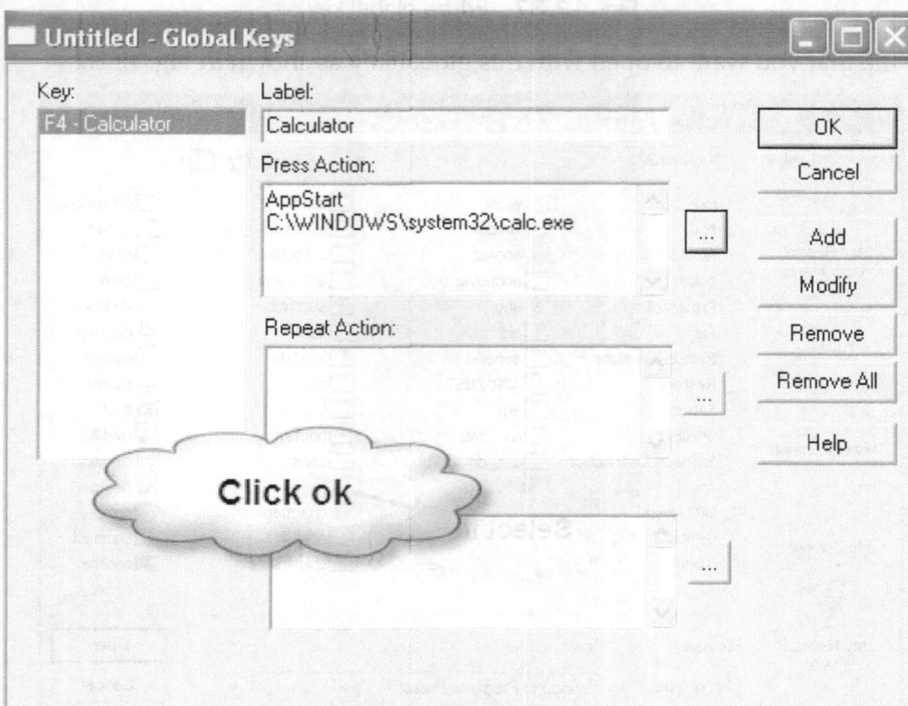

Fig. 12.40 Global key editor

Now save the display with some name *e.g.*, "Global". Then it will appear in the right pane of the project manager window as shown in Fig. 12.41.

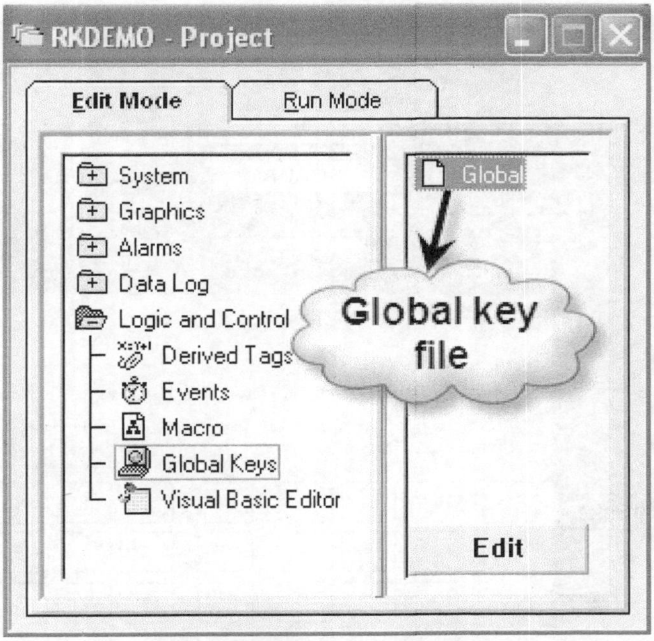

Fig. 12.41 Global key file

In startup window check the box against global key as shown in Fig. 12.42 and click OK.

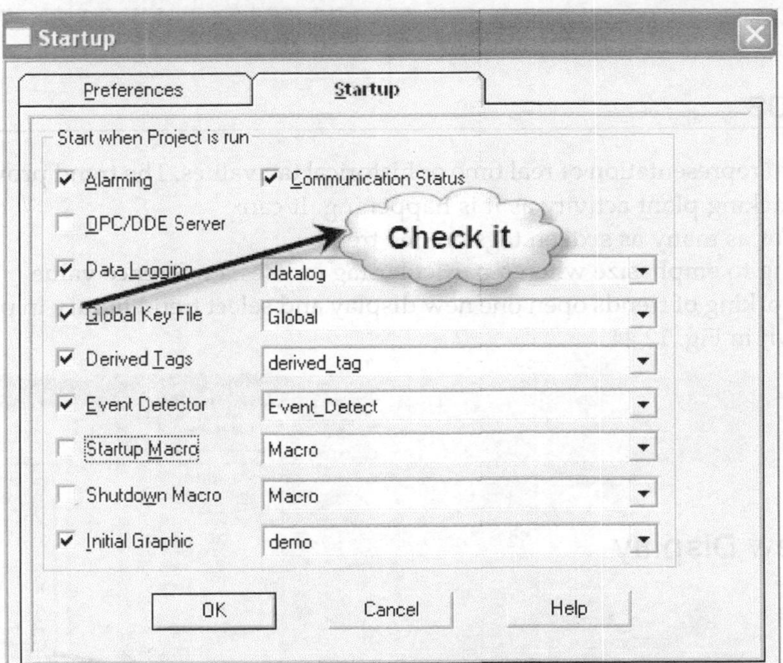

Fig. 12.42 Global key startup

Now run the display and press F4 key. Then a calculator will be opened on your display as shown in Fig. 12.43.

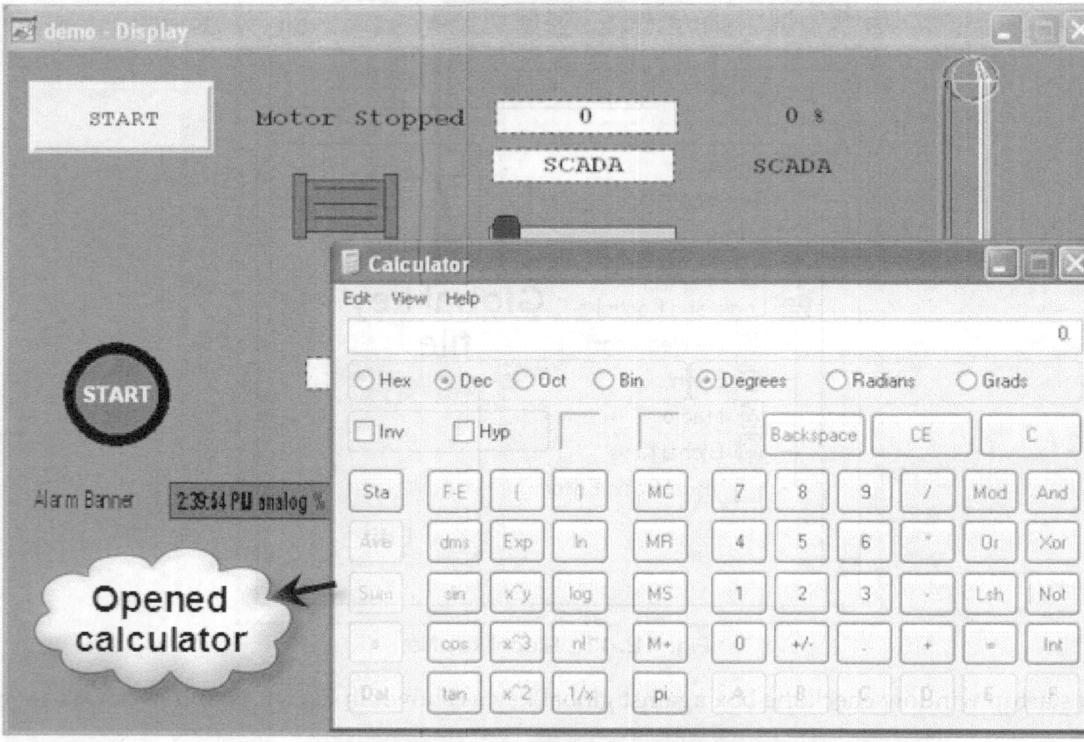

Fig. 12.43 Opened Calculator with global key

12.6 TRENDS

A trend is a visual representation of real time or historical tag values. The trend provides operators with a way of tracking plant activity as it is happening. It can:

(*a*) Plot data for as many as sixteen tags on one trend

(*b*) Use shading to emphasize when a particular tag crosses a reference value

To view the working of trends open one new display and select trend library in project manager window as shown in Fig. 12.44.

Fig. 12.44 Display and Trend library

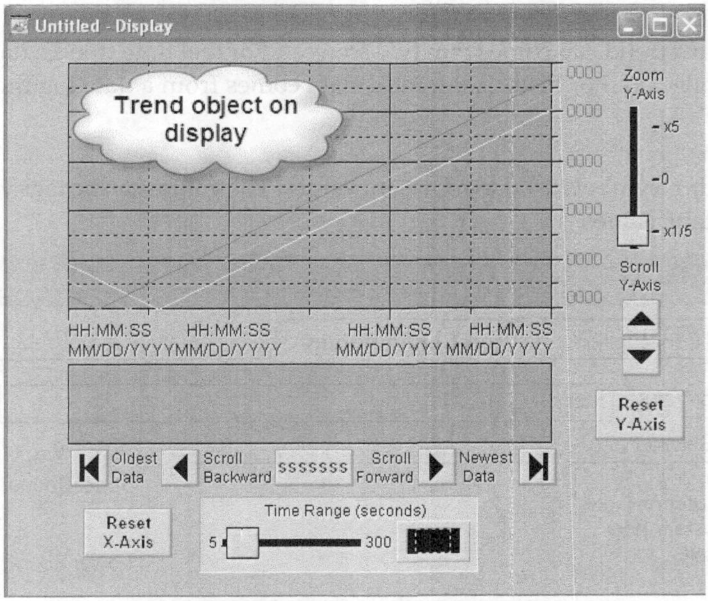

Fig. 12.45 Trend objects on display

Copy the required object from trend library and paste it on the display as shown in Fig. 12.47. Double click the trend to open the editor window as shown in Fig. 12.46.

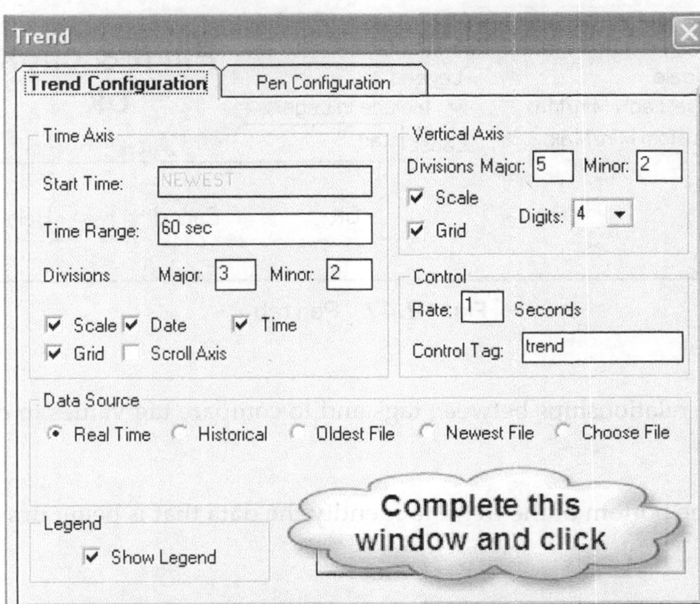

Fig. 12.46 Trend Editor

Trend Editor Box

The Trend editor box contains two tabs for configuring the appearance and content of the trend:

(*i*) Trend Configuration shown in Fig. 12.46.

(*ii*) Pen Configuration shown in Fig. 12.47.

Data Source

The data displayed in a trend can come from two sources. For real time trends, data comes from the value table as it is collected. For historical trends, data comes from a data log model's set of files.

Pens

Pens are the lines and/or symbols used to represent values. The values can be tags you are monitoring or they can be constant values.

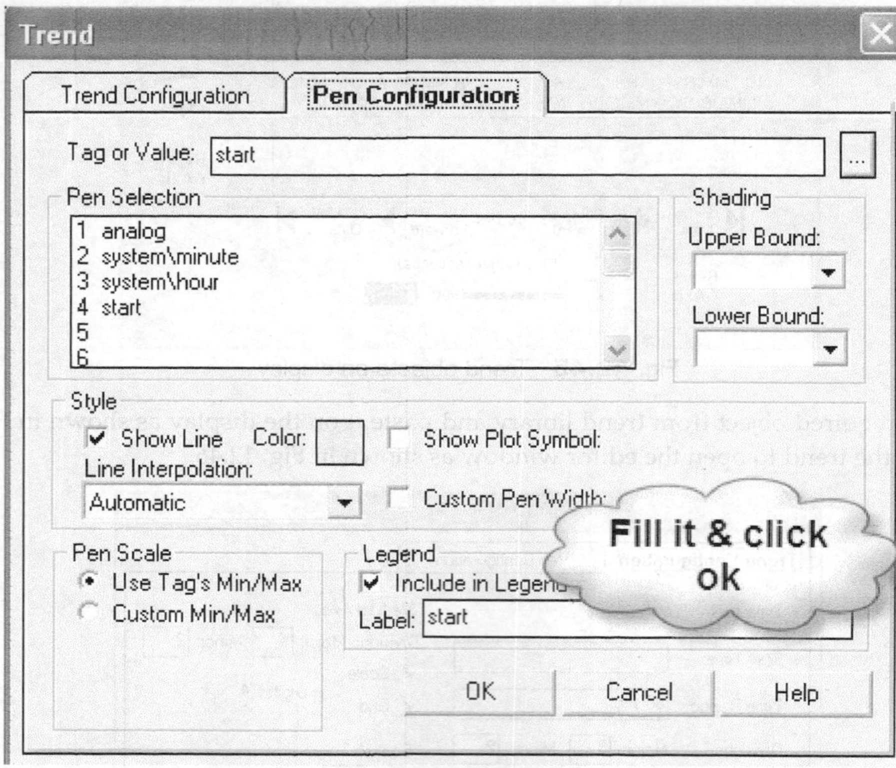

Fig. 12.47 Pen tab

Shading

Use shading to show relationships between tags and to compare tag values to constants.

Legend

Include a legend at the bottom of the trend to identify the data that is being drawn by each pen.

Control tags

Control tags are a type of memory tag that are created and updated by RSView32.

12.6.1 Trend Tab

I Time Axis

To configure the time axis, fill in the following fields as shown in Fig. 12.48.

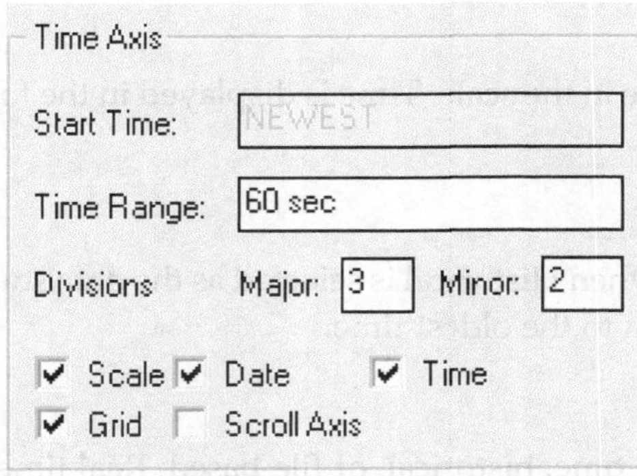

Fig. 12.48 Time axis

Start Time for a Real Time Trend

For a real time trend, NEWEST is shown. Data will start plotting as soon as the trend starts, with the newest data available.

Start Time for a Historical Trend

Start times for historical data can be relative or absolute. With relative time, plotting always starts relative to the newest or oldest data in the log model or log file. With absolute time, plotting always starts at the specified date and time.

To specify relative time select the word NEWEST or NEWEST minus (–) a time range or the word OLDEST or OLDEST plus (+) a time range. Specify the time range as a whole number followed by a unit. Valid units are: Sec, Min, Hour, Day, and Week.

Divisions

Specify the number of major and minor divisions required on the time axis. Keep the number of divisions small so they will be visible.

Scale

Check this box to have a time scale appear at the bottom of the trend. The scale can contain any or all of the following:

(*a*) The major and minor divisions,
(*b*) The date, and
(*c*) The time.

Grid

Check this box to have vertical grid lines extend from the major and minor divisions on the time axis.

Date

Check this box to show the date in the scale.

Scroll Axis

Check this box to have the grid for the time axis scroll across the trend. When this box is not checked, the grid for the time axis is fixed.

Time

Check this box to show the time in the scale. Time is displayed in the following format: HH:MM:SS (hour, minute, second).

Relative Time

This check box is visible only when Historical is selected as the data source. Check this box to show the time of the trend in relation to the oldest time.

II Data Source Selection

The data for trends can be real time, historical, or file-based. Real time data comes from the value table. Historical data comes from a data log model's set of files. File-based data comes from a single log file at a time.

For a Real Time Trend

Click the Real Time button and configure as shown in Fig. 12.49. The buffer for each real time trend pen can contain up to 2,000 tag values. A tag value is saved to the buffer whenever that value changes. When the buffer is full, the oldest value is overwritten with a new value.

Fig. 12.49 Data source selection

For a Local Historical Trend

To display historical data from a data log model in the running project, click Historical, then click Local and specify a data log model in the Model field.

For a Remote Historical Trend

To display historical data from a different project (from another location on the same computer or from another computer), click Historical, then click Remote. In the Model field, type the complete path.

For a File-based Trend

Click Oldest File to display data from the oldest file in the data model. For local data, specify a data log model in the Model field. For remote data, type the complete path.

III Legend

To include a legend, check the Show Legend box as shown in Fig. 12.50.

Fig. 12.50 Legend selection

12.6.2 Pen Tab

To configure the pens click the Pen Configuration tab in the trend toolbox as shown in Fig. 12.49. Select a pen and fill in the fields. The fields are described below. To move to another pen, select it in the Pen Selection field.

I Tag or Value

In the Tag or Value field, specify an analog or digital tag as shown in Fig. 12.51.

Fig. 12.51 Tag or value field

II Pen Style

Select a pen style as shown in Fig. 12.52.

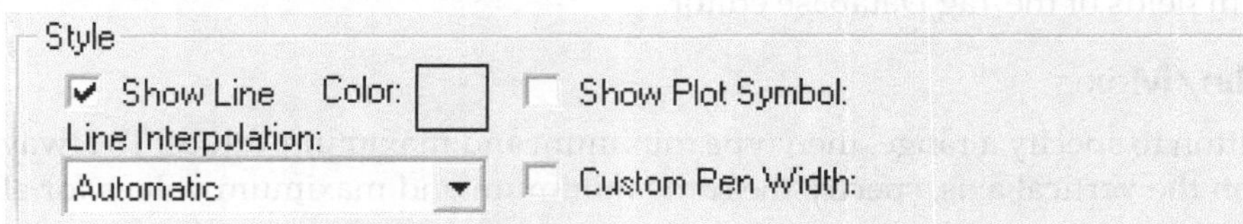

Fig. 12.52 Pen style selection

Show Line

Check this box if you want to draw a line between the plotted symbols. If you want *only* plot symbols visible, leave this box empty.

Color

Select a color for the pen. Do not choose the same color as the background of the trend object.

Show Plot Symbol

Check this box to plot a symbol. The plot symbol appears at each point that is plotted.

Line Interpolation

Line interpolation determines how a line is drawn. The choices are:

(*i*) **Automatic:** Draws the appropriate type of line based on the tag type. For example, for a digital tag, digital interpolation is automatically used. For an analog tag, linear interpolation is automatically used.

(*ii*) **Linear:** Draws a straight line between two points, which suggests a gradual change from the first point to the second point.

(*iii*) **Digital:** Draws a line along the time axis at the first value then, when the value changes, draws a vertical line to the second value.

(*iv*) **Full Width:** Draws the most recent value for a tag horizontally across the trend. For example, if the most recent value for a tag is 75, draws a line across the trend at 75. When the value changes, then it redraws the line at the new value.

(*v*) **Custom Pen Width:** The custom pen width affects the thickness of the line and the plot symbols. Check this box if you want to specify a different pen width.

III Pen Scale

The pen scale defines the range of numbers on the vertical axis. The vertical axis changes to reflect the minimum and maximum values for the selected pen's tag. It is shown in Fig. 12.53.

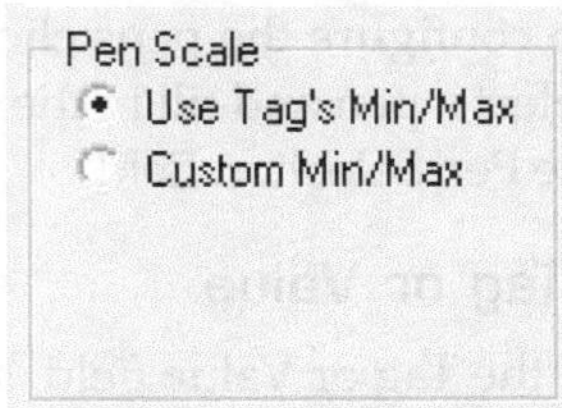

Fig. 12.53 Pen scale

The pen scale defines the range of numbers on the vertical axis. The vertical axis changes to reflect the minimum and maximum values for the selected pen's tag. For example, if Tag1 has a minimum value of 10 and a maximum value of 100, the range on the vertical axis will be 10 to 100 when the Tag 1 pen is selected. If Tag 2 has a minimum of –10 and a maximum of 50, the range on the vertical axis will change to –10 to 50 when the Tag 2 pen is selected. If you want all pens in the trend to use the same range on the vertical axis, specify the same range in the Custom Min/Max field.

Use Tag's Min/Max

Select this button to use the tag's minimum and maximum range as configured in the Minimum and Maximum fields of the Tag Database editor.

Custom Min/Max

Select this button to specify a range, then type minimum and maximum values. To always have the same range on the vertical axis, specify the same minimum and maximum values for all pens.

IV Shading

Use shading to compare the values of different pens. To have shading, at least two pens must be defined.

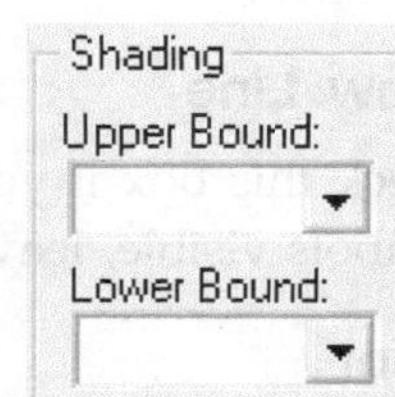

Fig. 12.54 Shading

In Fig. 12.54 for the upper and lower boundaries, select the pens with which to compare values. When the pen you are defining plots above the value of an upper boundary pen, or below the value of a lower boundary pen, the area between the pen you are defining and the boundary pen will be shaded. The shading color will be the color you select under Style when the configuration of trend display is complete the sane it with a name *e.g.*,"Trending" then it will appear in right pane of the project manager window as shown in Fig. 12.55.

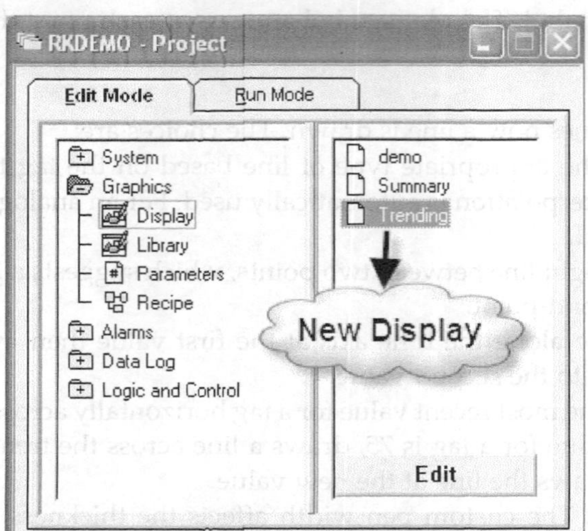

Fig. 12.55 Trending Display

Now run the project then you will monitor tags in the trend window as shown in Fig. 12.56.

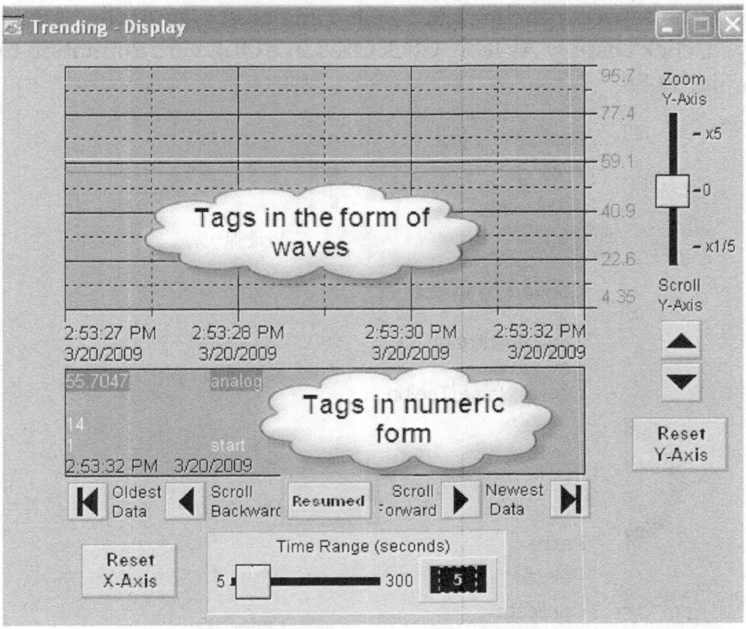

Fig. 12.56 Trend Output

12.7 OLE

OLE stands for Object Linking and Embedding. OLE objects are objects created in other Windows applications and linked to or embedded in an RSView32 graphic display. Click OLE Objects on the Objects menu or click the OLE in the Drawing Tools toolbox. Then Draw a box about the size you want for the object to open a window shown in Fig. 12.57 will appear. Select the file *e.g.*, "ofotonow" that you want to link with this object.

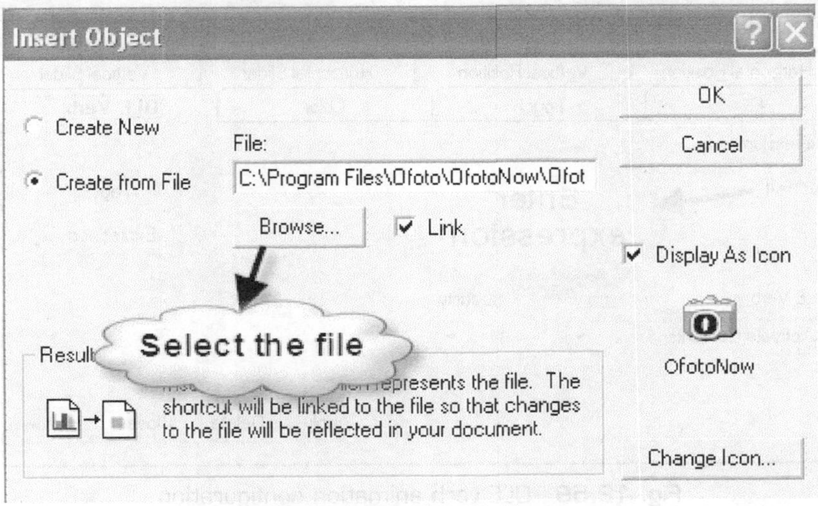

Fig. 12.57 File Linking

Click Create New to create and embed a new object. Click Create from File to insert an existing file. To create a link to the source file, click Link. Select an object type or file. If the embedded object icon is to be displayed, click Display As Icon. Click OK. Give OLE verb animation to the created OLE object as shown in Fig. 12.58.

Fig. 12.58 OLE verb animation selection

Configure OLE verb animation as per information given in Fig. 12.59.

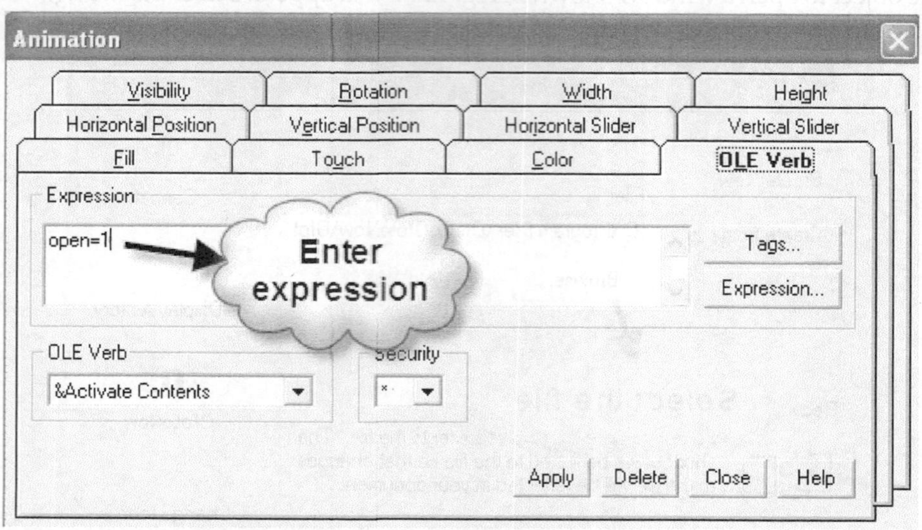

Fig. 12.59 OLE verb animation configuration

Select command line icon and create one command line on the display as shown in Fig. 12.60.

Fig. 12.60 Command line

Run the project and enter command as "set open 1" as shown in Fig. 12.61.

Fig. 12.61 Entering command

Then ofotonow will be opened as shown in Fig. 12.62.

Fig. 12.62 OLE opening ofotonow file

12.8 SECURITY

Security is used to prevent users from accessing certain parts of the system Security is based on a system of codes. Each code allows users with security privileges for that code to access the RSView32 features allowed by that code. Users can be assigned combinations of security codes, allowing each user to access a different set of features.

Example: Assigning security codes

The following security codes are set up:

(*a*) The DataLogOn command is assigned security code B

(*b*) A graphic display named demo is assigned security code D

(*c*) A tag named start is assigned security code E

In the User Account editor, the following security codes are assigned:

ME students are assigned security codes B, D, and E

STC participants are assigned security codes B and D as shown in Fig. 12.63.

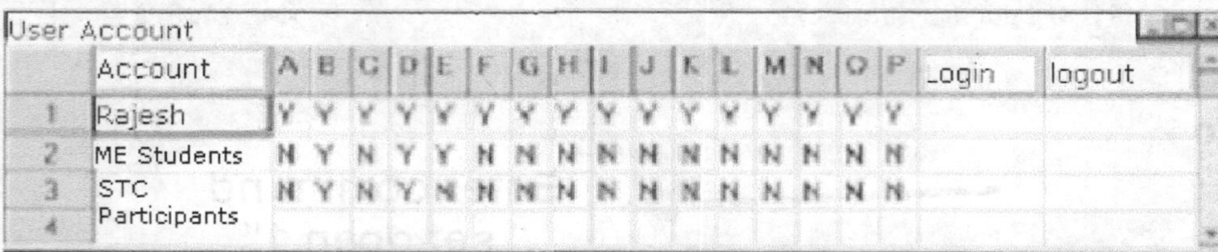

User Account

	Account	A	B	C	D	E	F	G	H	I	J	K	L	M	N	O	P	Login	logout
1	Rajesh	Y	Y	Y	Y	Y	Y	Y	Y	Y	Y	Y	Y	Y	Y	Y	Y		
2	ME Students	N	Y	N	Y	Y	N	N	N	N	N	N	N	N	N	N	N		
3	STC Participants	N	Y	N	Y	N	N	N	N	N	N	N	N	N	N	N	N		
4																			

Fig. 12.63 User account details

This means that ME students have access to the command, the display, and the tag. STC has access to the command and display, but not the tag.

12.8.1 Security Code Configuration

In project manager open security code editor as shown in Fig. 12.64.

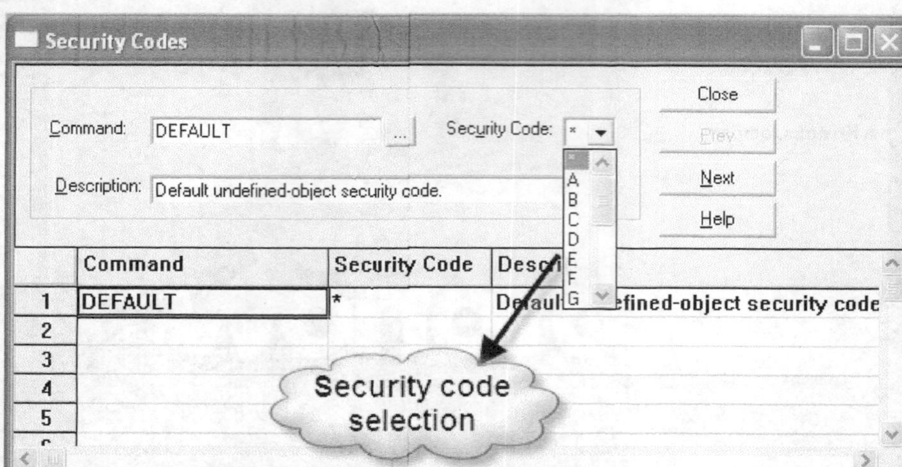

Fig. 12.64 Security code editor

The Default

The first record in the spreadsheet is called Default. Any commands or macros that are not listed in the spreadsheet use the security code selected for the default.

Security Codes

RSView32 has 17 security codes: an asterisk (*) and the letters A through P. The asterisk is for unlimited access, and the letters are for limited access. The letters are not hierarchical. All letters provide the same level of security.

(*a*) **Security by inclusion :** Security by inclusion means all RSView32 commands and all macros requiring security are listed in the Security Codes editor.

1. For Default, leave the security code as *.
2. In the Command field, type the command or macro you want to assign security to. If you assign security to the Login and Logout commands, be sure to give all users, including Default, access to these commands. Otherwise, users might be locked in or out of the system. It is recommended that Login and Logout keep the * security code.
3. In the Security Code field, select a code for the command or macro.
4. If required, type a descriptive remark in the Description field.
5. Repeat steps 2 through 4 for each command and macro.
6. Save the configuration by choosing Close.

(*b*) **Security by exclusion :** Security by exclusion means only the RSView32 commands and macros you want users who are assigned this code to have access to be listed in the Security Codes editor.

1. For Default, change the security code to any letter.
2. In the Command field, type a command or macro you want users who are assigned this code to have access to. Ensure you include the Login and Logout commands and assign them the * security code.
3. In the Security Code field, select a code for the command or macro.
4. If required, type a descriptive remark in the Description field.
5. Repeat steps 2 through 4 for each command and macro.
6. Save the configuration by choosing Close.

(*c*) **Preventing access to the Security Codes and User Accounts editors :** Both the Security Codes editor and the User Accounts editor have commands that can be used at runtime to open these editors. The Security command opens the Security Codes editor and the Account command opens the User Accounts editor. Be sure to secure the Security and Account commands in the Security Codes editor to prevent users from accessing these editors at runtime. If users open the editors, they can edit any information. Also be sure to restrict access to the ProjectShow command, as it allows access to these editors.

(*d*) **Using strict security :** When strict security is on, the system checks the security codes of commands and macros no matter where they are issued from. When strict security is off, the system checks the security codes of commands and macros only when they are issued from the command line.

When strict security is on, RSView32 checks security access to commands or macros issued from:

(*a*) Macros
(*b*) Command line
(*c*) Button objects
(*d*) Object, display, and global keys
(*e*) Objects configured with touch animation
(*f*) Alarm Identification field in the Tag Database editor

By default, strict security is on. For most projects, strict security is desirable.

(*e*) **Turning off strict security** : If you turn off strict security, security access to graphic displays, OLE verbs, and tags is still checked. As well, commands issued from the command line are checked.

1. Open the Security Codes editor.
2. On the menu bar, click Setup, and then click Strict Security then a new window will appear as shown in Fig. 12.65.

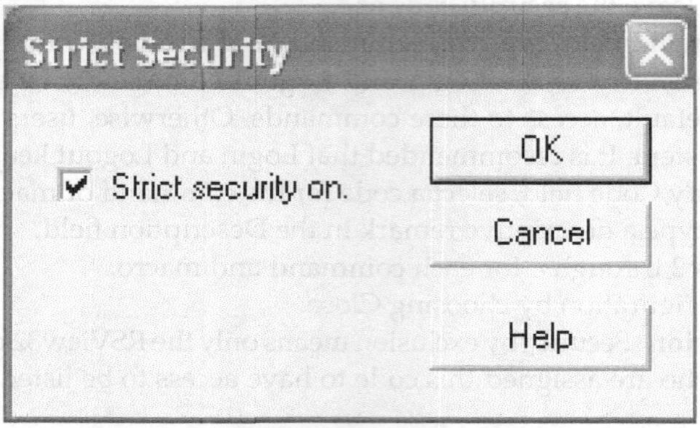

Fig. 12.65 Strict security

3. Clear the check box.
4. Click OK.

(*f*) **Graphic display security** : Security is assigned to a graphic display in the Graphic Display editor.

1. In the Graphic Display editor, open Display Settings by right-click of mouse button and then click Display Settings
2. Select a security code.
3. Click OK.

(*g*) **OLE object security** : Security is assigned to a graphic object with OLE verb animation in the Graphic Display editor.

1. Open the Graphic Display editor.
2. Select the OLE object.
3. Open the Animation dialog box.
4. Click the OLE Verb tab.
5. In the OLE Verb field, select the verb to secure.
6. Select a security code.
7. Click Apply

(*h*) **Tag security** : Security is assigned to tags in the Tag Database editor.

1. Open the Tag Database editor.
2. Select the tag that requires security.
3. Select a security code.
4. Click Accept.

12.8.2 User Accounts Creation

In the project manager window, select user accounts as shown in Fig. 12.66.

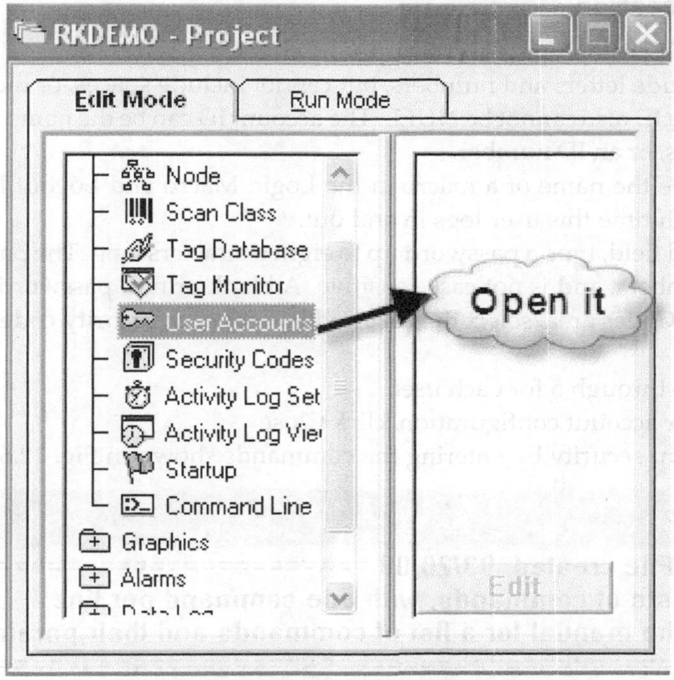

Fig. 12.66 User account selection

Then a new user account editor window will appear as shown in Fig. 12.67.

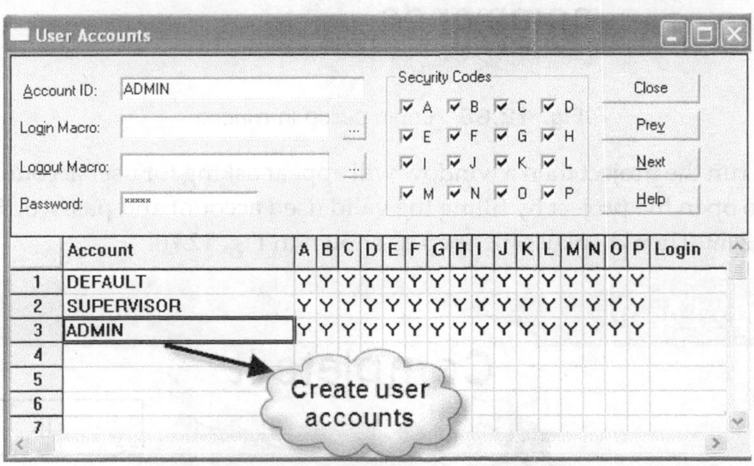

Fig. 12.67 User account editor window

About the Default User

The default user account is active when no user is logged into the system. When setting up security, keep the following in mind:

1. We cannot change the account ID.
2. We cannot add the password.

For a completely secure project, clear all security codes for the default user account. This ensures that the default user does not have any privileges.

User Accounts Creation

1. In the Account ID field, type a user ID up to 20 characters long.
   The ID can include letters and numbers, but cannot include spaces, or the characters / \ [ ] ; : | =,. +* ? <>. The ID also cannot be NULL. The account ID can be the name of a user, a name for a group of users, or an ID number.
2. If required, type the name of a macro in the Login Macro and Logout Macro fields. These macros run each time this user logs in and out.
3. In the Password field, type a password up to eight characters long. The password can include letters and numbers, and is not case sensitive. A blank or null password is acceptable.
4. In the Security Codes area, select the required box for each security code for user.
5. Click Accept.
6. Repeats steps 1 through 5 for each user.
7. To save the user account configuration, click Close

In macro file startup security by entering the commands shown in Fig. 12.68.

Fig. 12.68 Login setup in macro

Now when you run the project then a window will appear asking for user account and password. User will be able to open the project by filling the valid used account and password in this window as shown in Fig. 12.69. Then display will open as shown in Fig. 12.70.

Fig. 12.69 Login information

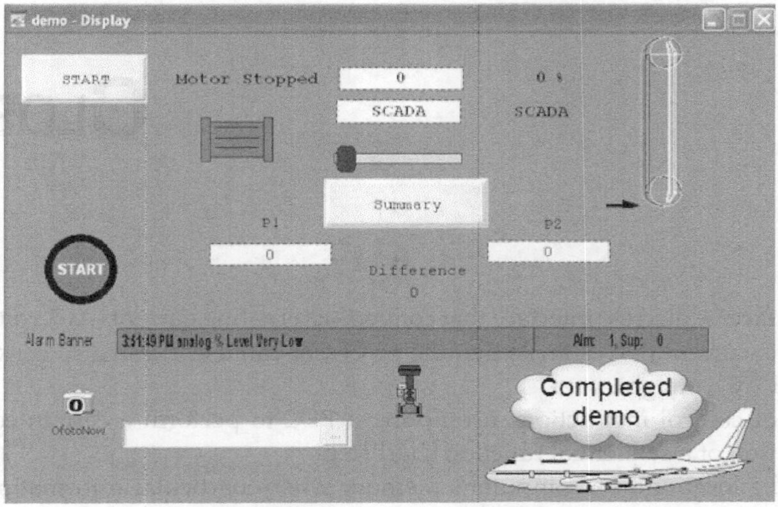

Fig. 12.70 Completed Demo

SUMMARY

Events are expressions that trigger actions. Expressions are equations containing tag values, mathematical operations, if-then-else logic, and other built-in RSView32 functions. A derived tag is a tag whose value is the result of an expression. The expression can be made up of mathematical operations, tag values from the value table, if-then-else logic, and other special functions. The derived tag is only evaluated when the expression changes. Derived tags can be device tags or memory tags. A macro is a series of commands stored in a macro file. The name of the macro file is then used like a command and can be used anywhere. Three different kinds of kys with RS 32 are : object keys, display keys and global keys.Object keys and display keys are different from global keys. Object keys and display keys are active only while a particular object or display has input focus. Global keys are active at all times. A trend is a visual representation of real time or historical tag values. The trend provides operators with a way of tracking plant activity as it is happening. OLE stands for Object Linking and Embedding. OLE objects are objects created in other Windows applications and linked to or embedded in an RSView32 graphic display. Security is used to prevent users from accessing certain parts of the system Security is based on a system of codes. Each code allows users with security privileges for that code to access the RSView32 features allowed by that code.

REVIEW QUESTIONS

1. Enlist and explain functions of different types of keys available with RSView32.
2. Explain the function of Trends in Automation.
3. What are the functions of various tales available in Trend editor box?
4. What are various fields available in Pen Tale? Explain each in detail.
5. What are various security codes available in RSView32?
6. Differentiate between security by inclusion and security by exclusion.
7. What ax MACROS? Enlist various steps involved in configuring Macros.

GLOSSARY

A

AC/DC I/O interface. A discrete interface that converts alternating current (AC) voltages from field devices into direct current (DC) signals that the processor can use. It can also convert DC signals into proportional AC voltages.

action. A set of control instructions prompting a PLC to perform a certain control function during the execution of a sequential function chart step.

address. (1) The location in a computer's memory where particular information is stored. (2) The alphanumeric value used to identify a specific I/O rack, module group, and terminal location.

addressability. The total number of devices that can be connected to a network.

address field. The sequence of eight (or any multiple of eight) bits immediately following the opening flag sequence of a frame, which identifies the secondary station that is sending (or is designated to receive) the frame.

algorithm. A set of procedures used to solve a problem.

alphanumeric code. A character string consisting of a combination of letters, numbers, and/or special characters used to represent text, commands, numbers, and/or code groups.

ambient temperature. The temperature of the air surrounding a device.

American National Standards Institute (ANSI). A clearinghouse and coordinating agency for voluntary standards in the United States.

American Wire Gauge (AWG). A standard system used to designate the size of electrical conductors. Gauge numbers have an inverse relationship to size; larger gauges have a smaller diameter.

analog device. An apparatus that measures continuous information signals (*i.e.*, signals that have an infinite number of values). The only limitation on resolution is the accuracy of the measuring device.

analog input interface. An input circuit that uses an analog-to-digital converter to translate a continuous analog signal, measured by an analog device, into a digital value that can be used by the processor.

analog output interface. An output circuit that uses a digital-to-analog converter to translate a digital value, sent from the processor, into an analog signal that can control a connected analog device.

analog signal. A continuous signal that changes smoothly over a given range, rather than switching suddenly between certain levels as discrete signals do.

analog-to-digital converter (A/D). A device that translates analog signals from field devices into binary numbers that can be read by the processor.

AND. A logical operator that requires all input conditions to be logic 1 for the output to be logic 1. If any input is logic 0, then the output will be logic 0.

application program. The set of instructions that provides control, data acquisition, and report generation capabilities for a specific process.

arithmetic instructions. Computer programming codes that give a PLC the ability to perform mathematical functions, such as addition, subtraction, multiplication, division, and square root, on data.

artificial intelligence (AI). A subfield of computer science dealing with the development of computer programs that solve tasks requiring extensive knowledge.

ASCII. For *American Standard Code for Information Interchange*. A seven-bit code with an optional parity bit used to represent alphanumeric, punctuation, and control characters.

ASCII I/O interface. A special function interface that transmits alphanumeric data between peripheral equipment and a PLC.

assembly language. A symbolic programming language that can be directly translated into machine language instructions.

asynchronous. Recurrent or repeated operations that occur in unrelated patterns over time.

B

back plane. A printed circuit board, located in the back of a chassis, that contains a data bus, power bus, and mating connectors for modules that will be inserted into the chassis.

backup. A device or system that is kept on hand to replace a device or system that fails.

backward chaining. A method of finding the causes of an outcome by analyzing its consequents to obtain its antecedents.

bandwidth. The range of frequencies expressed in Hertz over which a system is designed to operate.

base. The maximum number of digits used to represent values in a number system.

baseband coaxial cable. A communication medium that can send one transmission signal at a time at its original frequency.

BASIC module. An intelligent I/O interface capable of performing computational tasks without affecting the PLC processor's computing time.

battery backup. A battery or set of batteries that will provide power to the processor's memory in the event of a power outage.

baud. (1) The reciprocal of the shortest pulse width in a data communication stream. (2) The number of binary bits transmitted per second during a serial data transmission.

Baye's theorem. An equation that defines the probability of one event occurring based on the fact that another event has already occurred.

BCD. *See* **binary coded decimal**.

binary coded decimal (BCD). A binary number system in which each decimal digit from 0 to 9 is represented by four binary digits (bits). The four positions have a weighted value of 1, 2, 4, and 8, respectively, starting from the least significant (right-most) bit.

binary number system. A base 2 number system that uses only the numbers 0 and 1 to express all values. Each digit position of a binary number has a weighted value of 1, 2, 4, 8, 16, 32, 64, and so on, starting with the least significant (right-most) digit.

bit. For *binary digit*. The smallest unit of binary information. A bit can have a value of 1 or 0.

bit rate. *See* **baud**.

bit-wide bus network. An I/O bus network that interfaces with discrete devices that transmit less than 8 bits of data at a time.

blackboard architecture. The distribution of knowledge inferencing, as well as global and knowledge databases, in a control system through the use of several subsystems containing local, global, and knowledge databases that work independently of each other.

block. A group of words transmitted as a unit.

block check character (BCC). A character, placed at the end of a data block, that corresponds to the characteristics of the block.

block diagram. A schematic drawing.

block length. The total number of words transmitted at one time.

block transfer. A programming technique used to transfer up to 64 words of data to or from an intelligent I/O module.

Boolean action. A set of control instructions that assigns a discrete value to a variable during a sequential function chart step.

Boolean language. A PLC programming language, based primarily on the Boolean logic operators, that implements all of the functions of the basic ladder diagram instruction set.

Boolean operators. Logical operators, such as AND, OR, NAND, NOR, NOT, and exclusive-OR, that can be used singly or in combination to form logical statements that have output responses of TRUE or FALSE.

Boolean variable. A single-bit variable whose value is transmitted in the form of 1s and 0s.

Bourdon tube. A pressure transducer available in spiral, helical, twisted, and C-tube configurations that converts pressure measurements into displacement.

branch. A parallel logic path within a rung.

breadth-first search. A method of rule evaluation that evaluates each rule in the same level of a decision tree before proceeding downward.

bridge circuit. A mechanism found in transducer circuits that uses resistors to change the parameters (*e.g.*, voltage and current) of an incoming signal.

broadband coaxial cable. A communication medium that can transmit two or more transmission signals at one time via frequency division multiplexing.

burn-in procedure. The process of operating a device at an elevated temperature to identify early-failing parts.

bus. (1) A group of lines used for data transmission or control. (2) Power distribution conductors.

bus topology. A network configuration in which all stations are connected in parallel with the communication medium and all stations can receive information from any other station on the network.

bypass/control station. A device that allows a process to be switched to either PLC or manual control.

byte. A group of eight adjacent bits that are operated on as a unit, such as when moving data to or from memory.

byte-wide bus network. An I/O bus network, which interfaces with discrete and small analog devices, that can transmit between 1 and 50 or more bytes of data at a time.

C

cascade control. The use of two controllers to regulate a process so that the feedback loop of one controller is the set point of the other controller.

center of gravity method. A method of calculating the final output value of a fuzzy logic controller by finding the value that corresponds to the center of the mass under the control output curve.

centralized control. A PLC control system organization in which a central PLC controls several machines or processes.

central processing unit (CPU). The part of a programmable controller responsible for reading inputs, executing the control program, and updating outputs. Sometimes referred to as the *processor*, the CPU consists of the arithmetic logic unit, timing/control circuitry, accumulator, scratch pad memory, program counter, address stack, and instruction register.

centroid. The point in a geometrical figure whose coordinates equal the average of all the other points comprising the figure.

channel. A designated path for a signal.

channel capacity. The amount of information that can be transmitted per second on a given communication channel depending on the medium, line length, and modulation rate.

character. One symbol of a set of elementary symbols, such as a letter of the alphabet or a number.

chassis. A hardware assembly that houses PLC devices, such as I/O modules, adapter modules, processor modules, power supplies, and processors.

checksum. A transmission verification algorithm that adds the binary values of all the characters in a data block and places the sum in the block check character position.

chip. A very small piece of semiconductor material that holds electronic components. Chips are normally made of silicon and are typically less than 1/4 inch square and 1/100 inch thick.

closed loop. A control system that uses feedback from the process to maintain outputs at a desired level.

coaxial cable. A transmission medium, consisting of a central conductor surrounded by dielectric materials and an external conductor, that possesses a predictable characteristic impedance.

code. (1) A binary representation of numbers, letters, or symbols that have some meaning. (2) A set of programmed instructions.

coil. A ladder diagram symbol that represents an output instruction.

cold junction compensation. A compensation factor that allows a thermocouple to operate as though it has an ice-point reference.

collision detection (CSMA/CD). A network access method in which each node waits until there is no traffic on the network then transmits its message. If the node detects another transmission on the network, it will disable its transmitter and wait until the network clears before retransmitting the message.

combined error. *See* **propagation error.**

common bus topology. A network configuration in which individual PLCs connect to a main trunkline in a multidrop fashion.

compatibility. (1) The ability of various specified units to replace one another with little or no reduction in capability. (2) The ability of units to be interconnected and used without modification.

complement. A logical operation that inverts a signal or bit.

conditional probability inferencing. The conditional probability of an event happening in an artificial intelligence system.

constant voltage transformer. A transformer that maintains a steady output voltage (secondary) regardless of input voltage (primary) fluctuations.

contact. A ladder diagram symbol that represents an input condition.

contact output interface. A discrete interface, which does not require an external power source, that is triggered by the change in state of a normally open or normally closed contact.

contact symbology. A set of symbols used to express a control program through conventional relay symbols (*e.g.*, normally open contacts, normally closed contacts, etc.).

continuous-mode controller. A process controller that sends an analog signal to a process control field device.

control element. The output field device that regulates the actual control variable level in a process control system.

control logic. The control plan for a given system.

control loop. The method of adjusting the control variable in a process control system by analyzing process variable data and then comparing it to the set point to determine the amount of error in the system.

control panel. A panel that contains instruments used to control devices.

control program checkout. A final review of a PLC's control program prior to starting up the system.

control program printout. A hard copy of the control logic program stored in a PLC's memory.

control strategy. The sequence of steps that must occur during a process or PLC program to produce the desired output control.

control task. The desired results of a control program.

control variable. The independent variable in a process control system that is used to adjust the dependent variable, the process variable.

convergence. A point in a sequential function chart where many elements flow into one element.

counter. An electromechanical device that counts the number of times an event occurs.

counter instructions. Computer programming codes that allow a PLC to perform the counting functions (count up, count down, counter reset) of a hardware counter.

critically damped response. A second-order control system response in which the damping coefficient equals 1, causing the response to overshoot the set point and then quickly settle back to it.

current loop. A two-wire communication link in which the presence of a 20 milliamp current level indicates a binary 1 (mark) and its absence indicates no data, a binary 0 (space).

cyclic exclusive-OR checksum (CX-ORC). An error detection method in which the words in the data block are exclusive-ORed with the checksum word and then rotated to the left. This action is repeated until all of the words in the block have been operated on.

cyclic message. A scheduled message transmission.

cyclic redundancy check (CRC). An error detection method in which all the bits in a block are divided by a predetermined binary number. The remainder becomes the block check character.

D

data manipulation instructions. Computer codes that provide a PLC with the ability to compare, convert, shift, examine, and operate on data in multiple registers.

data table. The part of a processor's memory, containing I/O values and files, where data is monitored, manipulated, and changed for control purposes.

data transfer instructions. Computer codes that allow a PLC to move numerical data within a controller, either in single register units or in blocks of registers.

DC I/O interface. A discrete module that links a processor with direct current field devices.

dead time. The delay between the time a control system's control variable changes and the time the process variable begins to respond to the change.

debouncing. The act of removing intermediate noise from a mechanical switch.

decimal number system. A base 10 number system that uses ten numbers-0, 1, 2, 3, 4, 5, 6, 7, 8, and 9—to represent all values. Each digit position has a weighted value of 1, 10, 100, 1000, and so on, beginning with the least significant (rightmost) digit.

device bus network. A network that allows low-level input/output devices that transmit relatively small amounts of information to communicate directly with a PLC.

diagnostic AI system. The lowest level of artificial intelligence system. This type of system primarily detects faults within an application but does not provide information about possible solutions.

diagnostics. The detection and isolation of an error or malfunction.

differential input/output. A signal transmission system where inputs and outputs have individual return lines for each channel, as opposed to all data running through one line.

digital device. A device that processes and sends discrete (two-state) electrical signals.

digital signal. A noncontinuous signal that has a finite number of values.

digital-to-analog converter (D/A). A device that translates binary numbers from a processor into analog signals that field devices can understand.

direct-acting controller. A closed-loop controller whose control variable output increases in response to an increase in the process variable.

direct action I/O interface. A special I/O interface that detects, preprocesses, and transmits low-level and fast-speed signals.

discrete input interface. An input circuit that allows a PLC to receive data from digital field devices.

discrete-mode controller. A controller that sends a noncontinuous signal to the field device controlling a process.

discrete output interface. An output circuit that allows a PLC to send data to digital field equipment.

displacement transducer. A device that measures the movement of an object.

distributed control. A PLC control system organization in which factory or machine control is divided into several subsystems, each managed by a separate PLC, yet all interconnected to form a single entity.

distributed I/O processing. The allocation of various control tasks to several intelligent I/O interfaces.

divergence. A point in a sequential function chart where one element flows into many elements.

E

electrically alterable read-only memory (EAROM). A type of nonvolatile, programmable, read-only memory that can be erased completely by applying the proper voltage to the memory chip.

electrically erasable programmable read-only memory (EEPROM). A type of nonvolatile, programmable, read-only memory that can be erased by electrical pulses.

Electronic Industries Association (EIA). An agency that sets electrical/electronic standards.

encoder/counter module. An interface, which is used in positioning applications, that links encoders and high-speed counter devices with programmable logic controllers.

enhanced ladder language. A PLC language that implements basic ladder language instructions, as well as more sophisticated functional block instructions, which can perform multiple operations in a single instruction.

error. The difference between the set point and the process variable in a control system.

error-correcting code. A code in which each acceptable expression conforms to specific rules of construction that also define one or more nonacceptable expressions, so that if certain errors occur, they can be detected and corrected.

error deadband. The amount that the process variable can fluctuate from the set point before the control system provides corrective action.

error-detecting code. A code in which each expression conforms to specific rules of construction, so that if an error occurs in an expression, it can be detected.

exclusive-OR (XOR). A logical operation, which has only two inputs, that yields a logic 1 output if only one of the two inputs is logic 1 and a logic 0 output if both inputs are the same, either logic 1 or logic 0.

execute. To perform a specific operation by processing either one instruction, a series of instructions, or a complete program.

execution time. The time required to perform one specific instruction, a series of instructions, or a complete program.

executive memory. The part of the system memory that permanently stores a system's supervisory programs, as well as instruction software. This area of memory is not accessible to the user.

expert AI system. The highest level of AI systems. This type of system detects process faults, provides information about possible causes of the faults, and makes complex decisions about resulting actions based on statistical analysis.

F

FALSE. As related to PLC instructions, a reset logic state associated with a binary 0.

fast-input interface. An intelligent I/O module that functions as a pulse stretcher, detecting very fast input pulses that regular I/O modules cannot read.

fast-response interface. A special I/O module designed to detect fast inputs and respond with an output.

fiber-optic cable. A communication medium composed of thin fibers of glass or plastic enclosed in a material with low refraction.

first-order response. A process response to a rapid change in the control variable characterized by one lag time and a process response curve that slowly approaches the set point.

floating-point math. A data manipulation format, which is used to express a number by expressing the power of the base, that usually involves the use of two sets of digits. For example, in a floating decimal notation where the base is 10, the number 8,700,000 would be expressed as 8.7(10)6 or 8.7E6.

flowchart. A graphical representation of the definition or solution of a task or problem.

flowcharting. A method of pictorially representing the operation of a process in a sequential manner.

flow transducer. A device that measures the amount of solid, liquid, or gaseous materials flowing through a process by measuring either weight, differential pressure, or fluid motion.

forward chaining. A method for determining all possible outcomes for a given set of inputs.

full-duplex line. A communication line used to simultaneously transmit data in two directions.

function block diagram (FBD). A graphical PLC programming language in which instructions are programmed as blocks that are then used as needed to control process elements.

fuzzy processing. The interpretation of fuzzy input data to determine an appropriate outcome based on user-programmed IF…THEN rules.

fuzzy set. A group of membership functions.

G

gate. A circuit having two or more input terminals and one output terminal, where an output is present only when the prescribed inputs are present.

gateway. A device or pair of devices that connects two or more communication networks. This device may act as a host to each network and may transfer messages between the networks by translating their protocols.

global database. The section of an AI system that stores data measurements from the controlled process.

grade. A measure of how well a value fits into a given membership function.

Grafcet. A PLC programming language that uses an object-oriented, flowchart-like framework, along with steps, transitions, and actions, to define the control program.

Gray code. A cyclic code, similar to a binary code, in which only one bit changes as the counting number increases.

gross error. An error resulting from human miscalculation.

ground loop. A condition in which two or more electrical paths exist within a ground line.

guarantee error. A value of error derived from a known specification that defines the amount that a product or material will arithmetically deviate from the mean.

H

half-duplex line. A communication line that can transmit data in two directions, but in only one direction at a time.

Hamming code. An error-detecting code that combines parity and data bits to generate a byte containing a value that identifies the erroneous bit.

hard copy. A printed document.

hardware. All the physical components of a programmable controller, including peripherals, as opposed to the software components that control its operation.

hardwired logic. Logic control functions that are determined by the way a system's devices are physically interconnected.

hexadecimal number system (hex). A base 16 number system that uses the numbers 0, 1, 2, 3, 4, 5, 6, 7, 8, and 9 and the letters A, B, C, D, E, and F to represent numbers and codes.

host. A central computer in a network system.

I

IEC 1131 programming standard. A standardized set of PLC programming guidelines, set forth by the International Electrotechnical Commission, thatincludes general PLC information, equipment and test requirements, programming languages, user guidelines, and communication standards.

IEEE 802. A family of standards specified by the Institute of Electrical and Electronic Engineers for data communication over local and metropolitan area networks.

image table. An area in a PLC's memory dedicated to I/O data where 1s and 0s represent ON and OFF conditions, respectively.

individual control. A PLC control system organization in which a PLC controls a single machine or process.

inference engine. The section of an AI system where all decisions are made using the knowledge stored in the knowledge database.

input. Information sent to the processor from connected devices.

input device. Any connected equipment, such as control devices (*e.g.*, switches, buttons, and sensors) or peripheral devices (*e.g.*, cathode ray tubes and manual programmers), that supply information to the central processing unit. Each type of input device has a unique interface to the processor.

input/output system. A collection of plug-in modules that transmit control data between a PLC and field devices.

input table. The area of a PLC's memory where information about the status of input devices is stored.

instruction list (IL). A low-level, text-based PLC programming language that uses assembly language-like mnemonics to represent the control program.

integer variable. A nondiscrete variable whose value is transmitted in the form of a whole number.

integral controller. A continuous-mode controller whose output to the control field device changes according to how the error signal changes over time.

integral of time and absolute error open-loop tuning method (ITAE). A method used to determine the proper tuning constants for a controller based on the minimization of the integral of time and the absolute error of the response.

integral windup. The situation in which the control variable in a system remains at its maximum level even though the amount of error in the system starts to decrease.

intelligent I/O interface. A microprocessor-based module that can perform sophisticated processing functions independently of the central processing unit.

interface. A circuit that permits communication between a central processing unit and a field input or output device. Different devices require different interfaces.

interlock. A device actuated by the operation of another device to which it is linked to govern the succeeding operation of the same or allied devices.

I/O address. A unique number, assigned to each input/output device, that corresponds to the device's location in the rack enclosure. The address number is used when programming, monitoring, or modifying a specific input or output.

I/O address assignment document. A document that identifies every field device by address, type of input/output module, type of field device, and the function the field device performs.

I/O bus network. A network that lets input and output devices communicate directly to a PLC through digital communication.

I/O bus network scanner. A device connected to a PLC that reads and writes to field devices connected to an I/O bus network, as well as decodes the data in the network information packet.

I/O module. A plug-in assembly, containing two or more identical input or output circuits, that provides the connection between a processor and connected devices. Normal I/O module capacities are 2, 4, 8, and 16 circuits.

I/O scan time. The time required to update all local and remote I/O.

I/O update scan. The process of revising the bits in a PLC's I/O tables based on the latest results from reading the inputs and processing the outputs according to the control program.

I/O wiring connection diagram. A drawing that shows the actual connections of the field I/O devices to a PLC, including power supplies and subsystem connections.

isolated I/O interface. An input module in which each input has a separate return line. Isolated I/O interfaces can connect field devices powered from different sources to one module.

isolation transformer. A transformer that protects its connected devices from surrounding electromagnetic interference.

ITAE. *See* **integral of time and absolute error open-loop tuning method.**

K

K. 210. Used to denote memory size in either bits, bytes, or words.

knowledge AI system. A mid-level AI system that detects faults based on resident knowledge and also makes decisions about the cause of the fault and ensuing process actions.

knowledge database. The section of an AI system that stores information extracted from the expert.

knowledge inference. A decision-making methodology used to gather and analyze process data in order to draw conclusions.

knowledge representation. The way an artificial intelligence system strategy is organized.

L

label. A name given to a membership function.

ladder diagram. An industry standard for representing relay logic control systems.

ladder diagram language (LD). A graphical set of instructions that implements basic relay ladder functions in a PLC.

ladder relay instructions. Computer codes that implement relay coils and contacts and their corresponding functions in a PLC.

ladder rung matrix. A rectangular array that defines the maximum number of contacts that can be programmed in a ladder rung, along with the maximum number of parallel branches allowed in the rung.

lag time. The delay between the initial response of the process variable to a change in the control variable and the process variable's optimal response to it.

LAN. *See* **local area network.**

language. A set of symbols and rules for representing and communicating information between people and machines.

Laplace transform. A mathematical function used to convert differential equations from the time domain into the frequency domain so that they become easy-tomanage algebraic equations.

LCD. *See* **liquid crystal display.**

LD. *See* **ladder diagram language.**

lead resistance compensation. A factor that compensates for signal loss due to resistance present in electrical wires.

least significant bit (LSB). The bit representing the smallest value in a nibble, byte, or word.

least significant digit (LSD). The digit representing the smallest value in a byte or word.

LED. *See* **light-emitting diode.**

light-emitting diode (LED). A semiconductor diode whose junction emits light when current passes through it in a forward direction.

limit switch. An electrical switch actuated by the motion of a machine or equipment.

linear variable differential transformer (LVDT). An electromechanical mechanism that provides a voltage reference that is proportional to the movement or displacement of a core inside a coil.

liquid crystal display (LCD). A display device consisting of a liquid crystal hermetically sealed between two glass plates.

load. The power used by a machine or apparatus.

load cell. A force or weight transducer that is based on a direct application of a bonded strain gauge.

local area network (LAN). An ensemble of interconnected processing elements (nodes), which are typically located within a few miles of each other.

local rack. An enclosure, placed in the same area as the master rack, that contains a local I/O processor, which sends data to and from the central processing unit.

location. A storage position or register in memory identified by a unique address.

logic. The process of solving complex problems through the use of simple functions that can be either true or false.

logic diagram. A drawing that uses interconnected AND, OR, and NOT logic symbols to graphically describe a system's operation or control.

longitudinal redundancy check (LRC). An error-checking technique based on an accumulated exclusive-OR of transmitted characters. LRC characters are accumulated at both the sending and receiving stations.

loop tuning. The process of determining the proportional, integral, and derivative constants that will allow a PID controller to perform optimally.

LRC. *See* **longitudinal redundancy check.**

LSB. *See* **least significant bit.**

LSD. *See* **least significant digit.**

LVDT. *See* **linear variable differential transformer.**

M

mask. A logical function used to set certain bits in a word to an established state.

master. A device used to control other devices.

master control relay (MCR). A hardwired or softwired relay instruction that will de-energize its associated I/O devices when the instruction is de-energized.

master rack. The enclosure containing the CPU or processor module.

master/slave bus topology. A network configuration in which one master controller manages several slave controllers.

maximum value method. A method of calculating the final output value of a fuzzy logic controller by finding the rule output value with the highest membership function grade.

MCR. *See* **master control relay.**

mean. The average value of a set of data readings.

mean-time-between-failures study. A study, which contains data about the average time between equipment failures, that provides information about the reliability of a product.

median. The middle value of a set of data readings organized in ascending order.

medium access control (MAC). A technique that ensures that only one device is transmitting on a network at any given time.

membership function. A group of fuzzy logic rules used to divide input data into sets, which are then analyzed to provide reasoned control of a field device.

memory. The part of a programmable controller that stores data, instructions, and the control program either temporarily or semipermanently.

memory map. A diagram showing a system's memory addresses, as well as which programs and data are assigned to each section of memory.

message. A group of data and control bits transferred as an entity from a data source.

microprocessor. A digital, electronic logic package (usually on a single chip) capable of performing the program execution, control, and data-processing functions of a central processing unit. A microprocessor usually contains an arithmetic logic unit, temporary storage registers, instruction decoder circuitry, a program counter, and bus interface circuitry.

miniprogrammer. A portable device used for programming, changing, and monitoring a PLC's control logic.

mode. The most frequently occurring value in a set of data readings.

module. An interchangeable, plug-in item containing electronic components.

most significant bit (MSB). The bit representing the greatest value of a nibble, byte, or word.

most significant digit (MSD). The digit representing the greatest value of a byte or word.

MSB. *See* **most significant bit.**

MSD. *See* **most significant digit.**

multidrop link. A cable that terminates at more than one point.

multiplexing. The act of channeling two or more signals to one source using the same channel.

multiprocessing. Concurrent execution of two or more tasks residing in memory.

N

NAND. A logical operator that yields a logic 1 output if any input is logic 0 and a logic 0 output if all inputs are logic 1. This operator is a negated AND function, the result of negating the output of an AND gate by following it with a NOT symbol.

negative logic. The use of binary logic so that logic 0 represents the voltage level normally associated with logic 1 (*i.e.*, logic 0 = +5 V, logic 1 = 0 V).

network. A series of points (or devices) connected by some type of communication medium.

network communications instructions. Computer codes that allow a PLC to share data with other PLCs connected to a local area network.

network interface module. A special function interface that allows PLCs and other intelligent devices to communicate and transfer data over a high-speed local area communication network.

network layer. Layer 3 of the OSI protocol. This layer routes information in the network.

nibble. A group of four bits.

node. A station, such as a personal computer or a PLC, that is connected to a network and can thereby send and receive messages through the network.

nonreturn to zero invert on ones (NRZI). A self-clocking pulse code used to establish reliable synchronous transmission.

nonvolatile memory. A type of memory whose contents are not lost or disturbed if operating power is lost.

NOR. A logical operator that yields a logic 1 output if all inputs are logic 0 and a logic 0 output if any input is logic 1. This operator is a negated OR function, the result of negating the output of an OR gate by following it with a NOT symbol.

normal action. A set of IEC 1131-3 instructions that is executed continuously for the duration of an SFC step's activity.

normally closed contact. (1) A relay contact pair that is closed when the coil of the relay is not activated and open when the coil is activated. (2) A ladder program symbol that allows logic continuity (flow) if the referenced input is logic 0 when evaluated.

normally open contact. (1) A relay contact pair that is open when the coil of the relay is not activated and closed when the coil is activated. (2) A ladder program symbol that allows logic continuity (flow) if the referenced input is logic 1 when evaluated.

NOT. A logical operator that yields a logic 1 output if a logic 0 is entered at the input and a logic 0 output if a logic 1 is entered at the input. The NOT function, also called an *inverter*, is normally used in conjunction with AND and OR functions.

NRZI. *See* **nonreturn to zero invert on ones.**

O

octal number system. A base 8 number system that uses eight numbers-0, 1, 2, 3, 4, 5, 6, and 7-to represent all values.

off-line. The state of not being in continuous direct communication with the processor.

one's complement. An operation that represents the negative value of a binary word by assigning the most significant bit of the word with a value equal to its normal value minus one.

one shot. A programming technique that sets a storage bit or output to a certain state for only one scan.

on-line. The state of being in continuous communication with the processor.

open loop. A control system that does not receive process feedback in order to perform self-correcting actions.

optical coupler. A device that couples signals from one circuit to another by means of electromagnetic radiation.

OR. A logical operator that yields a logic 1 output if any input is logic 1 and a logic 0 output if all inputs are logic 0.

orifice plate. A transducer that measures fluid flow by measuring the pressure differential between two points.

OSI model. A description of network communications functions organized in seven layers to promote open system interconnections.

output. Information sent from the processor to connected field devices.

output device. Any connected equipment, such as control devices (*e.g.*, motors, solenoids, and alarms) or peripheral devices (*e.g.*, line printers, disk drives, and color displays), that receives information or instructions from the central processing unit. Each type of output device has a unique interface to the processor.

output table. The area of a PLC's memory where information about the status of output devices is stored.

overdamped response. A second-order control system response in which the damping coefficient is greater than 1, causing the response to overshoot the set point and then slowly settle back to it.

P

packet. Data and sequences of control bits arranged in a specified format and transferred as an entity during data transmission.

panel enclosure. The physical enclosure that houses a PLC's hardware and components.

parallel circuit. A circuit in which two or more of the connected components or contact symbols in a ladder program are connected to the same pair of terminals so that current may flow through all the branches.

parity. The even or odd characteristic of the number of 1s in a byte or word of memory.

parity bit. A bit added to a memory word as a means of error detection.

parity check. A check for a certain number of 1s and 0s in a memory word to ensure data integrity.

peripherals. External devices, such as line printers, disk drives, recorders, etc., that are connected to a PLC.

PID interface. *See* **proportional-integral-derivative interface.**

PLC. *See* **programmable logic controller.**

polling. A network access method where a master controller manages the communication process by interrogating each slave controller under it to determine whether the slave has any information to send.

positive logic. The conventional use of binary logic in which logic 1 represents a positive logic level (*e.g.*, logic 1 = +5 V, logic 0 = 0 V).

potentiometer. A simple transducer that measures displacement based on resistance changes due to the movement of a wiper arm.

power supply. The unit that supplies the necessary voltage and current to a system's circuitry.

presentation layer. Layer 6 of the OSI protocol. This layer communicates data while resolving syntax differences between network devices.

pressure transducer. A transducer that measures pressure by transforming exerted force into an electrical signal.

process. (1) Continuous and regular production executed in a definite, uninterrupted manner. (2) One or more entities threaded together to perform a requested service.

process bus network. A network that allows high-level analog input/output devices that transmit large amounts of information to communicate directly with a PLC.

process control. The regulation of process parameters to within specified target parameters through the manipulation of the control variable.

process gain. The ratio between a process's output and its input. In an ideal process control situation, the process gain equals one.

process variable. A process control system's dependent variable, which is controlled by its independent variable, the control variable.

program. A planned set of instructions stored in memory and executed in an orderly fashion by the central processing unit.

program coding. The process of translating a logic or relay diagram into PLC ladder program form.

program/flow control instructions. Computer codes that give a PLC the ability to direct the flow of operation and alter the order of execution of a control program.

programmable logic controller (PLC). A solid-state control device that can be programmed to control process or machine operations. It consists of five basic components: the processor, memory, input/output modules, the power supply, and the programming device.

programmable read-only memory (PROM). A read-only memory that can be programmed once and never altered again.

programming device. A device that is used to enter the control program into memory and make changes to the stored program.

program scan. The time required by the processor to evaluate and execute the control logic. This time does not include the I/O update time. The program scan repeats continuously while the processor is in the run mode.

PROM. *See* **programmable read-only memory.**

propagation error. A combined error caused by the interaction of two or more independent variables, each causing a different error.

proportional controller. A continuous-mode controller whose output to the control field device in proportional to the change in error.

proportional-derivative controller. A continuous-mode controller that uses both proportional and derivative actions to determine the control variable output based on both the amount of error and its rate of change.

proportional-integral controller. A continuous-mode controller that uses both proportional and integral actions to determine the control variable output based on the amount of error and its change over time.

proportional-integral-derivative controller. A continuous-mode controller that uses proportional, integral, and derivative actions to determine the control variable output based on the amount of error, its change over time, and its rate of change. This type of controller provides the optimum type of control in most process applications.

proportional-integral-derivative (PID) interface. An intelligent I/O module that provides automatic, closed-loop control of multiple, continuous-process control loops.

protocol. A formal definition of how communication will occur in a network.

pulse action. A set of IEC 1131-3 instructions that is executed only once after a step becomes active.

R

rack enclosure. The location in a PLC that physically houses plug-in devices, such as I/O modules and supplementary power supplies.

RAM. *See* **random-access memory.**

random-access memory (RAM). A volatile, alterable memory that provides storage for the application program and data.

random error. An error resulting from an unexpected action in a process line.

read. (1) To acquire data from a storage device. (2) The transfer of data between devices, such as a peripheral device and a computer.

read-only memory (ROM). A type of memory that permanently stores an unalterable program or set of instructions.

real variable. A nondiscrete variable whose value is transmitted in the form of fractional and floating-point data.

register. A temporary storage device for data and information (*e.g.*, timer/counter preset values). A PLC register is normally 16 bits wide.

register/BCD I/O interface. A multibit module that uses thumbwheel switches to interface between discrete devices and a programmable controller.

relay. An electrically operated device that mechanically switches electrical circuits.

relay logic. The representation of a control program or other logic in relay form (*i.e.*, using electrically operated devices to mechanically switch electrical circuits).

remote I/O subsystem. A system where some or all of the I/O racks are mounted away from the PLC.

remote rack. An enclosure, containing I/O modules and a remote I/O processor, located away from the CPU.

resistance temperature detector (RTD). A temperature transducer composed of conductive wire elements typically made of platinum, nickel, copper, or nickeliron.

resistance temperature detector (RTD) interface. An intelligent I/O module that interprets temperature information from RTD devices.

resolution. The smallest detectable increment of measurement.

response time. The time, including terminal delay, network delay, and service node delay, between the transmission of the last character of a network node's message and its receipt of the first character of the reply.

reverse-acting controller. A closed-loop controller whose control variable output decreases in response to an increase in the process variable.

ring topology. A network architecture where signals from one node are relayed through all the other nodes in the network.

RTD interface. *See* **resistance temperature detector interface.**

rule. An algorithm consisting of IF conditions and THEN actions that a fuzzy logic module uses to interpret input data and respond with a corresponding output value.

rule-based knowledge representation. A method of expressing an expert's knowledge in an AI system using IF…THEN rules that determine the actions and decisions to be made.

rung. A ladder program term that refers to the programmed instructions that drive one output. A complete control program may have several rungs.

S

safety control relay (SCR). A hardwired or softwired relay instruction that will deenergize its associated I/O devices when de-energized.

scaling. Changing analog output data to reflect engineering units.

scan. The process of reading all inputs, executing the control program, and updating all outputs.

scan time. The time required to complete the scan. Effectively, this is the time required to activate an output that is controlled by programmed logic.

SCR. *See* **safety control relay.**

scratch pad memory. A temporary storage area used by the CPU to store a relatively small amount of data used for interim calculations or control. Data that is needed quickly is stored in this area to avoid the extra access time involved in retrieving data from the main memory.

second-order response. A process response to a rapid change in the control variable characterized by two lag time and a process response curve that either oscillates around the set point or overshoots the set point before settling to it.

sequential function charts (SFC). An object-oriented programming framework that organizes actions written in IEC 1131-3 programming languages (ladder diagram, instruction list, function block diagram, and structured text) into a unified sequential control program.

series circuit. A circuit in which the components or contact symbols are connected end to end. All components must be closed to permit current flow.

servo motor interface. An intelligent I/O module used in applications requiring position control via a servo drive controller, which translates the rotational movement of a servo motor into linear displacement.

set point. The target process variable value in a process control system.

SFC. *See* **sequential function charts.**

SFC action. A set of IEC 1131-3 instructions, organized as an SFC program, that is activated when a certain step in the main SFC program becomes active.

single-ended input/output. An analog I/O connection in which the commons are electrically tied together resulting in only one return line.

single-precision arithmetic. Arithmetic instructions that use one register each to hold the operands and one or two registers to hold the result of the operation.

sinking configuration. An electrical configuration that causes a device to receive current when the device is ON.

slave. A remote system or terminal whose function is controlled by a master device.

software. The programs that control the processing of data in a system.

solenoid. A transducer that converts a current into linear motion through the use of one or more electromagnets that move a metal plunger.

solid-state. Circuitry designed using only integrated circuits, transistors, diodes, etc., without any electromechanical devices, such as relays.

sourcing configuration. An electrical configuration that causes a device to provide current when the device is ON.

special function instructions. Computer codes that allow a PLC to perform special operations, such as sequencing, diagnostics, and PID control.

ST. *See* **structured text**.

stand-alone action. A set of IEC 1131-3 programming instructions, not attached to the SFC program itself, that directs the program to jump to a particular step when the action's logical conditions are satisfied.

standard deviation. A measure of the dispersion of a set of data readings about the mean.

star-shaped ring topology. A network architecture in which signals from one node are relayed through all the other nodes in the network, yet a node can be bypassed in the event of its failure to avoid a break in the ring.

star topology. A network architecture in which all network nodes are connected to a central device that routes the nodes' messages.

static input wiring check. A procedure performed with power applied to the PLC and input devices that verifies that each input device is connected to the proper input terminal and is operating properly.

static output wiring check. A procedure performed with power applied to the PLC and output devices that verifies that each output device is connected to the proper output terminal and is operating properly.

steady state. The situation in which the error in a process control system is at zero or within the error deadband.

step. A stage in a control process as defined by the process's sequential function chart.

stepper motor interface. A positioning interface that controls a stepper motor, which translates incoming pulses into mechanical motion, by generating a pulse train indicating distance, rate, and direction commands to the motor.

step response. The process variable's response to a sudden change in the process input (*i.e.*, the control variable).

step test. A forced, sudden change in the control variable used to elicit a response from the process.

storage area. The area of a PLC's memory that stores blocks of input/output data, as well as data about the status of internal bits.

storage register assignment document. A document that lists the storage registers used in a control program, including their contents and a description of their function.

strain gauge. A mechanical transducer that measures body deformation (or strain) due to the force applied to a rigid body.

structured text (ST). A high-level, text-based PLC programming language, resembling the BASIC and PASCAL computer languages, that allows a control program or any other complex task to be broken down into smaller tasks.

subprogram. A semi-independent program, embedded in a larger, main control program, that executes a specialized control sequence when activated by the main program.

subroutine. A program segment in a ladder diagram that performs a separate task.

subsystem. A part of a larger system having the properties of a system in its own right.

sum-of-the-weights method. A method of changing values from other number systems into their decimal equivalents by multiplying each digit by the weighted value of its position and then summing the results.

synchronous. A type of serial transmission that maintains a constant time interval between successive events.

syntax. Rules governing the structure of a language.

system. A set of one or more PLCs, I/O devices and modules, computers, associated software, peripherals, terminals, and communication networks that together provide a means of performing information processing to control a machine or process.

system abstract. A definition of the process to be controlled including a clear statement of the control problem, a description of the design strategy, and a statement of objectives.

system configuration diagram. A drawing of the PLC control system that shows the location, simplified connections, and minimum details of the system's major hardware components.

system error. An error resulting from an instrument or from the environment.

system layout. The planned approach to placing and connecting PLC components to satisfy the control strategy and to provide system reliability and ease of maintenance.

T

tap. A device that provides mechanical and electrical connections to a trunk cable. A tap allows the signals on the trunk to be passed to a station and the signals transmitted by the stations to be passed to the trunk.

task. A set of instructions, data, and control information capable of being executed by a CPU to accomplish a specific purpose.

TCP/IP. *See* **transmission control protocol/internet protocol.**

termination. (1) The load connected to the output end of a transmission line. (2) A provision for ending a transmission line and connecting to a bus bar or other terminating device.

thermal transducer. A device that measures changes in temperature.

thermistor. A temperature transducer made of semiconductor material, such as oxides of cobalt, nickel, manganese, iron, and titanium, that exhibits changes in internal resistance proportional to changes in temperature.

thermocouple. A bimetallic temperature transducer that provides a temperature value by measuring the voltage differential caused by joining together two different metals at different temperatures.

thermocouple input module. A module that amplifies, digitizes, and converts the input signal from a thermocouple into a digital signal equivalent to the temperature reading.

thermopile. The connection of several thermocouples in series to enhance their resolution.

three-position controller. A discrete-mode controller that provides three output levels-ON, 50% ON, and OFF.

throughput. The speed at which an application or part of an application is performed. Throughput depends on the transmission speed, medium, protocol, packet size, and amount of data handled by a network.

thumbwheel switch. A rotating switch used to input numeric information into a controller.

time base. A unit of time generated by the system clock and used by software timer instructions. Typical time bases are 0.01, 0.1, and 1.0 seconds.

timer instructions. Computer codes that allow a PLC to perform the timing functions (ON-delay energize/de-energize, OFF-delay energize/de-energize, reset) of a hardware timer.

token. (1) A signal that grants bus transmission rights to a node on a network. (2) A signal that enables a transition or action in a sequential function chart.

token passing. A network transmission technique in which a token is passed along the bus and each node has a set amount of time to receive it and respond to it.

topology. The way in which a network or system is physically structured.

transducer. A device used to convert physical parameters, such as temperature, pressure, and weight, into electrical signals.

transfer function. The unique characteristics of a process that determine its output due to changes over time.

transient response. The behavioral response of a process.

transistor-transistor logic (TTL). A semiconductor logic family characterized by high speed and medium power dissipation in which the basic logic element is a multiple-emitter transistor.

transition. A variable input, action result, conditional statement, or other program element that signals a sequential function chart to progress from one step to another.

transmission control protocol/internet protocol (TCP/IP). A network protocol developed by the U.S. Department of Defense.

transmission medium. The physical device used to transfer data in a transmission system (*e.g.*, coaxial cable, fiber-optic cable, etc.).

transmitter. A device that amplifies a voltage signal.

tree topology. A network architecture in which the network has many nodes located in many branches of the network.

triac. A semiconductor device that functions as an electrically controlled switch for AC loads.

TRUE. As related to PLC instructions, a set logic state associated with a binary 1.

truth table. A table that shows the state of a given output as a function of all possible input combinations.

TTL. *See* **transistor-transistor logic.**

TTL I/O interface. A discrete interface that allows a controller to accept signals from TTL field devices, which are 5 VDC-level semiconductor devices.

turbine flow meter. A flow transducer that measures fluid flow by measuring the fluid's motion through the meter's multibladed rotor.

twisted-pair conductor. A communication medium used mainly for point-to-point applications that can transmit data up to 4000 feet at transmission rates as high as 250 kbaud.

two-position controller. A discrete-mode controller that provides two output levels-ON and OFF.

two's complement. A numbering system, used to express negative binary numbers, in which all numbers from right to left are inverted after the first 1 is detected.

U

underdamped response. A second-order control system response in which the damping coefficient is less than 1, causing the response to oscillate around the set point before settling to it.

user program memory. The memory section where the application control program is stored.

V

variable. A factor that can be altered, measured, and controlled.

Venturi tube. A transducer that measures fluid flow by measuring the pressure differential between two points.

vertical redundancy check (VRC). An error-detecting method in which a parity bit is added to each character in a message so that the number of bits in each character, including the parity bit, is either odd or even.

vibration transducer. A device that measures the vibration of a body by measuring its displacement, velocity, or acceleration.

volatile memory. A type of memory whose contents are irretrievable after operating power is lost.

VRC. *See* **vertical redundancy check**.

W

watchdog timer. A timer that monitors the logic circuits controlling a PLC. If a watchdog timer ever times out, it will disconnect the processor from the process because it will assume that the processor is faulty.

weighted value. The numerical value assigned to any single bit as a function of its position in a word.

weight input module. A special analog interface designed to read data from load cells, which convert force and weight values into electrical signals.

wire bundling. The technique of grouping an I/O module's wires according to their characteristics (*e.g.*, input, output, power).

wire input module. A special input interface designed to detect short-circuit or open-circuit connections between a module and its input devices.

word. The number of bits that the central processing unit operates on at one time when it is performing an instruction or operating on data. A word is usually composed of a fixed number of bits.

write. The process of putting information into a storage location. the control variable output in an open-loop system.

Z

Ziegler-Nichols closed-loop tuning method. A method for determining a controller's tuning constants by finding the value of the proportional gain that will cause the control loop to oscillate indefinitely at a constant amplitude when it is in a closed-loop system.

Ziegler-Nichols open-loop tuning method. A method for determining the tuning constants for a controller by testing the process variable's response to a change in gain, at constant amplitude when it is in a open loop system.

INDEX